# *Decommissioning and Radioactive Waste Management*

## Dr A. Rahman CRadP MSRP

**Principal Consultant, Atkins Limited**

**WHITTLES PUBLISHING**

CRC Press
Taylor & Francis Group

Published by
**Whittles Publishing**,
Dunbeath,
Caithness KW6 6EY,
Scotland, UK
www.whittlespublishing.com

Distributed in North America by
CRC Press LLC,
Taylor and Francis Group,
6000 Broken Sound Parkway NW, Suite 300,
Boca Raton, FL 33487, USA

© 2008 A. Rahman
ISBN 978-1904445-45-6
USA ISBN 978-1-4200-7348-5

Printed by MPG Books Ltd, Bodmin, Cornwall

*In memory of my loving mother
who instilled in me the importance of education and
knowledge, whatever the situation. I hope that this book will
inspire readers to acquire knowledge and skill wherever they
may be and whatever their circumstances.*

# Contents

Preface ................................................................................................................. xi

Acknowledgements ........................................................................................... xii

PART I RADIATION SCIENCE

**Chapter 1: Radiation** .................................................................................... 1
   1.1 Types of radiation ................................................................................ 1
   1.2 Radiation quantities ............................................................................. 4
   1.3 Interaction of radiation with matter ..................................................... 8

**Chapter 2: Biological effects of radiation** ................................................. 28
   2.1 Modes of exposure .............................................................................. 28
   2.2 Effects of radiation ............................................................................. 28
   2.3 Dose–response characteristics ............................................................ 29
   2.4 Internal irradiation .............................................................................. 33
   2.5 Radiotoxicity ....................................................................................... 35
   2.6 Mechanisms of biological damage ..................................................... 36

**Chapter 3: Radiological protection** ........................................................... 43
   3.1 The aim of radiological protection ..................................................... 43
   3.2 Dosimetric quantities ......................................................................... 44
   3.3 System of radiological protection ....................................................... 49
   3.4 Radiological protection in practice ..................................................... 50
   3.5 UK MoD requirements ....................................................................... 53
   3.6 MoD standards ................................................................................... 53

**Chapter 4: Statistical methods** .................................................................. 57
   4.1 Basic statistical quantities .................................................................. 57
   4.2 Probability distributions ..................................................................... 62
   4.3 Sampling design ................................................................................. 70
   4.4 Test of hypotheses ............................................................................. 73

PART II DECOMMISSIONING

**Chapter 5: Decommissioning of nuclear facilities** ................................... 79
   5.1 Introduction ........................................................................................ 79
   5.2 Facilities to be decommissioned ........................................................ 80
   5.3 Scale of the decommissioning problem .............................................. 81
   5.4 Reasons for decommissioning ........................................................... 82
   5.5 Overall decommissioning strategies at the
       end of the useful life ........................................................................ 82
   5.6 Decommissioning operations ............................................................. 83
   5.7 Nuclear decommissioning authority ................................................... 85
   5.8 Decommissioning of nuclear facilities ............................................... 86
   5.9 Decommissioning of nuclear submarines ........................................... 92

**Chapter 6: Regulatory aspects in decommissioning** ...................................96
  6.1 Introduction ....................................................................................96
  6.2 International organisations ...........................................................97
  6.3 UK national regulations ..............................................................102
  6.4 Regulatory controls ....................................................................113
  6.5 UK government policy .................................................................114
  6.6 MoD regulatory regime ..............................................................115

**Chapter 7: Safety aspects in decommissioning** ..............................122
  7.1 Introduction ................................................................................122
  7.2 Safety objectives ........................................................................123
  7.3 Strategy for achieving objectives ..............................................123
  7.4 Technical safety ..........................................................................124
  7.5 Radiological protection ...............................................................128
  7.6 Application of the ALARA/ALARP principle .............................131
  7.7 Safety from chemical hazards ....................................................140
  7.8 Safety from industrial hazards ...................................................145
  7.9 Safety documentation .................................................................146
  7.10 Quality assurance .....................................................................148

**Chapter 8: Financial aspects of decommissioning** ........................151
  8.1 Introduction ................................................................................151
  8.2 Overall decommissioning cost estimation ................................152
  8.3 Methodologies and techniques for
      decommissioning cost estimation ..............................................153
  8.4 Factors influencing decommissioning cost estimate .................155
  8.5 Cost estimation guidelines .........................................................157
  8.6 European cost estimate methodology .........................................158
  8.7 Discounting technique .................................................................163
  8.8 Funding mechanisms ..................................................................167
  8.9 Fund Management .......................................................................168

**Chapter 9: Project management** ...................................................170
  9.1 Introduction ................................................................................170
  9.2 General project management ......................................................171
  9.3 Nuclear decommissioning project management .........................172
  9.4 Decommissioning strategy ..........................................................173
  9.5 Project management plan ............................................................173
  9.6 Schedule management .................................................................175
  9.7 Cost management ........................................................................178
  9.8 Qualitymanagement ....................................................................180
  9.9 Human resource management .....................................................181
  9.10 Information management system ...............................................183

9.11  Risk management  ................................................184
9.12  Contract and procurement management  ..................184

**Chapter 10: Planning for decommissioning**  ...............191
10.1  Introduction  ..................................................191
10.2  Decommissioning options  ................................191
10.3  Detailed planning description  .........................192

**Chapter 11: Site/facility characterisation**  ................200
11.1  Introduction  ..................................................200
11.2  Characterisation objectives  ............................201
11.3  Process of characterisation  .............................201
11.4  Methods and techniques of characterisation  ......203
11.5  Instrumentation  .............................................204
11.6  Direct measurement  .......................................208
11.7  Scanning  ......................................................209
11.8  Sampling  ......................................................211
11.9  Statistical evaluation  ......................................213
11.10  Computer calculations  ..................................214
11.11  BR 3 characterisation  ...................................214

**Chapter 12: Environmental impact assessment
             and best practicable environmental option**  ........218
12.1  Introduction  ..................................................218
12.2  Purpose of the EIA process  ............................219
12.3  EIA: a phased process  ....................................220
12.4  Implementation of environmental impact assessment  ..............227
12.5  ES preparation  ..............................................233
12.6  Best practicable environmental option  ..............235

**Chapter 13: Decontamination Techniques**  ...............240
13.1  Introduction  ..................................................240
13.2  Chemical decontamination  ..............................240
13.3  Mechanical decontamination  ...........................244
13.4  Other decontamination techniques  ...................249
13.5  Emerging technologies  ...................................249

**Chapter 14: Dismantling techniques**  ......................252
14.1  Introduction  ..................................................252
14.2  Mechanical cutting tools  .................................252
14.3  Thermal cutting tools  .....................................256
14.4  Abrasive water jet cutting  ...............................257
14.5  Electrical cutting techniques  ............................262

14.6 Emerging technologies .................................................262
14.7 Mechanical demolition techniques .............................265
14.8 Developments in the EU ........................................266

**Chapter 15: Case histories and lessons learnt** ..............268
15.1 Introduction .........................................................268
15.2 EU decommissioning pilot projects .........................268
15.3 KGR Greifswald, Germany ...................................277
15.4 Calder Hall in the UK ..........................................282

PART III RADIOACTIVE WASTE MANAGEMENT
**Chapter 16: Radioactive waste classification and inventory** .............288
16.1 Introduction .........................................................288
16.2 Radioactive substances and radioactive waste ...........288
16.3 Classification of radioactive waste .........................289
16.4 Waste inventory in the UK ...................................295
16.5 Partitioning and transmutation ..............................304
16.6 Waste inventory produced by the CoRWM ...............305

**Chapter 17: Management of radioactive waste** ...............308
17.1 Introduction .........................................................308
17.2 General principles of waste management .................308
17.3 Regulatory issues of waste management ...................310
17.4 Exemption and clearance levels .............................317
17.5 Environmental discharge .......................................328

**Chapter 18: Treatment and Conditioning of Radioactive Waste** ..........331
18.1 Introduction .........................................................331
18.2 Operations Preceding Treatment and Conditioning .....332
18.3 Treatment of Waste ...............................................333
18.4 Conditioning of Waste ...........................................345
18.5 Characterisation of Conditioned Waste ...................356

**Chapter 19: Storage and transportation of radioactive waste** ..............358
19.1 Storage ................................................................358
19.2 Safety Aspects of Radioactive Material Transport ......359
19.3 Transportation of radioactive materials ...................361
19.4 Waste packages ....................................................366
19.5 Transportation of waste .........................................371

**Chapter 20: Disposal of radioactive waste** ....................378
20.1 Introduction .........................................................378

20.2 Pre-disposal management ...................................................................379
20.3 Safety objectives ..............................................................................379
20.4 Disposal concepts and systems ..........................................................380
20.5 Multi-barrier concept .......................................................................381
20.6 Waste emplacement options ..............................................................388
20.7 Disposal practices ............................................................................390
20.8 Performance assessment of disposal system ......................................394
20.9 Disposal options for nuclear submarines ..........................................397

**Appendix 1: Abbreviations and acronyms** ...................................................401
**Appendix 2: Physical quantities, symbols and units** ..............................407
  Table A2.1  Physical quantities, symbols and units (SI) ..........................407
  Table A2.2  Physical constants, symbols and units (SI) ...........................407
  Table A2.3  SI prefixes ..........................................................................408

**Appendix 3: Properties of elements and radionuclides** ..........................409
  Table  A3.1  Properties of elements .......................................................409
  Table  A3.2  Half-lives and specific activities of some significant
                    radionuclides ...................................................................413
**Appendix 4: NDA management of decommissioning
                    activities in the UK** ...................................................420
  A4.1  NDA operating regime .................................................................420
  A4.2  NDA operating model .................................................................421
  A4.3  Supply chain ...............................................................................423

**Appendix 5: Institutional framework and regulatory standards in
                    decommissioning in selected EU Member States** ..............426
  Table A5.1  Overview of decommissioning regulatory issues
                    in selected EU Member States ........................................427

**Appendix 6: Multi-attribute utility analysis:
                    A major decision-aid technique** .........................................430
  A6.1  Introduction ................................................................................430
  A6.2  MUA methodology .......................................................................431
  A6.3  Steps in the MUA ........................................................................432
  A6.4  Outcome of the analysis ..............................................................437
  A6.5  Discussion and conclusions .........................................................438
  A6.6  Glossary of terms ........................................................................440

**Index** ..........................................................................................................445

# PREFACE

More than a decade ago I was requested by the Director of the MoD's nuclear training establishment (Royal Naval College/HMS SULTAN) to offer a training course on 'Nuclear Decommissioning and Radioactive Waste Management' at graduate and post-graduate level. In preparing for the training course, I acutely felt the need for a standard text book on the subject which was lacking. Subsequently, during the period of my work in the nuclear industry, I prepared a training course on the same subject, in association with the British Nuclear energy Society, which was offered to the British nuclear industry. As a partner in a consortium of European nuclear decommissioning organisations under the 6th Framework Programme of the European Commission, I was involved as the editor-in-chief in offering training courses to the European nuclear industry. It became apparent that an up-to-date book dealing with all aspects of decommissioning, both in the UK and the EU, was needed.

The subject matter is certainly approached from a practical perspective. As nuclear decommissioning and radioactive waste management are safety oriented and regulatory driven, safety issues form the underlying theme of the book. Decommissioning activities covering radiological protection, nuclear and non-nuclear safety, national and international legislation, project management as well as practical issues such as decontamination, dismantling and waste management all require fundamental understanding before undertaking practical work. This book deals with all of these issues maintaining a balance between academic knowledge and industrial experience to help workers perform decommissioning activities safely, securely and cost effectively.

In order to cover all aspects of nuclear decommissioning, radioactive waste management and associated issues, the book has been divided into three parts. Part One deals with radiation science, which is the foundation for the rest of the book, and includes the basic concepts of radiation, the biological effects of radiation, radiological protection and statistical methods.

Part Two introduces decommissioning, with its multifarious activities, detailing the extent of decommissioning activities both in the UK and abroad; UK, European and international legislation; the implementation of safety procedures against radiation hazards, as well as the industrial, chemical and biological hazards. The financial and environmental considerations, project management, decontamination and dismantling techniques are also discussed comprehensively. This part of the book concludes with case studies of decommissioning activities and presents lessons learnt − of importance to the future planning of decommissioning projects.

Part Three deals specifically with radioactive waste management from regulatory requirements and implementation of these requirements to the treatment, storage, transportation, and finally, the methods of disposal of the various categories of radioactive wastes.

One major impediment to the revival of the nuclear industry is the lack of trained workers. During the past two to three decades, when the nuclear industry was, at best, stagnating, there was no influx of new blood into the industry. Now that an anticipated upturn, both in decommissioning activities and in new build, is underway, there is an urgent need to address this skill shortage. This book is intended to provide a comprehensive overview for those currently working in this vital sector or studying with the intent to join the industry.

If the issues associated with nuclear waste disposal can be resolved in a safe, secure and cost-effective manner, public concerns regarding nuclear matters will be allayed and the nuclear industry will surge forward. This is indeed a timely book in view of the current climate concerns and the ever-growing demand for energy supplies.

*Dr A Rahman*

# Acknowledgments

In the course of writing this book, I have been greatly helped by a number of people as well as organisations over a number of years. Particular mention must be made of my daughter, Zara, who has scrupulously and meticulously carried out checking and cross checking of the book at various stages of production. I am also indebted to the International Atomic Energy Agency (IAEA) for permission to use their material in the book. I must also acknowledge my gratitude to the Nuclear Decommissioning Authority (NDA) for their review of the NDA-related material used in the book. I should also mention of the support given by Atkins Limited by sponsoring the book. Last but not least, mention should be made of the publisher whose steadfast support and optimism helped to steer the book to successful completion.

# 1

# RADIATION

## 1.1 Types of Radiation

The phenomenon of emission of energy in the form of waves or particles and its transmission through air or vacuum is generally called radiation. Radiation can take the form of electromagnetic waves such as X-rays, γ-rays (gamma rays), ultraviolet rays, radio waves etc. It can also be a stream of sub-atomic particles, such as α-particles (alpha particles), β-particles (beta particles) and neutrons, which are collectively known as the particulate radiation. All of these radiations can be either ionising or non-ionising.

### 1.1.1 Electromagnetic Radiation

The electromagnetic radiation consists of waves which are characterised by a wavelength, $\lambda$, and a frequency, $v$ ($c = \lambda v$, where c is the velocity of light $= 3 \times 10^8$ m.s$^{-1}$). The quantum of an electromagnetic radiation is known as a **photon** and its energy is given by

$$E = hv \qquad (1.1)$$

where h is the Planck constant ($= 6.626 \times 10^{-34}$ J.s) and $v$ is the frequency of the wave (Hz).

X-rays and γ-rays are electromagnetic waves of very high frequencies ($> 10^{18}$ Hz) and consequently their photon energies are also high: $\sim 6.6 \times 10^{-16}$ J = 4 keV; ($1.6 \times 10^{-19}$ J = 1 eV). If we now compare this energy with the photon energy of visible light whose frequency is around $10^{15}$ Hz; we obtain $E = hv = (6.6 \times 10^{-34}$ J.s $\times 10^{15}$ s$^{-1}$)/($1.6 \times 10^{-19}$ J.eV$^{-1}$) = 4 eV. This amount of energy is lower than the binding energy of an electron ($\sim 34$ eV) in an atom of an air particle and consequently the energy of the visible light cannot cause any ionisation (The process of ionisation is explained in Section 1.1.3.). The electromagnetic radiation spectrum is shown in Figure 1.1.

The photon of electromagnetic waves can be viewed as an energy packet or as a particle carrying an energy $E$ (wave–particle duality). If the energy $E$ of the photon is high enough to cause ionisation in the target atom, then that radiation is **ionising,** otherwise it is non-ionising.

Nucleons in an atom, like electrons, have energy levels. However, the energy levels of the nucleons are much higher than those of electrons. When

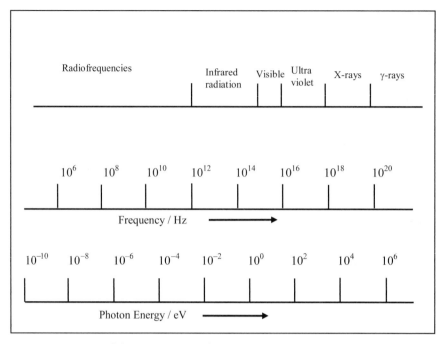

**Fig. 1.1** *Spectrum of electromagnetic radiation*

a nucleon moves from a higher energy level to a lower energy level, γ-rays are emitted. The γ-rays that are emitted are of the order of keV or more. X-rays, on the other hand, originate from the transition of atomic electrons from a higher energy state to a lower energy state and hence the X-ray energy is lower than the γ-ray energy.

## 1.1.2 Particulate Radiation

The particulate radiation consists of sub-atomic particles such as α- and β-particles, which are charged, or neutrons which are uncharged.

An α-particle, consisting of two protons and two neutrons bound together very strongly as a single entity, is a massive particle in nuclear dimensions. For a given energy, its large mass means a low velocity which, when coupled with its double charge (from two protons), means that it has a high ionisation capacity and consequently a short range.

A β-particle, on the other hand, is a singly charged particle and is much lighter than an α-particle. It can be either negatively charged or positively charged. When it is negatively charged it is called an electron and when it is positively charged it is called a positron. The β-particles arising from the disintegration of a radionuclide are emitted with a continuous spectrum having a characteristic end-point energy, $E_{max}$. For a given energy, the low mass of the β-particle results in a high velocity and low ionisation capacity and hence a long range.

**Table 1.1** *Characteristics of particulate radiation*

| Radiation | Origin | Atomic mass unit | Electronic charge | Range | |
|---|---|---|---|---|---|
| | | | | Air (m) | Tissue (m) |
| Alpha | Radioactive decay | 4 | 2 | $3 \times 10^{-2}$ | $4 \times 10^{-5}$ |
| Beta | Radioactive decay | $5.5 \times 10^{-4}$ | $\pm 1$ | 3.0 | $1 \times 10^{-2}$ |
| Fast neutron | Fission | 1 | 0 | $(L)_{water} = 6 \times 10^{-2}(*)$ | |
| Thermal neutron | Moderation | 1 | 0 | $(L)_{water} = 3 \times 10^{-2}(**)$ | |

\* This length represents the straight line distance between the point of introduction of a fast neutron in a medium and its thermalisation. It is known as the fast-diffusion length

\*\* This is the thermal diffusion length which is evaluated as $1/\sqrt{6}$ of the average distance from thermalisation to absorption of neutrons

The neutrons are mostly produced from the fission of fissile nuclides and carry a continuous spectrum of energy up to about a few MeV. The neutrons of higher energy may, however, be produced by the fusion process. As neutrons are uncharged particles, they do not react electrically with the electron cloud of the atom. They undergo physical collisions with atomic nuclei losing energy in every encounter. The phenomenon of a neutron losing energy as a result of scattering by atomic nuclei is commonly referred to as thermalisation. Following collisions, the neutrons may be absorbed by the target nucleus. The thermalised as well as non-thermalised neutrons, when absorbed, produce compound nuclei which may eventually break up giving rise to nuclear radia-tion in the form of γ-photons, neutrons, charged particles and recoil nuclei. Each of these radiations can cause ionisation in target materials. Neutrons, because of their various modes of interactions with target atoms and conse-quent generation of various types of secondary radiation, are difficult particles to shield and hence they are considered significant from the radiological point of view. Table 1.1 shows the origin and properties of various types of particu-late radiation.

## 1.1.3 Ionising and Non-ionising Radiation

The absorption of radiation, either particulate or electromagnetic, involves the transfer of energy from the incoming radiation to the target atom. In the case of particulate radiation, the energy transferred is its kinetic energy whereas

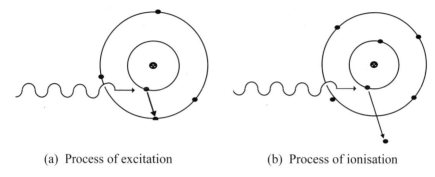

(a) Process of excitation          (b) Process of ionisation

**Fig. 1.2** *Processes of excitation and ionisation*

in the case of electromagnetic radiation, it is the energy of photons which are scattered and/or absorbed by the target atoms. If the energy of the incident radiation is low, the transferred energy will also be low. This may result in the electrons of the target atom only moving from a lower energy state (inner shell) to a higher energy state (outer shell) while remaining bound to the atom. This process is called **excitation**. On the other hand, if the incident energy and hence the transferred energy is high, the electrons may be ejected from the atomic orbits leaving the initial atom positively charged. This process is known as **ionisation**. Figure 1.2 shows the processes of excitation and ionisation.

Ionisation may cause the break-up of the atomic and molecular structure of the target material. In the case of organic material, ionisation can cause the break-up of the cellular structure which may result in malfunctioning of the cell leading to adverse biological effects. This is precisely the reason why radiological protection is needed to protect living cells from the harmful effects of ionising radiation. If the radiation is non-ionising, such as that encountered in ordinary visible light or low frequency electromagnetic waves, the biological damage arises from different mechanisms and the consequences to the cells are very much smaller.

## 1.2 Radiation Quantities

### 1.2.1 Nuclides and Isotopes

The word **nuclide** is a generic term used to specify a nuclear species with a specific atomic number, Z and a specific mass number, A and which is in a defined nuclear state. A nuclide can be either stable or unstable. If the nuclide happens to be unstable, it is radioactive and it is referred to as a **radionuclide**. At present there are known to be about 2500 nuclides, of which about 800 have been found to be radioactive.

Although all the atoms of a particular element contain the same number of protons, they may occur with different numbers of neutrons. The atoms of the same element (same number of protons) but with different number of neutrons (hence different mass numbers) are called **isotopes**. For example, phosphorus-31 with the mass number of 31 and phosphorus-32 with a mass number of 32 are Phosphorus isotopes. Similarly C-12, C-13 and C-14 are carbon isotopes.

## 1.2.2 Radioactivity

The **radioactivity** (in short, **activity**), $A$ of a radioactive material is defined as the rate of disintegration of its nuclei. Thus the activity, $A$ is given by

$$A = \frac{dN}{dt} = N\lambda \tag{1.2}$$

where $N$ is the number of atoms or nuclei and $\lambda$ is the decay constant, which is defined as the probability of decay per unit time of a radionuclide. This decay constant, $\lambda$ is the fundamental characteristic of a radionuclide. The unit of $\lambda$ is the reciprocal of time. From the above equation, one can write

$$-dN = N\lambda\,dt$$

where the negative sign has been inserted in order to show explicitly that a number, $dN$, of atoms decay in the time interval, $dt$. From this equation it can be shown that

$$N = N_0 \exp(-\lambda t) \tag{1.3a}$$

Multiplying both sides by the decay constant, $\lambda$, one obtains

$$A = A_0 \exp(-\lambda t) \tag{1.3b}$$

where $A_0$ is the activity at $t = 0$.

Thus the activity of a radionuclide follows an exponential decay law which is characterised by the decay constant. The decay constant, $\lambda$ is related to the half-life, $t_{1/2}$ of a radionuclide by

$$\lambda = \frac{\ln 2}{t_{1/2}} = \frac{0.693}{t_{1/2}} \tag{1.4}$$

The unit of activity is becquerel (Bq), which is equal to one disintegration per second. The old unit of activity was curie (Ci), defined as the activity of one gram of radium (Ra-226), which decays at the rate of $3.7 \times 10^{10}$ per second. Thus

$$1\,\text{Ci} = 3.7 \times 10^{10}\ \text{Bq} \tag{1.5}$$

The specific activity of a material is defined as the activity per unit mass and hence its unit is Bq.kg$^{-1}$. This is also referred to as the mass activity concentration. When there are a number of elements in the material, the mass activity concentration (Bq.kg$^{-1}$) based on the total activity divided by the total mass is relevant.

The other definition of specific activity is, according to the definition given by the International Commission on Radiation Units and Measurements (ICRU), the activity of a radionuclide per unit mass ($Bq.kg^{-1}$) calculated from atomic or molecular considerations. It is given by

$$\text{specific activity} = \frac{\left(\text{Avogadro's constant}\right) \times \lambda}{\text{mole}} = \qquad (1.6)$$

where Avogadro's constant (also called Avogadro number), $N_A = 6.023 \times 10^{23}$ per mole.

This quantity is useful when a specific radionuclide is under consideration. The mole is the atomic or molecular mass in grams. For example, a mole of C-12 is 12 g or 0.012 kg. The value of $\lambda$ can be calculated using equation (1.4). The specific activities of various radionuclides are given in Table A3.2 of Appendix 3. The basic physical quantities and fundamental constants, their symbols and values used in this book are given in Appendix 2. The detailed description and definition of physical quantities can be found in [1–3].

### 1.2.3 Particle Fluence Rate or Flux

For a mono-directional beam of particles, the particle fluence rate, $\phi$ is the number of particles crossing a unit area normal to the beam in unit time. It has the dimension of $m^{-2}.s^{-1}$. The fluence, $\Phi$ is the time integral of fluence rate and has the dimension of $m^{-2}$.

For multi-directional particles, the fluence rate, $\phi$ at a point is the number of particles crossing the surface of a sphere of unit diametral area in unit time. Thus the particle fluence rate, $\phi$ is

$$\phi = \frac{dN}{dt.da} = \frac{dN}{dt.4\pi r^2} \qquad (1.7)$$

where $r$ is the radius of the sphere (m), and $da$ is the surface area of the sphere ($m^2$). The particle fluence, $\Phi$ for multi-directional particles is given by

$$\Phi = \frac{dN}{4\pi r^2} \qquad (1.8)$$

In many textbooks, particle fluence rate and flux are used synonymously. But the ICRU defines **flux** as the number of particles incident on a target material in unit time [4].

### 1.2.4  Energy Fluence Rate

The energy fluence rate, $\psi$ is defined as the quotient of energy of particles crossing a sphere of unit diametral area in unit time. Thus

$$\psi = \frac{dE}{dt.da} \qquad (1.9)$$

where $da$ is the surface area of the sphere (m²). So the unit of $\psi$ is J.m⁻².s⁻¹ = W.m⁻².

The energy fluence, $\Psi$ is

$$\psi = \frac{d E}{da} \tag{1.10}$$

The unit of energy fluence is J.m⁻².

## 1.2.5 Specific Ionisation and Stopping Power

These two parameters, **specific ionisation and stopping power,** are characteristics of directly ionising radiation. When the incoming radiation carries a sufficiently high energy to cause ionisation in the target atoms, then that radiation is termed a directly ionising radiation. This radiation loses energy by interacting directly with the target atoms. This energy loss goes to the target atoms in the form of absorbed energy. Normally charged particles are the directly ionising radiation.

The stopping power, $S$ of a material for the charged particle is defined as the energy loss per unit length within the medium. A charged particle travelling at a high velocity will lose energy not only due to collision but also due to radiative loss such as bremsstrahlung. The bremsstrahlung (the German word meaning 'breaking radiation' or 'deceleration radiation') is the electromagnetic radiation which is produced when a fast moving charged particle such as an electron loses energy upon changing velocity and is deflected by the electric field surrounding a positively charged atomic nucleus. So $S$ can be written as [1]

$$S = \left[\frac{dE}{dl}\right]_{col} + \left[\frac{dE}{dl}\right]_{rad} \tag{1.11}$$

The unit of $S$ is J.m⁻¹.

The specific ionisation, $I$ of a particle is defined as the number of ion pairs produced per unit path length of the particle in a medium. If $W$ is the average energy required to produce an ion pair, then the specific ionisation, $I$ is given by

$$I = \frac{\left[\frac{dE}{dl}\right]_{col}}{W} \frac{\frac{J}{m}}{\frac{J}{\text{ion pair}}} \frac{\left[\frac{dE}{dl}\right]_{col}}{W} \text{ ion pairs}/m \tag{1.12}$$

Conversely, one can say that the energy loss per unit path length due to collision is given by the product of $I$ and $W$. The specific ionisation of α-particles of 1 MeV energy in air at normal temperature and pressure is about 4000 ion pairs/mm, whereas for β-particles of the same energy in air it is about 10 ion pairs/mm.

## 1.2.6 Linear Energy Transfer

When an ionising radiation traverses a medium, the charged particles (primary and/or secondary) bounce from one site of interaction to another and leave a trail of ionised particles. The extent of the ionisation depends on the type and energy of the incoming particle. The **Linear Energy Transfer (LET)** of a particle is defined as the average energy deposited along the track of the particle per unit length. Thus the LET is directly related to the degree of ionisation and hence to the biological effectiveness. The LET, $L_\infty$ is given by

$$L_\infty = \frac{dE}{dl} \tag{1.13}$$

where $dE$ is the energy (J) lost by a charged particle locally and $dl$ is the distance (m) traversed by the charged particle. Thus the unit of LET is $J.m^{-1}$.

It should be noted that the LET is almost equal to the stopping power defined in equation (1.11) except that the LET does not include any radiative loss from the incoming particle.

The magnitude of LET increases rapidly with the mass and energy of the incoming particle. The LET of $\alpha$-particles is considerably larger than that of electrons of the same energy. For example, the LET of a 1 MeV $\alpha$-particle in water is about 90 $keV.\mu m^{-1}$, while it is only 0.19 $keV.\mu m^{-1}$ for a 1 MeV electron. For this reason, $\alpha$-particles and other multiply charged particles are referred to as high LET radiations, whereas electrons and other sparsely ionising radiation such as $\gamma$-rays and X-rays are low LET radiations. As neutron interactions with materials lead to the generation of heavy, high LET-charged particles, so neutrons are also considered to be high LET radiation.

## 1.3 Interaction of Radiation with Matter

When a radiation (particulate or electromagnetic) strikes a medium, it initiates a variety of physical processes which depend on the type of radiation, its energy, the type of target material etc. The charged particles such as $\alpha$- and $\beta$-particles, Auger and internal conversion electrons, fission fragments etc. cause ionisation (as well as excitation) in the atoms of the target material by direct electrical interaction and hence they are called the **directly ionising radiation**. On the other hand, neutrons and photons (except in photoelectric effect) do not ionise target atoms directly but give rise to particles which do and hence they are referred to as **indirectly ionising radiation**.

When an indirectly ionising radiation strikes a medium, it may penetrate the medium, if the medium is thin enough, without any collision or interaction and hence suffers little degradation in energy. However, it may undergo some sort of interaction such as scattering and hence becomes somewhat degraded in energy. In the case of electromagnetic radiation, the scattered photons would have lower energies and hence lower frequencies whereas in the case of neutrons, the scattered neutrons would carry lower kinetic energies. It should be noted that scattering of neutrons may also accompany γ-rays. The mechanisms of the interaction of γ-rays with matter are quite different from those of neutron interactions. Hence these two processes will be discussed separately.

## 1.3.1 Processes of γ-Ray Interactions with Matter

The term 'γ-ray' is used here to include both the γ-rays originating from the nuclei and the X-rays originating from the atom due to the transition of atomic electrons. Both of these radiations are electromagnetic in nature, they differ only in their energies and hence frequencies [1].

The **photoelectric effect, Compton effect** and **pair production** are the three processes by which γ-rays interact with matter. In the photoelectric process, the incident γ-ray interacts with the entire atom, the γ-photon disappears, the atom recoils and one of the atomic electrons is ejected from the atom. The ejected electron is called the photoelectron. This process dominates at low photon energies. In the Compton effect, the incident γ-photon undergoes an elastic scattering by an electron; the photon does not disappear, it only loses part of its energy. Due to this loss of energy, the frequency of the photon decreases and the electron recoils. This process becomes significant when the incident photon energies range from a few hundred keV to about 1 MeV. In pair production, the incident γ-photon having a minimum energy of 1.02 MeV disappears and an electron–positron pair is produced. The excess energy of the photon, i.e. over and above 1.02 MeV, is carried away by the electron–positron pair as their kinetic energies. Thus when an electron and a positron annihilate each other, two photons, each of 0.511 MeV, are generated. Pair production dominates at high energies of the photons (>1.02 MeV).

The probabilities of these processes taking place per atom are quantified by their respective cross-sections. The total cross-section, $\sigma$ (m$^2$) is the sum of these respective cross-sections.

$$\sigma = \left( \sigma_{pe} + \sigma_{ce} + \sigma_{pp} \right) \qquad (1.14)$$

where $\sigma_{pe}$, $\sigma_{ce}$ and $\sigma_{pp}$ are, respectively, the photoelectric, Compton and pair production cross-sections (m$^2$).

When these cross-section are multiplied by the atom density, $N$ (the number of atoms per unit volume) of the target material, a quantity called the attenuation coefficient is obtained. The **total attenuation coefficient,** $\mu$ giving the probability of removal of a photon from the beam by any of the above-mentioned interaction processes is given by

$$\mu = N\sigma = N\left( \sigma_{pe} + \sigma_{ce} + \sigma_{pp} \right) = \mu_{pe} + \mu_{ce} + \mu_{pp} \qquad (1.15)$$

where $\mu$ is the total attenuation coefficient (m$^{-1}$) and $\mu_{pe}$, $\mu_{ce}$ and $\mu_{pp}$ are the attenuation coefficients of the photoelectric, Compton and pair production effects respectively.

The parameter, $\mu$ may be shown mathematically to quantify the fraction of the total energy in the beam that is removed in a unit distance of the attenuating material. The ratio of the total attenuation coefficient, $\mu$ to the density, $\rho$ of the attenuating material is called the **mass attenuation coefficient,** $\mu/\rho$ (m$^2$.kg$^{-1}$).

The energy of the incoming photon that is deposited to the absorber is the kinetic energy of the photoelectron following the photoelectric effect, the Compton energy that is not scattered away and the kinetic energies of the electron–positron pair following pair production. The energy generated by the annihilation of the electron–positron pair (1.02 MeV) may not be absorbed within the medium. The fraction of energy that is absorbed within the medium per unit distance is given by $\mu_a$.

$$\mu_a = \mu_{pe} + \mu_{ce}(1-x) + \mu_{pp}(h\nu - 1.02)/h\nu \qquad (1.16)$$

where $\mu_a$ is the total absorption coefficient (m$^{-1}$), $\mu_{pe}$, $\mu_{ce}$ and $\mu_{pp}$ are the attenuation coefficients from the three processes and $x$ is the fraction of the Compton energy that is scattered.

This quantity, $\mu_a$ is called the **total absorption coefficient.** When the absorption coefficient $\mu_a$ is divided by the density, $\rho$ of the material, a quantity, called the **mass absorption coefficient,** $\mu_a/\rho$ (m$^2$.kg$^{-1}$) is obtained, where $\rho$ is the density of the absorbing material (kg.m$^{-3}$).

Thus, it is obvious from equations (1.15) and (1.16) that the total absorption coefficient is smaller than the total attenuation coefficient. (For brevity, the

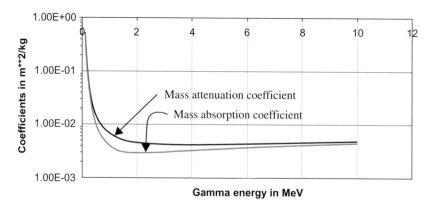

**Fig. 1.3** *Mass attenuation and mass absorption coefficients for Pb*

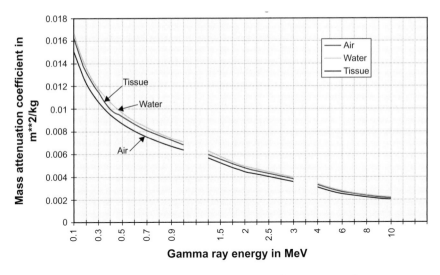

**Fig. 1.4** *Mass attenuation coefficients for air, water and tissue material*

word 'total' will be omitted from these two quantities). In other words, all the attenuated energy, i.e. the energy that is removed from the incoming γ-rays is not all absorbed within the thickness of the target material. The variation in the mass attenuation and mass absorption coefficients for Pb as a function of energy is shown in Figure 1.3.

Graphical presentations of the mass attenuation and mass absorption coefficients for air, water and tissue materials are shown in Figures 1.4 and 1.5, respectively.

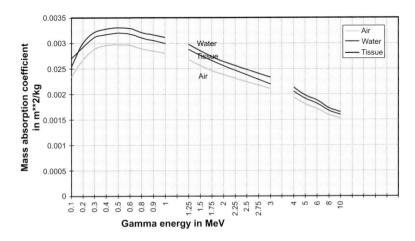

**Fig. 1.5** *Mass absorption coefficients for air, water and tissue materials*

## 1.3.2 Dose Calculations from γ-Rays

### 1.3.2.1 Free Space Dose Calculation

If $E$ is the energy of the γ-rays (J) and $\phi$ is the mono-energetic γ-ray fluence rate (number of photons.m$^{-2}$.s$^{-1}$), then the energy deposition rate per unit mass, (Gy.s$^{-1}$) is given by:

$$\bar{D} = E\,\phi \left( \frac{\mu_a}{\rho} \right) \text{Gy.s}^{-1} \tag{1.17}$$

where $\mu_a/\rho$ is the mass absorption coefficient (m$^2$.kg$^{-1}$ ).

If $\mu_a/\rho$ for air is used in the above equation, then the absorption dose rate represents the free space dose rate without any attenuation.

It should be noted that equation (1.17) only applies to a mono-energetic beam. If γ-rays have a distribution of energies, it is necessary to integrate equation (1.17) over the whole spectrum if it is continuous, or to sum over the spectrum if it is discrete. For a discrete spectrum the dose rate is

$$\bar{D} = \sum_i E_i\,\phi_i \left( \frac{\mu_a}{\rho} \right)_i \text{Gy.s}^{-1} \tag{1.18}$$

where $\phi$ is the fluence rate of γ-rays at energy $E_i$ and $(\mu_a/\rho)_i$ is the mass absorption coefficient at energy $E_i$ .

The total absorbed dose over a period, $T$ can be obtained by integrating the above equation with respect to time

$$D = \int_0^T \bar{D}\,dt = \int_0^T E \left( \frac{\mu_a}{\rho} \right) \phi(t)\,dt$$

If $E$ is constant in time, then

$$D = \int_0^T E \left( \frac{\mu_a}{\rho} \right) \phi(t)\,dt = E \left( \frac{\mu_a}{\rho} \right) \int_0^T \phi(t)\,dt$$

$$= E\,\Phi \left( \frac{\mu_a}{\rho} \right) \tag{1.19}$$

where $\Phi = \int_0^T \phi(t)\,dt$

is called the γ-ray fluence and has the dimension of photons.m$^{-2}$. The term $\Phi E$ is called the energy fluence.

## Example 1.1

A radioactive waste steel container contains an amount of activity which gives $3 \times 10^{15}$ photons.m$^{-2}$.s$^{-1}$ at the surface, with an average energy of 0.8 MeV. What is the energy deposition rate at the surface of the container?

Given $\left[ \dfrac{\mu_a}{\rho} = 0.00274 \ \text{m}^2 \, \text{kg}^{-1} \right]$

*Solution*

$$E = 0.8 \ \text{MeV} = 0.8 \times 1.6 \times 10^{-13} \ \text{J} = 1.28 \times 10^{-13} \ \text{J}$$
$$\phi = 3 \times 10^{15} \ \text{photons.m}^{-2}.\text{s}^{-1}$$
$$\mu_a / \rho = 0.00274 \ \text{m}^2.\text{kg}^{-1}$$

Now using equation (1.17)

$$\bar{D} = 1.28 \times 10^{-13} \ \text{J} \times 3 \times 10^{15} \ \text{photons} / \left( \text{m}^2.\text{s} \right) \times 0.00274 \ \text{m}^2.\text{kg}^{-1}$$

$$= 1.05 \ \text{J.kg}^{-1}.\text{s}^{-1} = 1.05 \ \text{Gy.s}^{-1}$$

## Example 1.2

Calculate the absorbed dose rate at a distance of 3 m from a 240 MBq Co-60 point source.

*Solution*

The energy, $E$ of the $\gamma$-radiation from Co-60 is

$$E = 1.17 + 1.33 = 2.5 \ \text{MeV} = 2.5 \times 1.6 \times 10^{-13} \ \text{J} = 4.0 \times 10^{-13} \ \text{J}$$

$$\phi = \frac{240 \times 10^6}{4\pi \, r^2} \ \text{Bq} = \frac{240 \times 10^6}{4\pi \, 3^2} \ \text{Bq.m}^{-2}$$

$$= 2.12 \times 10^6 \ \text{m}^{-2}.\text{s}^{-1}$$

From the standard table, $\mu_a / \rho$ for air at 2.5 MeV = 0.0023 m$^2$.kg$^{-1}$

So $\bar{D} = 4.0 \times 10^{-13} \ \text{J} \times 2.12 \times 10^6 \ \text{m}^{-2}.\text{s}^{-1} \times 0.0023 \ \text{m}^2.\text{kg}^{-1}$

$= 1.95 \times 10^{-9} \ \text{J.kg}^{-1}.\text{s}^{-1}$

$= 1.95 \times 10^{-3} \ \mu\text{Gy.s}^{-1}$

### 1.3.2.2 Dose Calculation from Attenuated γ-Rays

The γ-ray dose rate due to attenuation when it passes through a material of thickness, $t$ and total attenuation coefficient, $\mu$ is given by

$$\bar{D} = \bar{D}_0 \exp(-\mu t) \tag{1.20}$$

where $\bar{D}$ is the dose rate after the thickness $t$ and $\bar{D}_0$ is the dose rate without the material (Gy.s$^{-1}$), which is calculated using equation (1.17).

A quantity called the **half-value thickness** (HVT) or **half-value layer** (HVL) for a particular material is defined as the thickness of the material which would reduce the initial dose rate due to radiation to half its initial value. Putting $t_{1/2}$ as the HVL, one obtains

$$\frac{\bar{D}}{\bar{D}_0} = 0.5 = \exp(-\mu\, t_{1/2})$$

$$t_{1/2} = \frac{\ln 2}{\mu} = \frac{0.693}{\mu} \tag{1.21}$$

---

## Example 1.3

A Co-60 point source gives a dose rate of 400 μGy.h$^{-1}$ at 1 m. Find: (i) at what distance from the source should a barrier be placed if the dose rate is to be lower than or equal to 10 μGy.h$^{-1}$? (ii) what thickness of Pb would give the same level of protection at 1 m? Given: HVL of Pb for Co-60 is 12.5 mm.

*Solution*

(i) As dose rates from point sources are inversely proportional to the squares of distances i.e., $D_1 r_1^2 = D_2 r_2^2$

$$400 \times 1^2 = 10 \times r^2$$

$$r^2 = 40 \;\therefore\; r = 6.32 \text{ m}$$

(ii) As $t_{1/2} = 12.5 \text{ mm} = \dfrac{0.693}{\mu} \;\therefore\; \mu = \dfrac{0.693}{12.5} \text{ mm}^{-1}$

Using equation (1.20)

$$10 = 400 \exp(-\mu t) = 400 \exp\left(-\frac{0.693}{12.5} t\right)$$

or $\dfrac{10}{400} = \exp\left(-\dfrac{0.693}{12.5} t\right)$

Taking natural logarithms on both sides, $t = 66.5$ mm.

---

*(continued)*

It should be noted that the scattered γ-rays in the form of the build-up factor have not been included here. Inclusion of the build-up factor will increase the dose rate.

Obviously the value of the HVL depends on the material of the medium and on the energy of the incoming radiation. Another related quantity, called the **tenth-value layer**, $t_{1/10}$, is defined as the thickness required to reduce the dose to one-tenth of its initial value. Table 1.2 gives approximate values of $t_{1/2}$ and $t_{1/10}$, for Pb and water.

**Table 1.2** *Values of $t_{1/2}$ and $t_{1/10}$*

| Photon energy (MeV) | Lead (mm) $t_{1/2}$ | $t_{1/10}$ | Water (mm) $t_{1/2}$ | $t_{1/10}$ |
|---|---|---|---|---|
| 0.5 | 4 | 13 | 72 | 239 |
| 1.0 | 9 | 30 | 98 | 326 |
| 1.5 | 12 | 40 | 120 | 400 |
| 2.0 | 13 | 43 | 140 | 465 |
| 3.0 | 14.5 | 48 | 175 | 581 |
| 5.0 | 14 | 46.5 | 230 | 764 |
| 10.0 | 12.5 | 41.5 | 316 | 1050 |

### 1.3.2.3 Dose Calculation with Build-up Factors

Gamma-ray shields are used to reduce the doses to individuals or to reduce the γ-ray intensity exposing individuals. This is generally known as the **biological shield**. The dose calculation for the biological shield will be presented here.

It would be easy to calculate the γ-ray fluence rate if every time a photon interacted with matter it disappeared. Then $\phi$ would simply be the uncollided fluence rate which would be $\phi_0 \exp(-\mu t)$. Unfortunately γ-rays do not disappear at each interaction. In the Compton effect, they are merely scattered with some loss of energy. In the photoelectric effect, X-rays are quite often produced subsequent to this interaction. In pair-production, annihilation radiation which

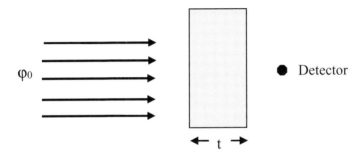

**Fig. 1.6** *Mono-directional beam of γ-rays incident on a slab shield*

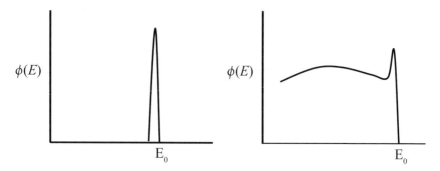

**Fig. 1.7** *Energy spectrum of incident γ-ray beam*

**Fig. 1.8** *Energy spectrum of γ-rays emerging from a shield*

produces two photons, each of 0.511 MeV, inevitably follows (see Section 1.3.1). A mono-energetic mono-directional beam of photons incident on a shield (see Figure 1.6), with an energy spectrum (see Figure 1.7) emerges from the shield with a continuous spectrum (see Figure 1.8). The peak at $E_0$ in Figure 1.8 is the unscattered photon fluence rate exponentially reduced from the fluence rate of Figure 1.7.

The fluence rate, $\phi(E)$ at any point is thus dependent on the energy spectrum of the incoming radiation as well as on the property of the shielding material. The dose rate can thus be written, similarly to equation (1.17), as

$$\bar{D} = \int_0^{E_0} E\, \phi(E) \left(\frac{\mu_a}{\rho}\right) dE$$

$$\bar{D} = \bar{D}_0 \, \exp(-\mu t)\, B_m(\mu t) \tag{1.22}$$

where $\bar{D}_0$ is the dose rate in the absence of the shield and $B_m(\mu t)$ is the exposure build-up factor for a mono-energetic beam in the presence of the shield.

The build-up factor for the fluence rate for a mono-energetic beam can be defined as

$$B_m(\mu t) = \frac{\text{total fluence rate at the detector}}{\text{uncollided fluence rate at the detector}} \tag{1.23}$$

Hence the total fluence rate (also known as build-up fluence rate) is the product of the uncollided fluence rate and $B_m(\mu t)$. So

$$\phi = \phi_u B_m(\mu t) = \phi_0 \, \exp(-\mu t) B_m(\mu t) \tag{1.24}$$

The values of $B_m(\mu t)$ are usually tabulated as the exposure build-up factor as a function of energy, as shown in Table 1.3 [1]. The quantity $\mu t$ (= $t/l$, where $l$ is the mean free path, i.e. the average photon travel distance between interactions) is quoted as the thickness of the shield material in terms of the number of mean free paths.

**Table 1.3** *Exposure buildup factor for plane mono-directional beam [1]*

| Material | $E_0$ / MeV | $\mu t$ | | | | |
|---|---|---|---|---|---|---|
| | | 1 | 2 | 4 | 7 | 10 |
| Water | 0.5 | 2.63 | 4.29 | 9.05 | 20.0 | 35.9 |
| | 1.0 | 2.26 | 3.39 | 6.27 | 11.5 | 18.0 |
| | 2.0 | 1.84 | 2.63 | 4.28 | 6.96 | 9.87 |
| | 3.0 | 1.69 | 2.31 | 3.57 | 5.51 | 7.48 |
| | 4.0 | 1.58 | 2.10 | 3.12 | 4.63 | 6.19 |
| | 6.0 | 1.45 | 1.86 | 2.63 | 3.76 | 4.86 |
| | 8.0 | 1.36 | 1.69 | 2.30 | 3.16 | 4.00 |
| Iron | 0.5 | 2.07 | 2.94 | 4.87 | 8.31 | 12.4 |
| | 1.0 | 1.92 | 2.74 | 4.57 | 7.81 | 11.6 |
| | 2.0 | 1.69 | 2.35 | 3.76 | 6.11 | 8.78 |
| | 3.0 | 1.58 | 2.13 | 3.32 | 5.26 | 7.41 |
| | 4.0 | 1.48 | 1.90 | 2.95 | 4.61 | 6.46 |
| | 6.0 | 1.35 | 1.71 | 2.48 | 3.81 | 5.35 |
| | 8.0 | 1.27 | 1.55 | 2.17 | 3.27 | 4.58 |
| Lead | 0.5 | 1.24 | 1.39 | 1.63 | 1.87 | 2.08 |
| | 1.0 | 1.38 | 1.68 | 2.18 | 2.80 | 3.40 |
| | 2.0 | 1.40 | 1.76 | 2.41 | 3.36 | 4.35 |
| | 3.0 | 1.36 | 1.71 | 2.42 | 3.55 | 4.82 |
| | 4.0 | 1.28 | 1.56 | 2.18 | 3.29 | 4.69 |
| | 6.0 | 1.19 | 1.40 | 1.87 | 2.97 | 4.69 |
| | 8.0 | 1.14 | 1.30 | 1.69 | 2.61 | 4.18 |

## Example 1.4

A mono-directional beam of 2 MeV γ-rays of intensity $10^{10}$ photons.m$^{-2}$.s$^{-1}$ is incident on a Pb slab shield of 0.1 m thickness. What would the dose rate at the outer surface of the shield be?

### Solution
From the standard attenuation coefficient table one obtains

$$\frac{\mu_a}{\rho} = 0.00238 \text{ m}^2\text{kg}^{-1}$$

$\mu/\rho = 0.00457$ m$^2$.kg$^{-1}$ for Pb at 2 MeV

As $\rho$ for Pb is $11.34 \times 10^3$ kg.m$^{-3}$,

$\therefore \mu = 0.00457 \times 11.34 \times 10^3$ m$^{-1}$

$\quad = 5.18 \times 10^1$ m$^{-1}$

and $\mu t = 5.18 \times 10^1 \times 0.1 = 5.18$

Using the exposure build-up factor of Table 1.3 for a mono-directional source of 2 MeV energy, for Pb one obtains, $B_m(\mu t = 4) = 2.41$ and $B_m(\mu t = 7) = 3.36$. Interpolating between these two values of $\mu t$, $B_m$ for $\mu t = 5.18$ becomes 2.78. Now using equation (1.22)

$$\bar{D} = E\phi \frac{\mu_a}{\rho} B_m(\mu t) \exp(-\mu t)$$

$\quad = 2 \times 1.6 \times 10^{-13} \text{ J} \times 10^{10} \text{ m}^{-2}.\text{s}^{-1} \times 0.00238 \text{ m}^2.\text{kg}^{-1} \times 2.78 \times \exp(-5.18)$

$= 1.19 \times 10^{-1} \mu\text{Gy.s}^{-1}$

$= 0.43 \text{ mGy.h}^{-1}$

Equation (1.22) relates exclusively to a mono-directional beam incident on a slab shield. The build-up factors can be computed for other types of sources. If a point isotropic source emitting $S$ photons per unit time is surrounded by a spherical shield of radius $r$, then the dose rate with build-up factor can be written as

$$\bar{D} = \bar{D}_0 \exp(-\mu r) B_p(\mu r) \tag{1.25}$$

where $\bar{D}_0$ is the dose rate in the absence of the shield and $B_p(\mu r)$ is the point isotropic exposure build-up factor.

The build-up fluence rate for a point source can then be written as

$$\phi = \phi_u B_P(\mu r) = \phi_0 \exp(-\mu r) B_P(\mu r) \tag{1.26}$$

If a point source emitting $S$ photons per second is considered, then the dose rate, $\bar{D}$ at the outer surface of a spherical shield of radius $R$ is given by

$$\bar{D} = \frac{S}{4 \pi R^2} E \frac{\mu_a}{\rho} \exp(-\mu R) B_p(\mu R) \tag{1.27}$$

The values of $B_p(\mu R)$ as a function of energy are shown in Table 1.4.

Table 1.4 *Exposure build-up factor for isotropic point source*

| Material | $E_0$ (MeV) | $\mu_0 r$ | | | | |
|---|---|---|---|---|---|---|
| | | 1 | 2 | 4 | 7 | 10 |
| Water | 0.5 | 2.52 | 5.14 | 14.3 | 38.8 | 77.6 |
| | 1.0 | 2.13 | 3.71 | 7.68 | 16.2 | 27.1 |
| | 2.0 | 1.83 | 2.77 | 4.88 | 8.46 | 12.4 |
| | 3.0 | 1.69 | 2.42 | 3.91 | 6.23 | 8.63 |
| | 4.0 | 1.58 | 2.17 | 3.34 | 5.13 | 6.94 |
| | 6.0 | 1.46 | 1.91 | 2.76 | 3.99 | 5.18 |
| | 8.0 | 1.38 | 1.74 | 2.40 | 3.34 | 4.25 |
| Iron | 0.5 | 1.98 | 3.09 | 5.98 | 11.7 | 19.2 |
| | 1.0 | 1.87 | 2.89 | 5.39 | 10.2 | 16.2 |
| | 2.0 | 1.76 | 2.43 | 4.13 | 7.25 | 10.9 |
| | 3.0 | 1.55 | 2.15 | 3.51 | 5.85 | 8.51 |
| | 4.0 | 1.45 | 1.94 | 3.03 | 4.91 | 7.11 |
| | 6.0 | 1.34 | 1.72 | 2.58 | 4.14 | 6.02 |
| | 8.0 | 1.27 | 1.56 | 2.23 | 3.49 | 5.07 |
| Lead | 0.5 | 1.24 | 1.42 | 1.69 | 2.00 | 2.27 |
| | 1.0 | 1.37 | 1.69 | 2.26 | 3.02 | 3.74 |
| | 2.0 | 1.39 | 1.76 | 2.51 | 3.66 | 4.84 |
| | 3.0 | 1.34 | 1.68 | 2.43 | 3.75 | 5.30 |
| | 4.0 | 1.27 | 1.56 | 2.25 | 3.61 | 5.44 |
| | 6.0 | 1.18 | 1.40 | 1.97 | 3.34 | 5.69 |
| | 8.0 | 1.14 | 1.30 | 1.74 | 2.89 | 5.07 |

## Example 1.5

An isotropic point source emits $10^8$ $\gamma$-rays per second with a photon energy of 1 MeV. The source is to be shielded with a spherical iron shield. What must be the radius of the shield if the exposure rate at the surface is to be 10 $\mu$Gy.h$^{-1}$?

*Solution*

Here the final dose rate, $\bar{D}$ of equation (1.27) is given as 10 $\mu$Gy.h$^{-1}$. Substituting the value of $(\mu_a/\rho)$ in air from the standard mass absorption coefficient table in equation (1.27), one obtains

$$\frac{10 \times 10^{-6}}{3600} \frac{\text{Gy}}{\text{s}} = \frac{10^8 \, \text{s}^{-1}}{4\pi r^2 \, \text{m}^2} \times 1.0 \times 1.6 \times 10^{-13} \, J \times 0.0028 \frac{\text{m}^2}{\text{kg}} \exp(-\mu r) B_p(\mu r)$$

$$1 = 1.28 \times \frac{\exp(-\mu r) B_p(\mu r)}{r^2} \tag{1.28}$$

This equation can be solved graphically by plotting the RHS as a function of $\mu r$ and then taking the value of $\mu r$ for which RHS becomes 1. First, the RHS must be made a function of $\mu r$. Multiplying both numerator and denominator of the equation by $\mu^2$, one obtains

$$1 = \frac{1.28 \times (\mu)^2 \exp(-\mu r) B_p(\mu r)}{(\mu r)^2} \tag{1.29}$$

Using a standard mass attenuation coefficient table, one obtains $\mu/\rho$ = 0.00595 m$^2$ kg$^{-1}$ for Fe at 1 MeV. Taking $\rho$ for Fe as 7.86 $\times$ $10^3$ kg.m$^{-3}$, $\mu$ = $0.00595 \times 7.86 \times 10^3 = 4.68 \times 10^1$ m$^{-1}$. Substituting the value of $\mu$ in equation (1.29)

$$1 = \frac{1.28 \times (4.68 \times 10^1)^2 \exp(-\mu r) B_p(\mu r)}{(\mu r)^2}$$

$$= \frac{2.80 \times 10^3 \exp(-\mu r) B_p(\mu r)}{(\mu r)^2} \tag{1.30}$$

The RHS can now be plotted on a semi-log paper as shown in Figure 1.9. When the RHS = 1, $(\mu r)$ is equal to 6.4 and so $r$ = 6.40/46.8 = 0.137 m = 13.7 cm.

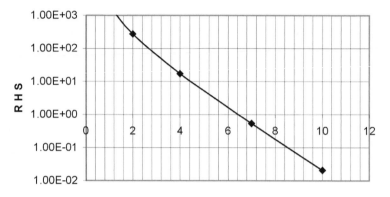

**Fig. 1.9** *RHS of equation (1.30) as a function of μr*

For practical solution of shielding problems, the exposure build-up factor for point sources can be conveniently expressed as

$$B_P(\mu r) = A_1 \exp(-\alpha_1.\mu r) + A_2 \exp(-\alpha_2.\mu r) = \Sigma A_n \exp(-\alpha_n.\mu r) \quad (1.31)$$

where $A_1$, $A_2$, $\alpha_1$ and $\alpha_2$ are parametric constants which depend on the incoming radiation energy and on the properties of the shielding materials.

The above equation is known as the Taylor form of the build-up factor. As $r$ tends to zero, $B_p(0)$ must approach unity as there can be no build-up of scattered radiation in a shield of zero thickness. So

$$A_1 + A_2 = 1. \qquad \therefore \quad A_2 = 1 - A_1$$

So equation (1.31) becomes

$$B_P = A_1 \exp(-\alpha_1.\mu r) + (1 - A_1) \exp(-\alpha_2.\mu r)$$

Changing $A_1$ to $A$, one obtains

$$B_P = A \exp(-\alpha_1.\mu r) + (1 - A) \exp(-\alpha_2.\mu r) \qquad (1.32)$$

Some of the values of $A$, $\alpha_1$ and $\alpha_2$ for more commonly used materials are given in Table 1.5.

A more accurate but somewhat more difficult expression for $B_p$ is the Berger form which is

$$B_P = 1 + C\mu r \exp(-\beta\mu r) \qquad (1.33)$$

where $C$ and $\beta$ are parametric constants which depend on the γ-energy and the shield material.

**Table 1.5.** *Parameters in the Taylor form of exposure buildup factor for point isotropic source.*

| Material | $E_0$/MeV | A | $\alpha_1$ | $\alpha_2$ |
|----------|-----------|-----|-----------|-----------|
| **Water** | 0.5 | 100.84 | −0.12687 | −0.10925 |
| | 1.0 | 19.60 | −0.09037 | −0.02522 |
| | 2.0 | 12.612 | −0.05320 | 0.01932 |
| | 4.0 | 11.163 | −0.02543 | 0.03025 |
| | 8.0 | 4.635 | −0.02633 | 0.07097 |
| **Concrete** | 0.5 | 38.225 | −0.14824 | −0.10579 |
| | 1.0 | 25.507 | −0.07230 | −0.01843 |
| | 2.0 | 18.089 | −0.04250 | 0.00849 |
| | 4.0 | 11.460 | −0.02600 | 0.02450 |
| | 8.0 | 8.972 | −0.01300 | 0.02979 |
| **Lead** | 0.5 | 1.677 | −0.03084 | 0.30941 |
| | 1.0 | 2.984 | −0.03503 | 0.13486 |
| | 2.0 | 5.421 | −0.03482 | 0.04379 |
| | 4.0 | 3.897 | −0.08468 | −0.02383 |
| | 8.0 | 0.368 | −0.23691 | −0.05864 |

## 1.3.3 Neutron Interactions

The ways in which neutrons interact with matter are varied and complex. There are a number of mechanisms for neutron interaction with target atoms. There are three modes in which the neutrons interact with the atomic nuclei. These are

**Elastic scattering:** In this process, the neutron strikes the nucleus which is normally in the ground state; the neutron reappears and the nucleus is left in its ground state. This is denoted by the (n, n) reaction.

**Inelastic scattering:** This process is identical to elastic scattering except that the nucleus is left in an excited state. This reaction is denoted by (n, n'). The excited nucleus decays by the emission of γ-rays, called inelastic γ-rays. Inelastic scattering occurs with energetic neutrons.

**Radiative capture:** In this process, the neutron is captured by the nucleus and following this capture, one or more γ-rays, termed capture γ-rays, are emitted. This is an exothermic interaction and is denoted by (n,γ). Since the original neutron is absorbed, this process is also known as an absorption reaction. This reaction is most probable at resonance and the thermal energies of the neutrons.

The interaction of neutrons with a target material is characterised by the **microscopic cross-section**, $\sigma$ of the target material. The probability of the interaction of a neutron with a target nucleus is effectively specified by the cross-sectional area of a target nucleus. The unit of $\sigma$ is the barn (1 b = $10^{-28}$ m²). The total cross-section, $\sigma_t$ is given by

$$\sigma_t = \sigma_s + \sigma_i + \sigma_\gamma + \sigma_f \qquad (1.34)$$

where $\sigma_s$, $\sigma_i$, $\sigma_\gamma$ and $\sigma_f$ are the cross-sections for elastic scattering, inelastic scattering, radiative capture and fission process respectively.

The product $\sigma N$, where $N$ is the atom density (number of atoms per unit volume) of target material, is called the **macroscopic cross-section**, $\Sigma$ whose unit is $m^{-1}$. The fluence rate of neutrons after passage through a thickness of material is thus given by

$$\phi = \phi_0 \exp\left(-\Sigma_t t\right) \tag{1.35}$$

where $\Sigma_t$ is the total macroscopic cross-section ($m^{-1}$), $\phi_0$ is the initial fluence rate of neutrons (no.$m^{-2}.s^{-1}$) and $t$ is the thickness of the material (m).

For shielding purposes, fast neutrons generated within the reactor core must be slowed down to thermal neutrons. This is done because the absorption cross-sections of the fast neutrons in most materials are very small whereas those of the thermal neutrons are quite high, often some orders of magnitude higher. It can be shown that a neutron, on average, loses 50% of its energy in an elastic collision with a hydrogen nucleus, more than with any other nucleus. When fast neutrons have thus been reduced in energy by successive scattering by hydrogeneous materials, they can be absorbed by some absorption reaction. It may, however, be noted that when thermal neutrons are captured in water by a $^1$H(n,$\gamma$)$^2$H reaction, the emitted $\gamma$-energy is 2.2 MeV. On the other hand, if they are captured by iron, a 7.6 MeV and a 9.3 MeV $\gamma$-ray are emitted. To reduce the intensity of such $\gamma$-rays, boron is added to the reactor shield. Boron (B) has a high thermal absorption cross-section (~759 b) and undergoes the $^{10}$B(n,$\alpha$)$^7$Li reaction, the accompanying $\gamma$-ray energy is only 0.5 MeV.

## Example 1.6

What fraction of 10 MeV neutrons would be transmitted through a 1 cm thick Pb shield? (Given that the atomic weight of Pb is 207.21, density is 11.3 $\times$ $10^3$ kg.$m^{-3}$ and $\sigma$ at 10 MeV is 5.1 b).

*Solution*
The atomic density of Pb is

$$N = \rho N_A / M$$

$$= \frac{11.3 \times 10^3 \, (kg/m^3) \times 6.026 \times 10^{23} \, (atoms/mole)}{207.21 \times 10^{-3} \, kg/mole}$$

$$= 3.29 \times 10^{28} \, atoms.m^{-3}$$

$$\sigma N = 5.1 \times 10^{-28} \, (m^2) \times 3.29 \times 10^{28} \, (m^{-3})$$

$$= 1.68 \times 10^1 \, m^{-1}$$

*(continued)*

$$\frac{\phi}{\phi_0} = \exp\left(-1.68 \times 10^1 \times 10^{-2}\right)$$

$$= 0.856$$

So the fraction of unscattered neutrons transmitted through a 1 cm thick Pb shield is 0.856.

### 1.3.3.1 Sources of Neutrons

The most significant source of neutrons is the operating reactor. Large numbers of neutrons are generated in the fission process. The average number of neutrons per fission is 2.42 ($v = 2.42$ from U-235 fission). For the purposes of a biological shield, these neutrons are divided into two categories.

(i) **Prompt fission neutrons**: These neutrons are produced at the instant when fission occurs. They are most numerous and energetic ($E > 1$ MeV).

(ii) **Delayed fission neutrons**: These neutrons are produced at a slightly later to the fission process. The average delay time is about 12 s in U-235 fission. They are less energetic (~400 keV) and far less numerous than the prompt neutrons and consequently less significant from the radiological point of view. However, they become relatively more significant in a shut down reactor because of the delayed nature of their generation.

## 1.3.4 Neutron Dose Calculation

For neutron dose calculations, neutron energies are conveniently divided into fast and thermal ranges and consequently the dose calculations are separated into: (i) fast neutron dose calculation; and (ii) thermal neutron dose calculation.

### 1.3.4.1 Fast Neutron Dose Calculation

The dose absorbed by a tissue from a beam of fast neutrons may be calculated by considering the energy absorbed by each of the tissue elements. The main mechanism of energy absorption is the elastic scattering of the incident neutrons. The scattered nuclei dissipate their energies in the immediate vicinity of the primary neutron interaction. The neutron absorbed dose rate is

$$\bar{D} = E\phi(E)\sum_i \sigma_{si} N_i f_i \qquad (1.36)$$

where $E$ is the neutron energy in joules, $\phi(E)$ is the neutron fluence rate of energy $E$ in $m^{-2}.s^{-1}$. $N_i$ is the atom density per kg of the $i$th element in the target ($N_A$ (Avogadro's constant) × mass fraction/molecular mass). The values of the molecular mass are given in Appendix 3.

$\sigma_{si}$ is the scattering cross-section of the $i$th element for neutrons of energy $E$ in m$^2$ and $f_i$ is the mean fractional energy transferred from neutrons to scattered nuclei.

For isotropic scattering, the average fraction of neutron energy transferred to a nucleus of mass number $A$ is

$$f = 2A/(A+1)^2 \qquad (1.37)$$

The composition of soft tissue is given in Table 1.6. The calculated values of $f$ and $N$ are also given in Table 1.6.

**Table 1.6** *Tissue composition and elemental values*

| Element | Mass fraction of element | $f_i (= 2A/(A+1)^2)$ | $N_i$ (atom.kg$^{-1}$) |
|---------|--------------------------|----------------------|------------------------|
| Oxygen | 0.7139 | 0.111 | 2.69E+25 |
| Carbon | 0.1489 | 0.142 | 7.48E+24 |
| Hydrogen | 0.1000 | 0.500 | 5.98E+25 |
| Nitrogen | 0.0347 | 0.124 | 1.49E+24 |
| Sodium | 0.0015 | 0.080 | 3.93E+22 |
| Chlorine | 0.0010 | 0.053 | 1.70E+22 |

## Example 1.7

What is the absorbed dose rate to a soft tissue from a beam of 5 MeV neutrons with a fluence rate of $2 \times 10^7$ neutrons m$^{-2}$.s$^{-1}$?
(The scattering cross-sections for 5 MeV neutrons in oxygen, carbon, hydrogen, nitrogen, sodium and chlorine are respectively 1.55, 1.65, 1.50, 1.00, 2.3 and 2.8 barns).

*Solution*
The scattering cross-section for 5 MeV neutrons in tissue is calculated as

| Element | $\sigma_{si}$ (m$^2$) | $N_i \, \sigma_{si} f_i$ (m$^2$.kg$^{-1}$) |
|---------|----------------------|--------------------------------------------|
| Oxygen | 1.55E -28 | 4.63E -04 |
| Carbon | 1.65E -28 | 1.75E -04 |
| Hydrogen | 1.50E -28 | 4.49E -03 |
| Nitrogen | 1.00E -28 | 1.85E -05 |
| Sodium | 2.30E -28 | 7.23E -07 |
| Chlorine | 2.80E -28 | 2.52E -07 |
| Total | | 5.14E -03 |

*(continued)*

Now substituting these values in equation (1.36), the absorbed dose rate is

$$\bar{D} = 5 \times 1.6 \times 10^{-13} \, (\text{J}) \times 2 \times 10^{7} \left( \text{n.m}^{-2}.\text{s} \right) \times 5.14 \times 10^{-3} \left( \text{m}^{2}.\text{kg}^{-1} \right)$$

$$= 8.22 \times 10^{-8} \, \text{Gy.h}^{-1}$$

$$= 2.96 \times 10^{-4} \, \text{Gy.h}^{-1}$$

### 1.3.4.2 Thermal Neutron Dose Calculation

The doses to the soft tissues from thermal neutrons are received via: (i) the transmutation reaction, $^{14}\text{N}(\text{n,p})\,^{14}\text{C}$; and (ii) the radiative capture reaction, $^{1}\text{H}(\text{n},\gamma)\,^{2}\text{H}$. Both of these are absorption reactions. In the case of the N-14 reaction, the dose rate may be calculated as

$$\bar{D} = \phi E (N\sigma) \tag{1.38}$$

where $\phi$ is the thermal neutron fluence rate (no. $\text{m}^{-2}.\text{s}^{-1}$); $E$ is the energy release per reaction (= 0.63 MeV = $1.01 \times 10^{-13}$ J); $N$ is the number of nitrogen atoms per unit mass of tissue (= $1.49 \times 10^{24}$ atoms.kg$^{-1}$); and $\sigma$ is the absorption cross-section in nitrogen for thermal neutrons (= $1.75 \times 10^{-28}$ m$^{2}$)

In the case of the hydrogen reaction, each reaction generates a $\gamma$-photon of 2.23 MeV. The specific activity, i.e. the number of reactions per unit mass per second is given by

$$A = \phi N\sigma \tag{1.39}$$

where $\phi$ is the thermal neutron fluence rate (n m$^{-2}.\text{s}^{-1}$); $N$ is the number of hydrogen atoms per kg of tissue (= $5.98 \times 10^{25}$ atoms.kg$^{-1}$); and $\sigma$ is the absorption cross-section in hydrogen (= $3.3 \times 10^{-29}$ m$^{2}$).

When the specific activity has been calculated, the absorbed dose rate due to $\gamma$-rays in the tissue may be calculated using the following equation:

$$\bar{D} = EA(\text{AF}) \tag{1.40}$$

where $E$ is the energy of the emitted photon per unit activity (J); $A$ is the calculated specific activity; and (AF) is the absorbed fraction for 2.23 MeV $\gamma$-ray.

# Revision Questions

1. What is meant by radiation? What are the main types of radiation? List the names of the components of these main types of radiation and describe them briefly.

2. Why some radiations are ionising, while others are not? Illustrate your argument with a numerical example showing the distinction between ionising and non-ionising radiations.

3. What are the various types of particulate radiation encountered in the nuclear industry? Give indicative ranges for these radiations in air as well as in body tissues.

4. Explain clearly the phenomena of ionisation and excitation. Show these two processes diagrammatically.

5. Deduce the radioactive decay law

$$N = N_0 \, \exp(-\lambda t)$$

where $N_0$ is the number of atoms at $t = 0$, $\lambda$ is the decay constant, and $N$ is the number of atoms after a time $t$.

6. Define the following terms and state units where applicable:
   (i)   radionuclide
   (ii)  isotope
   (iii) activity
   (iv) mass activity concentration
   (v)  specific activity (ICRU definition)
   (vi) particle fluence rate
   (vii) energy fluence rate
   (viii) specific ionisation

7. What are the various processes of $\gamma$-ray interaction with matter? Explain each of them briefly.

8. What is the basic difference between the mass attenuation coefficient and the mass absorption coefficient? Explain the difference clearly.

9. What is the significance of the half-value thickness of a material? How is it related to the total attenuation coefficient?

10. Describe the significance of the build-up factor in $\gamma$-ray dose calculations. Show diagrammatically the incident and emergent energy spectra to establish the need for the build-up factor.

11. What are the various modes of interaction of neutrons with matter? Briefly explain each of them.

12. Define the macroscopic cross-section of the interaction of a neutron with matter. How does the neutron fluence rate vary with the thickness of the material.

13. What are the neutron energies that are generated during the fission process in a nuclear reactor? Briefly describe these neutron types with their energy spectra.

14. On what basis are neutron dose calculations separated into into fast and slow neutron dose calculations?

## REFERENCES

[1] Lamarsh J.R., *Introduction to Nuclear Engineering*, 3rd edn., Addison-Wesley, Reading, MA, USA, 1975.

[2] Parker R.P., Smith P.H.S. and Taylor D.M., *Basic Science of Nuclear Medicine*, 2nd edn., Churchill-Livingstone, London, UK, 1984.

[3] Jenkins P.F., Harbison S.A. and Martin A., *An Introduction to Radiation Protection*, 5th edn., Hodder Arnold, London, UK, 2006.

[4] International Commission on Radiation Units and Measurements, ICRU Report 60, 1999.

# 2

# BIOLOGICAL EFFECTS OF RADIATION

## 2.1 Modes of Exposure

When human beings are exposed to ionising radiation from sources outside the body, it is known as **external exposure**. If the sources of radiation happen to be within the body, then the exposure is termed **internal exposure**.

The sources outside the body may emit radiations of various types: α- or β-particles, γ-rays or X-rays as well as neutrons. Even when α and β radiations are not shielded, they are quickly attenuated even in air as they are charged particles and hence their effects in terms of energy transfer to a human body would be small. On the other hand, electromagnetic radiations such as γ-rays and X-rays could travel significant distances in air and transfer significant amount of energy to human beings. In addition, radiation workers working in the vicinity of nuclear reactors may be exposed to neutron irradiation. Neutrons, being uncharged particulates, may travel significant distances in air. The calculations of doses from external exposures have been described in Chapter 1.

Internal exposure would arise when radioactive particles are inhaled or contaminated foodstuff is ingested or radioactivity enters the body through wounds in the body. Internal exposure is much more hazardous than external exposure, as the source of exposure remains within the body whereas one can move away from the source of external radiation. The effects of external or internal irradiation may be acute or chronic depending on the magnitude of the exposure. For the purposes of radiological protection, doses from both types of radiation need to be estimated and added together. This chapter deals with the effects of internal and external exposures on human bodies.

## 2.2 Effects of Radiation

The absorption of radiative energy by living organisms produces effects which are potentially harmful. The nature and extent of the harm depends not only on the amount of energy absorbed by the organism but also on the rate of absorption and the type of radiation that imparts that energy.

If the absorbed dose is high such that the harmful effects become apparent immediately or are clinically detectable within a short period of time, then the effect is termed as the **deterministic effect**. On the other hand, if the dose is low, the harmful effects may not appear for a long period of time and then may appear as a malignancy in the exposed person or as a genetic defect to an offspring. This is known as the **stochastic effect**, meaning that it is random or statistical in nature.

Radiation effects may also be categorised by the levels of exposure. If the level of exposure is high or very high, then it is called an **acute exposure**. Such exposures would invariably be for a short period of time, as nobody would normally be left at such high exposure levels for any length of time. The effects show up soon after exposure and hence they are also known as the **early effects**. Obviously acute exposures would lead to a deterministic effect. Acute exposures would arise from nuclear explosions or severe nuclear accidents such as criticality accidents etc. On the other hand, if level of exposure is small but extended over a long period, then it is known as **chronic exposure**. The effects of such small exposures take a long time to develop and hence they are known as **late effects.** Chronic exposures arise from day-to-day radiation work, radon exposures in the home and workplace etc. Chronic exposures are stochastic in nature. These effects take a long time to develop: cataracts may appear after 2 years, leukaemia may develop after quite a few years and lung cancer may take as long as 20 years to develop.

To understand the effects of radiation, one needs to look at the dose-response characteristic more closely and that is done in the following section.

## 2.3 Dose–Response Characteristics

The biological response to radiation doses can be separated into two broad categories: one arising from high doses or high dose rates and the other from low doses or low dose rates. The effects from exposure to high doses or high dose rates become apparent within a short period of time and can be clearly identified. This response is, therefore, deterministic. Examples of deterministic effects are: skin burns, cataracts, vomiting, general malaise and, in extreme cases, malfunctioning of the central nervous system. The body, which has a built-in repair mechanism always strives to overcome damage and the damage becomes apparent only when the capacity of the body's repair mechanism is exceeded. Thus, for the deterministic effect to take place, there must be a minimum level of dose below which the effect is indeterminate and that level of dose is the **threshold level** for the deterministic effect. With low doses or low dose rates, the body may overcome minor damage and no outward symptom of damage may be available immediately. Nonetheless, the body may suffer damage which may show up a long time after the incident. This period between the incident and the expression of damage is called the **latency period**. During the latency period, the body's repair mechanism tries to repair the damage. In a healthy human being, the radiation damage may be totally repaired and hence no lasting damage will result; whereas the same damage to a physically weak human being may not be totally repaired and after the latency period the damage shows up as a malignancy.

## 2.3.1 Stochastic Effects of Radiation

The stochastic effects of radiation apply when the dose levels are lower than the threshold value. These effects occur due to the existence of a finite probability of deleterious effects of radiation on cells, even when the dose level is very small. It will be shown in Section 2.6.4 that damage even to a single cell may ultimately lead to the development of a malignant condition after a prolonged and variable latency period. Although the multistage clonal growth is the presently assumed mode of development of the malignant condition, the initial phase of the whole process may start with a single track of radiation within the tissue. Consequently, it is prudent to assume that there is no level of radiation which may be considered harmless and safe. In other words, for the stochastic effect, it is considered that there is no threshold level and so the dose–response curve must go through the origin.

The next question is what would be the shape of the dose–response curve? The biological response to low or very low levels of dose without any outward symptom of damage is very difficult to quantify. Reliance has to be made on statistical studies. Quantitative information on the risk of cancer from exposures to radiation at low doses comes from epidemiological studies on a human population exposed to intermediate or high doses and dose rates. The three principal sources of information for epidemiological studies are: the survivors of the nuclear weapon explosions at Hiroshima and Nagasaki, patients exposed to medical radiation and the nuclear industry workers exposed to radiation. Such studies demonstrate that radiation damage increases with the increase in dose or dose rate (at least for low LET). For the purposes of radiological protection at low doses and low dose rates, a linear variation of the dose–response curve for the induction of cancer is assumed. This estimation

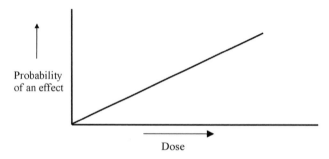

**Fig. 2.1** *Stochastic effects of radiation*

of radiation damage in the form of cancer as a function of low doses is based on the effects detectable at intermediate and high doses and high dose rates and then transposing such effects to low doses and low dose rates. This effectiveness parameter is called the Dose and Dose Rate Effectiveness Factor (DDREF) [1]. The dose–response curve for the stochastic effect is assumed to be linear and is shown in Figure 2.1 (see also Section 3.1).

So, the stochastic effect is that for which the probability of an effect occurring, rather than its severity, is regarded as a linear function of dose without threshold. This effect is also referred to as the Linear No-Threshold (LNT) effect.

## 2.3.2 Deterministic Effects of Radiation

As the radiation dose increases, there comes a point when a substantial number of cells has either been killed or irreparably damaged such that a clinically detectable impairment of the function of the tissue or organ becomes apparent. It had been found that if the dose level to an organ is roughly below 500 mGy, then it is clinically difficult to detect quickly and pathological disorders may not show up soon. This minimum level of dose at which the deterministic effect takes place is called the threshold value. It varies from organ to organ of an individual as well as from individual to individual. For the lenses of the eye, the threshold level is considered to be 150 mGy, as the eye is very radiosensitive. There are, however, subtle asymptomatic changes (e.g. lowering of blood cell counts and chromosomal aberrations) which occur at doses well below such thresholds and can be measured by modern techniques. But for the purposes of radiological protection, the above threshold levels are generally used.

The next question is, how does the response vary with dose beyond the threshold level? It has been found experimentally on cultured cells and animals that immediately beyond the threshold level, the severity of the damage varies slowly with the dose and, as the dose level increases, the severity per unit dose increases progressively and eventually the severity tends to level off at

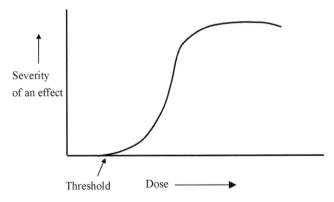

**Fig. 2.2** *Deterministic effects of radiation*

very high doses. If the damage is considered to be the death of cells, then at very high doses all of the cells are dead and any further increase in the dose would show no increase in damage and thus the response levels off. The dose-response of the deterministic effect is shown in Figure 2.2.

So, the deterministic effect is that for which the severity of an effect varies with dose and for which a threshold may therefore exist.

The shape of the population dose–response relationship for deterministic effects is sigmoid in character [2]. The formulation of the risk function, $R$ in terms of dose, $D$ is given by equations (2.1) and (2.2).

$$R = 1 - \exp(-H) \tag{2.1}$$

$$H = \ln(2)\left(\frac{D}{D_{50}}\right)^V \tag{2.2}$$

The function $H$ is known as the hazard function and $D_{50}$ is the dose at which the effect is detected in 50% of the population. $LD_{50}$ (median lethal dose) for lethal effects and $ED_{50}$ (median effective dose) for morbidity are used when applying equation (2.2). The shape factor, $V$, determines the steepness of the risk function. Examples of the risk function, $R$, are given in Figure 2.3 in terms of the normalised dose, $D/D_{50}$ for values of $V$ of 2, 5 and 10. To produce a sigmoid shape, $V$ must be greater than unity. The threshold dose is specified (by the ICRP based on examination of the relevant data) below which $H$ is estimated to be very small, approaching zero. The threshold value, when expressed as a function of $D_{50}$, generally lies in the range 0.2–0.5. In order to prevent unnecessary computation to estimate the risks at low levels of dose, the threshold value is specified as the dose which confers a risk of 1%. The threshold values can be computed from a knowledge of $D_{50}$ and $V$ (see Figure 2.3).

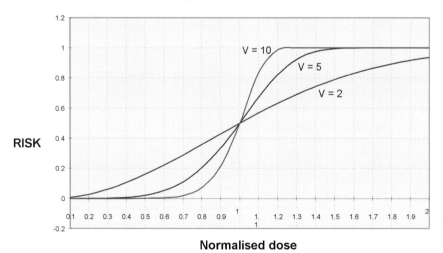

**Fig. 2.3** *Risk as a function of normalised dose*

## 2.4 Internal Irradiation

Internal irradiation is caused by the incorporation of radioactive materials within the body in various physical and chemical forms. The main routes by which radioactive contamination enters the body are: (i) inhalation, (ii) ingestion, and (iii) body contamination. The suspended radioactive particles in air may be inhaled by the human beings. These inhaled particles are then taken directly into the respiratory system. If the material is soluble in blood, a good fraction of it is likely to be absorbed in the blood stream and the remaining fraction is exhaled and swallowed into the digestive system. On the other hand, if the materials are insoluble, they are mostly exhaled but a small fraction may be deposited in the respiratory tract causing prolonged irradiation to the tracheo-bronchial regions. If the materials are ingested through food and water, they irradiate the various organs of the digestive system as well as being distributed around the body through the body fluids. Body contamination, leading to skin absorption offers a route of entries of suspended contamination into the blood stream.

The retention period of radioactive materials within the body is expressed by a quantity called the **effective half-life**. The activity of a radionuclide decays exponentially following the radioactive decay law. This decay law is valid regardless of the physical and chemical conditions under which the materials may find themselves, whether within or outside the body system. The rate of excretion of a substance from the body is also considered to follow an exponential behaviour as the excretion rate depends on the amount of material present. Thus, if $N_0$ is the initial number of radionuclides in the body, then the number of nuclides, $N$ present after time $t$ is given by

$$N = \left(N_0 \ \exp\left(-\lambda_r t\right)\right)\left(\exp\left(-\lambda_b t\right)\right) \tag{2.3}$$

where $\lambda_r$ is the radioactive decay constant and $\lambda_b$ is the biological decay constant. Thus

$$N = N_0 \ \exp\left(-\left(\lambda_r + \lambda_b\right)t\right) \tag{2.4}$$

The effective decay constant, $\lambda_{eff}$, is defined as

$$\lambda_{eff} = \lambda_r + \lambda_b \tag{2.5}$$

As $\lambda = \dfrac{0.693}{T}$, so $\dfrac{0.693}{T_{eff}} = \dfrac{0.693}{T_r} + \dfrac{0.693}{T_b}$

or, $T_{eff} = \dfrac{T_r \ T_b}{T_r + T_b}$ $\tag{2.6}$

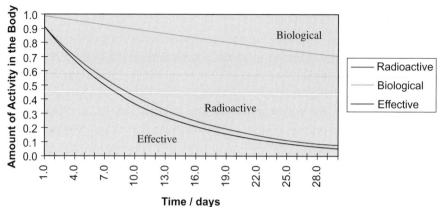

**Fig. 2.4** *Typical clearance curve of a radionuclide from the body*

The general decay pattern of a radionuclide within the body is shown in Figure 2.4.

The effective half-life is determined by measuring the amount of activity in an organ or in blood samples over a period of time and then plotting these values on a log-linear graph. The effective half-life is the time that it takes for the measured sample activity to fall to half its original value.

It is worth noting here that a small amount of radioactive material outside the body may give a small dose, but the same amount of radioactive material lodged within the body may offer a significant dose and cause serious biological damage. This is because whereas the dose reduction strategy such as 'time', 'distance' and 'shielding' may be effective against external radiations, there is no such preventive mechanisms against internal irradiation. Once the radioactive materials are within the body, they remain there until they are completely excreted by the body system following the effective half-life. The dose received by the whole body during this prolonged but variable period of time is called the **committed effective dose** (see also Section 3.2.6). The total dose received from internal irradiation is thus likely to be significant. The protective action against internal irradiation is to reduce airborne activity concentrations and ban contaminated food products from consumption. The airborne activity levels are therefore constantly monitored in workplaces by sensitive instruments. The only recourse following the intake of a significant amount of radioactive material is to induce excretion by invasive medical practices such as the application of Diethylene Triamine Pentaacetic Acid (DTPA).

The problem with internal contamination is that the dose levels cannot be measured directly and hence mathematical models representing the metabolic and dosimetric behaviours of radionuclides within the body are used. The International Commission on Radiological Protection (ICRP) had produced dose coefficients (dose per unit intake) for over 800 radionuclides from intake by inhalation and ingestion by workers as well as by the public [3]. That database offers a simple but accurate method of calculation of internal dose uptake.

## 2.5 Radiotoxicity

There is no precise definition of radiotoxicity. However, the term is used quite widely both in technical journals as well as in everyday language to indicate the toxicity arising from a radionuclide or its harmfulness to human beings.

A measure of the radiotoxicity of a radionuclide may be made in terms of the intrinsic hazard quantified by the committed effective dose coefficient, that is, the committed effective dose per unit intake of activity. But the problem is that the committed effective dose is dependent on the mode of intake e.g. ingestion and/or inhalation. Consequently, the radiotoxicity of a radionuclide would be different for different modes of intake. Table 2.1 gives the committed effective dose coefficients to adult members of the public from some of the significant radionuclides [4, 5]

The radiotoxicity may be defined as the product of the committed effective dose coefficient and the concentration of radioactivity in the material. So it is given by

$$R_{tx} = E(\tau)C \tag{2.7}$$

where $E(\tau)$ is the committed effective dose integrated over a period of 50 years in Sv and $C$ is the activity concentration in Bq.kg$^{-1}$.

Thus the radiotoxicity, $R_{tx}$ is given in terms of Sv.kg$^{-1}$.

**Table 2.1** *Dose coefficients to an adult member of the public*

| Radionuclide | Committed effective dose coefficient (Sv.Bq$^{-1}$) for an adult member of the public | |
| --- | --- | --- |
| | Inhalation | Ingestion |
| Co-60 (5.27 y) | 3.1 E−08 (S) | 3.4 E−09 |
| Sr-90 (29.12 y) | 1.6 E−07 (S) | 2.8 E−08 |
| Tc-99 (2.13E+05 y) | 1.3 E−08 (S) | 6.4 E−10 |
| I-129 (1.57E+07 y) | 3.6 E−08 (F) | 1.1E−07 |
| Cs-137 (30.0 y) | 3.9 E−08 (S) | 1.3 E−08 |
| U-235 (7.04E+08y) | 8.5 E−06 (S) | 4.7 E−08 |
| U-238 (4.47E+09 y) | 8.0 E−06 (S) | 4.5 E−08 |
| Pu-239 (2.41E+04 y) | 5.0 E−05 (M) | 2.5 E−07 |

Notes: (1) The half-lives of the radio-nuclides are given along side the name of the radio-nuclides in the 1ˢᵗ column.
(2) The letters within the parenthesis: F, M and S denote respectively fast, moderate and slow absorption rates by the body fluid following intake by inhalation. The absorption rates indicate the clearance rates from the lung (following inhalation) to the gastrointestinal tract [5].

## 2.6 Mechanisms of Biological Damage

When a target material is exposed to ionising radiation, energy is transferred from the incoming radiation to the material in question by the processes of ionisation and excitation. The deposition of energy on the target atoms is a random process. The fractionation of the incoming energy between the ionisation and excitation processes is dependent on the photon energy and the physical property of the target material. Generally, in body tissues, a large fraction of the deposited energy is used to excite atoms, whereas only a small fraction serves to ionise the atoms. It is the ionisation which changes the structure of the atom which, in turn, changes the chemical behaviour of the molecule containing the atom. As atoms and molecules are the ultimate units of the cells and cells are the ultimate building blocks of living creatures, radiation changes the chemical behaviour of the living organisms. So to understand the overall effect of radiation damage on biological systems, one needs to know about the cell, its structures and composition and the way in which cells function within the body system.

### 2.6.1 Cells

The cells are the basic units of life, the microscopic building blocks from which the body is constructed. An adult contains approximately 40 trillion ($\sim 4 \times 10^{13}$) cells [6] with an average cell diameter of about 10 μm. The cells in the body vary greatly in shape, size and detailed structure according to the functions they perform. Muscle cells, for example, are long and thin. Many nerve cells are also long and thin, and are designed to transmit electrical impulses, while the hexagonal cells of the liver are equipped to carry out a multitude of chemical processes. Doughnut-shaped red blood cells transport oxygen and carbon dioxide round the body, while spherical cells in the pancreas make and replace the hormone insulin.

Cells may be divided into two classes: **somatic cells** and **germ cells**. Somatic cells are the ordinary cells, which make up the organs, tissues and other structures of the body. Examples of somatic cells are: muscle cells, blood cells, brain cells etc. The germ cell is of two types: sperm cell of the male and the egg or ovum cell of the female. The germ cell only functions during reproduction. The sperm cells are produced in the testes of the male, whereas the ova cells are produced in the ovaries of the female.

### 2.6.2 Cell Structure

Despite significant variations in shape and size, all body cells have a similar structure and pattern. Around the outside of every cell is a boundary wall called the **cell membrane**, which encloses a jelly-like substance called **cytoplasm**. Embedded in the cytoplasm is the nucleus which contains **chromosomes**. The basic structure of a human cell is shown in Figure 2.5.

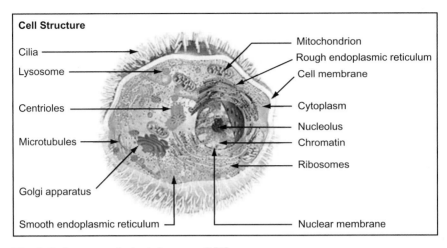

**Fig. 2.5** *Structure of a basic human cell* [7]

The **cell membrane** has a definite structure. It is approximately 7 nm (= $7 \times 10^{-9}$ m) thick and porous and comprises a double layer of protein and a layer of fat molecules — rather like a sandwich with the fat as the filling. As substances pass in or out of the cell, they are either dissolved in the fat or passed through the membrane. Some cells have hair-like projections called cilia on their membranes. The cell membrane is relatively insensitive to radiation.

The **cytoplasm** is a transparent, dilute mixture of water and various molecules and electrolytes. Organ-like structures called organelles, each of which performs specific functions for the cell as organs do in the body as a whole, are suspended in the cytoplasm. The cytoplasm of all cells contains microscopic sausage-shaped organs called **mitochondria** which carry out metabolic process within the cell e.g. they convert oxygen and nutrients into the energy needed for the activities of the cells. The metabolic functions such as protein synthesis, anaerobic glycolysis etc. which are essential for the survival and growth of the cells are carried out here. Cytoplasm is relatively insensitive to radiation damage.

The nucleus of the cell is embedded in the cytoplasm but it is separated from it by a membrane called nuclear membrane which allows the interchange of molecules between the nucleus and cytoplasm. There are 46 chromosomes arranged in 23 pairs (except in the sperm and ova cells which contain 23 chromosomes each) inside each nucleus. The chromosome, a threadlike structure, contains thousands of genes, each with enough information for the production of one type of protein. The genes are made up of complex molecules, the most important of which is the double helix of the Deoxyribo Nucleic Acid (DNA) molecule. The cross-links between the helices are the basis on which genetic information is coded. Damage to the DNA molecule can cause somatic as well as genetic defects.

[37]

### 2.6.3 Cell Reproduction

Cells reproduce to enable organisms to grow and to replenish the loss due to natural death at the end of the lifespan or loss due to other causes. The lifespan of a human cell can vary from approximately a day (lymphocytes) to a few years (bone cells). The reproduction of cells occurs in two ways: **mitosis** and **meiosis.**

The mitotic cells are the ordinary cells in the body. Ordinarily the chromosomes within the nucleus reside as an entangled mass called **chromatin**. As soon as the cell is about to divide, the chromosomes untangle themselves, becoming thicker and shorter. In this process the chromosomes duplicate in pairs by splitting lengthwise. These double chromosomes then move apart and go to opposite ends of the nucleus. Finally the cytoplasm is halved and new walls are formed round the two new cells, each of which has the normal number of 46 chromosomes. Thus the new cells, known as daughter cells, are an exact replica of the parent cells. The main features of the cell division are shown in Figure 2.6.

Meiosis is, on the other hand, a special type of cell division which occurs during the formation of the sexual reproduction cells (sperm in the male and ovum in the female). Each of the sperm and egg cells divides in two, as in mitosis, by splitting the chromosomes lengthwise and then pairing them up so that one chromosome from the mother and another chromosome from the father lie side by side. The embryo, and subsequently the offspring, develops from this single cell (the fertilised ovum).

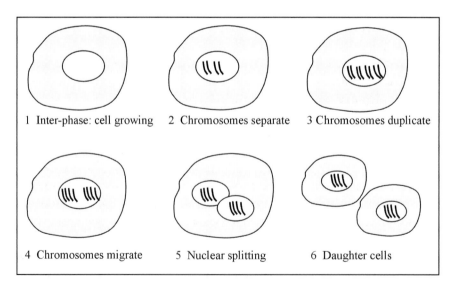

1 Inter-phase: cell growing    2 Chromosomes separate    3 Chromosomes duplicate

4 Chromosomes migrate    5 Nuclear splitting    6 Daughter cells

**Fig. 2.6** *Mitotic cell division*

## 2.6.4 Cell Damage

In all biological systems water is the most abundant molecule and radiation-induced splitting of the water molecule, known as the radiolysis of water, is a primary event in the initiation of biological damage. The cytoplasm is composed of 70–85% of water by volume. Due to the process of radiolysis of water, $H_2O^+$ and $e^-$ are produced which lead to the formation of free radicals such as $OH^0$ and $H^0$. These free radicals react with the organic molecules within the cell and may damage the structure and functions of these molecules.

There are four possible outcome following cellular damage caused by ionisation. The cells may be killed outright, they may be unable to reproduce due to DNA damage, they may remain viable but modified due to DNA modification and, of course, they may recover completely.

If the dose level is very high or dose rate is very high, cells may be killed outright within a short period of time. Generally, cells are very radio resistant during the inter-phase period and quite high doses of radiation are required to kill them (~10 Gy). The loss of a limited number of cells does not lead to serious biological damage, but the loss of a large number of cells does have immediate biological effects. Examples of the effects of cell death in tissues include skin burns, cataracts, sterility etc.

Even if the dose levels or dose rates are not sufficiently high to kill cells straightaway, they may damage the DNA molecules such that cells fail to reproduce and die at the end of their lifespan. Failure to reproduce by small number of cells would not affect the biological function of the organ or tissue badly. However, if a significant number are involved, there will be observable damage at the end of the cell lives which may be reflected by the loss of tissue function. This means that cells which divide frequently, due to their short lifespan, such as lymphocytes, are more vulnerable to radiation damage than those which divide less frequently, due to their long lifespan, such as nerve cells, bone cells etc.

These two effects of radiation, which can be determined or identified early, are the deterministic effects. As these effects require high doses, a minimum threshold level is required before the onset of such effects.

When the dose level is relatively low, lower than the threshold level, there may not be any identifiable cellular damage. In this case, the irradiated cell remains viable, it is functional but modified. Despite the existence of highly effective cellular repair mechanisms, the clone of cells resulting from the reproduction of a modified but viable somatic cells may result, after a prolonged and variable latency period, in the manifestation of a malignant condition, a cancer. This is the stochastic effect. If the damage occurs in a cell whose function is to transmit genetic information to later generations, the effect is genetic or hereditary.

It should, however, be noted that malignancy is not the inevitable outcome of radiation exposure, even when the dose level is high enough to cause cell modification. At the present level of understanding of radiation tumorigenesis, it is considered that tumour development involves multistage clonal growth. This hypothesis predicts that the evolution of the malignant state may be broadly divided into three phases: **initiation, promotion** and **progression** [1]. It should be noted that these three processes may not always be distinctive, nor can they be detected in all types of tumours. However, they represent the sequential stages in the development of tumorigenesis. This multistage tumour development and the variation between tumour types may explain the different latency periods seen in radiation-induced tumours. The initiation phase may begin with the exposure of even a single cell to radiation when the cellular DNA suffers mutations which affect the activity of a single gene or a limited number of genes. Such mutation may arise from the loss or inactivation of specific genes within the chromosome. Such somatic cell mutations are believed to provide the carrier cells with some form of growth or selective advantage such that they have the potential to evade normal tissue constraints and proliferate to form clones of pre-neoplastic cells. It is believed that the cells having stem-like properties are the initiators of the multistage process. This is called the initiation phase. There are non-genotoxic chemicals, which on their own exhibit little or no tumorigenic activity, but in concert with mutated cells are able to promote the development of tumours. They do this by stimulating the growth-enhancing genes of the mutated cells (the promotion phase). However, these induced pre-neoplastic cells do not automatically progress to malignancy unless additional, secondary chromosomal and gene mutations take place. These mutations may alter the cell cycle controls, respond to growth regulatory factors, become invasive and metastatic (the progression phase). A malignancy only occurs after all three of these phases.

Finally, if the dose levels or dose rates are very low, the body's immune system may repair the damage to the cells or to the cellular DNA perfectly. Following such a process, there would be no lasting damage and the cells would have recovered fully.

# Revision Questions

1. What are the various types of radiation that can cause external exposure? Why are some types of radiation more hazardous from the external exposure point of view than others? Explain briefly. What protective measures can be taken against each of these radiations?

2. Explain how internal exposure occurs. Explain why an internal exposure to a small amount of radioactive material is more hazardous than an external exposure to the same amount even at close proximity.

3. Briefly explain the following terms giving an example of each of these effects:
   - (i) deterministic effects
   - (ii) stochastic effects
   - (iii) acute exposure
   - (iv) chronic exposure
   - (v) early effect
   - (vi) late effect

4. Show diagrammatically the dose–response characteristic of the stochastic and deterministic effects, clearly indicating the parameters used in the axes. Why is it assumed that there is a threshold level of dose for the deterministic effect, but not for the stochastic effect? What is the significance of the latency period in biological damage?

5. For the deterministic effect, the risk function is basically sigmoidal in character. Can you explain why?

6. Explain the significance of the biological half-life, radioactive half-life and effective half-life. Show them diagrammatically.

7. Explain the following terms:
   - (i) effective dose
   - (ii) committed effective dose

8. What is radiotoxicity? How is it defined? Explain the definition of radiotoxicity giving the relevant unit.

9. Sketch the basic structure of a cell and label it accordingly.

10. Cells are divided into two classes: somatic cells and germ cells. Explain the functions of these cells with some examples.

11. What do you understand by the processes of mitosis and meiosis? Briefly explain the process of cell reproduction.

12. Which part of the cell is most vulnerable to radiation damage and why?

13. Explain the process of radiation damage to the biological system from a cellular point of view.

## REFERENCES

[1] International Commission on Radiological Protection, 1990 Recommendations of the International Commission on Radiological Protection, ICRP Publication 60, Vol. 21 No. 1–3.
[2] National Radiological Protection Board, Risk from Deterministic Effects of Ionising Radiation, Documents of the NRPB, Vol. 7, No. 3,1996.
[3] International Commission on Radiological Protection, The ICRP Database of Dose Coefficients: Workers and Members of the Public, Version 1.0, 2001.
[4] International Atomic Energy Agency, International Basic Safety Standards for Protection against Ionising Radiation and for the Safety of Radiation Sources, Jointly sponsored by FAO, IAEA, ILO, OECD/NEA, PAHO, WHO, IAEA Safety Series No. 115, Vienna, 1996.
[5] International Commission on Radiological Protection, Age Dependent Doses to Members of the Public from Intake of Radionuclides: Part 5, Compilation of Ingestion and Inhalation Dose Coefficients, ICRP Publication 72, 1996.
[6] J. R. Lamarsh, *Introduction to Nuclear Engineering*, Addison-Wesley series in nuclear science and engineering, 1975.
[7] Wikipedia, The Free Encyclopedia, http://en.wikipedia.org/wiki/Cytoplasm

# 3

# RADIOLOGICAL PROTECTION

## 3.1 The Aim of Radiological Protection

The subject matter of radiological protection deals with ways and means of protecting human beings and their descendants, both individually and collectively, as well as the environment from the harmful effects of ionising radiation. No amount of ionising radiation, however small, is considered harmless. The higher the dose, the higher is the probability of harm and this dose–response relationship is considered linear within the limit of the threshold value (see Section 2.3.1). Beyond the threshold level, there is a non-linear dose–response relationship (see Section 2.3.2). Radiological protection deals mainly with low doses and low dose rates where deterministic effects do not occur.

The principle of the Linear No-Threshold (LNT) effect of low doses is considered to be a fundamental plank of radiological protection. The question which immediately springs to mind is that what is then the effect of the natural background radiation to which every human being (and every living creature) is continuously exposed? There must be some discernible effects from such ubiquitous exposures.

Indeed, it had been found that over 20% of the human population die of cancer of one type or another arising from causes which cannot be attributed to man-made activity. This death rate can thus be attributed to ubiquitous background radiation doses and other cancer-causing agents. This background radiation level is not uniform over the surface of the earth. In the UK, the average dose that a person receives from the background radiation is about 2.6 mSv per year. Figure 3.1 shows this background radiation effect in the shaded block and the LNT hypothesis is applicable beyond that level. Radiological protection deals with man-made radiation exposures, including medical exposures and ways of minimising such exposures to reduce the stochastic effects to as low as is reasonably practicable.

But before going into the details of radiological protection principles and practices, it is essential to know the essential dosimetric quantities which are used in radiological protection. These dosimetric quantities follow the definitions of the ICRP recommendations [1]

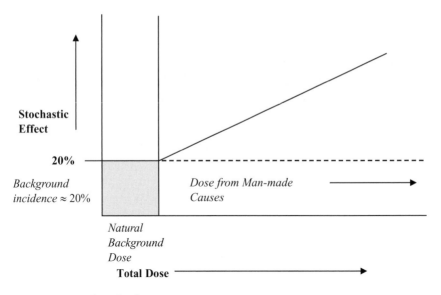

**Fig. 3.1** *LNT effect of radiation*

## 3.2 Dosimetric Quantities

### 3.2.1 Absorbed Dose

The fundamental dosimetric quantity in radiological protection is the absorbed dose, $D$. This is the mean energy deposited by a radiation in a unit mass of the medium. The absorbed dose can be stated mathematically as

$$D = \frac{dE}{dm} \tag{3.1}$$

where $dE$ is the mean energy imparted by the ionising radiation to the medium in a volume element (J), and $dm$ is the mass of the material in the volume element (kg).

The SI unit for the absorbed dose is called gray (Gy). Its units are joules per kilogram (J.kg$^{-1}$). The old unit is called the rad and is equal to 0.01 J.kg$^{-1}$. Thus 1 Gy = 100 rad.

It should be noted that although the absorbed energy may be in a small, microscopic volume or it may vary from point to point within the volume, for radiological protection purposes it is sensible as well as convenient to consider the whole of the absorbed energy averaged over the whole of the organ or tissue. When this is done, it becomes the tissue or organ absorbed dose. As an example, if a liver weighing 2 kg receives a radiation energy of 5 J at any part of the organ, the absorbed dose to the liver is estimated as 5/2 J kg$^{-1}$ = 2.5 J kg$^{-1}$ = 2.5 Gy.

## 3.2.2 Radiation Weighting Factor

The biological effectiveness in inducing stochastic effects by an ionising radiation depends not only on the absorbed dose but also on the type of the incoming radiation. The property associated with the type (and energy) of a radiation is characterised by a parameter called the **radiation weighting factor**, $w_R$. The values of this parameter have been chosen by the ICRP [1] to represent approximately the Relative Biological Effectiveness (RBE) in inducing stochastic effects at low doses. (Previously, in ICRP Publication 26 [2], the radiation weighting factor was associated with the absorbed dose at a point and was called the quality factor, $Q$). The relative biological effectiveness is related to the LET, a measure of the density of ionisation along the track of an ionising particle. A low LET radiation giving a low density of ionisation and hence low RBE would have a lower radiation weighting factor than the high LET radiation giving a high density of ionisation. The ICRP has provided a table of radiation weighting factors for various types and energies of radiation, as shown in Table 3.1.

The radiation weighting factors for neutrons of varying energies have been given discrete values. In practical dosimetric measurements and calculations, such discrete values would pose problems. So a smooth curve without discontinuity has been provided by the ICRP to represent the radiation weighting factors for neutrons (see Figure 3.2). The equation for this smooth curve is

$$w_R = 5 + 17\exp\left[-\left(\ln(2E)\right)^2/6\right]$$

**Table 3.1** *Radiation weighting factors*

| Radiation | Radiation weighting factor, $w_R$ |
|---|---|
| Photons, all energies | 1 |
| Electrons, all energies | 1 |
| Neutron energies: | |
| < 10 keV | 5 |
| 10–100 keV | 10 |
| > 100 keV–2 MeV | 20 |
| > 2–20 MeV | 10 |
| > 20 MeV | 5 |
| Protons, all energies | 5 |
| α-particles, heavy nuclei | 20 |

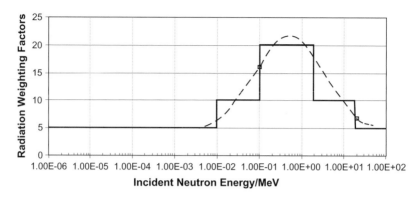

**Fig. 3.2** *Radiation weighting factors for neutrons. The smooth curve is an approximation [1]*

### 3.2.3 Equivalent Dose

The absorbed dose to an organ or tissue weighted by the radiation weighting factor is the parameter which is of interest in radiation protection. This parameter is called the equivalent dose and is given by

$$H_T = \sum_R w_R \, D_{T,R} \tag{3.2}$$

where the summation is carried over all types of radiation. $H_T$ is the equivalent dose in the tissue, $T$ (J.kg$^{-1}$); $w_R$ is the radiation weighting factor for the type of radiation, $R$; and $D_{T,R}$ is the absorbed dose averaged over tissue or organ, $T$ due to the radiation type, $R$.

Although the basic unit of the equivalent dose is J.kg$^{-1}$, it is given a special name, the sievert (Sv), to differentiate it from the absorbed dose. The old unit was the rem which is approximately equal to 0.01 J.kg$^{-1}$. Thus 1 Sv = 100 rem.

Equation (3.2) specifies that when a radiation field is composed of various types of radiation (or, for neutrons it is composed of various energies), the absorbed doses must be calculated for each type of radiation and weighted by the corresponding value of $w_R$ and then summed together to obtain the total equivalent dose.

### 3.2.4 Tissue Weighting Factor

To determine the total health detriment from non-uniform irradiation of the body, some mechanism needs to be found whereby the risk of a fatal cancer among the organs can be calculated. The United Nations Scientific Committee

**Table 3.2** *Grouping of tissues for tissue weighting factors*

| Organs | $w_T$ | Total $w_T$ |
|---|---|---|
| Bone surfaces | 0.01 | 0.02 |
| Skin | | |
| Bladder | 0.05 | 0.30 |
| Breast | | |
| Liver | | |
| Oesophagus | | |
| Thyroid | | |
| Remainder | | |
| Colon | 0.12 | 0.48 |
| Lung | | |
| Red bone marrow | | |
| Stomach | | |
| Gonads | 0.20 | 0.20 |
| Whole body | | 1.00 |

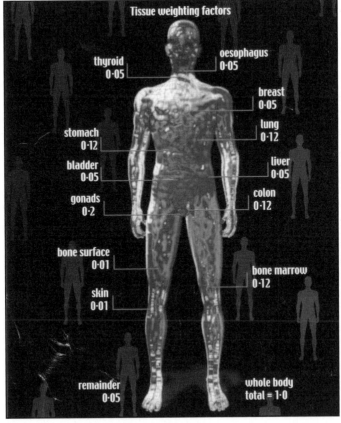

**Fig. 3.3** *Assigned values of tissue weighting factors*

on the Effects of Atomic Radiation (UNSCEAR) have derived a list of fatal cancer probabilities in organs using age average risk coefficients [3]. The ICRP used these UNSCEAR values but rounded them up in order to avoid giving the impression of biological precision in risk estimate analysis. The ICRP had placed the various organs and tissues in groups and assigned certain values which are shown in Table 3.2. These values are called the **tissue weighting factors**, $w_T$. The assigned values of tissue weighting factors to various organs and tissues are shown in Figure 3.3 [4].

The tissue weighting factors, $w_T$, represent the risks various organs and tissues suffer from uniform irradiation when normalised to the whole body irradiation risk of 1. So $w_T$ when multiplied by the equivalent dose to the tissue or organ represents the relative contribution of that organ or tissue to the total detriment resulting from uniform irradiation of the whole body.

### 3.2.5 Effective Dose

The effective dose, $E$ is defined as the sum of the products of equivalent doses to various tissues with their respective tissue weighting factors. As the equivalent dose itself is the weighted absorbed dose, the effective dose becomes the doubly weighted absorbed dose. Thus the effective dose, $E$ is

$$E = \sum_T w_T H_T = \sum_T w_T \sum_R w_R D_{T,R} \tag{3.3}$$

The SI unit of this quantity is also the sievert, Sv.

### 3.2.6 Committed Equivalent Dose and Committed Effective Dose

When a radionuclide is incorporated within the body, that radionuclide may be distributed around the body or may preferentially irradiate specific organs or tissues depending on the biokinetic behaviour of that nuclide. There would be a period during which the radionuclide would give rise to equivalent doses to the tissues or organs of the body at varying rates. The time integral of the equivalent dose rate is called the **committed equivalent dose**, $H(t)$ where $t$ is the integration time in years. If the time $t$ is not specified, then it is taken to be 50 years for adults and 70 years for children. Thus the committed equivalent dose is

$$H_T(t) = \int_0^t \overline{H_T} \, dt \tag{3.4}$$

When the committed equivalent doses are multiplied by the respective tissue weighting factors and then summed, the **committed effective dose**, $E(t)$ is obtained. The committed effective dose can also be obtained by the time integral of effective dose rate over a period of 50 years for adults and 70 years for children. Thus the committed effective dose, $E(t)$ is

$$E(t) = \sum_T H_T(t).w_T = \int_0^t \bar{E} \, dt \qquad (3.5)$$

## 3.2.7 Collective Equivalent Dose and Collective Effective Dose

When a group of people is exposed to a source of radiation which affects a specific organ or tissue, $T$, a quantity called the **collective equivalent dose**, $S_T$, to the organ is evaluated. This is given by

$$S_T = \int_0^\alpha H_T \frac{dN}{dH_T} \, dH_T \qquad (3.6)$$

where $(dN/dH_T)$ is the number of individuals receiving an equivalent dose between $H_T$ and $H_T + dH_T$, and $\alpha$ is the upper limit of the equivalent dose.

When a measure of the whole body radiation exposure in a population is desired, the **collective effective dose**, $S$, can be calculated. This is given by

$$S = \int_0^\alpha E \frac{dN}{dE} \, dE \qquad (3.7)$$

where $E$ is the effective dose to an individual.

The SI unit for both the collective equivalent dose and the collective effective dose is the man-Sv. It should be noted that no time period is specified for either of these two quantities. Therefore, the time period and population over which they are calculated should be specified.

## 3.3 System of Radiological Protection

Radiological protection aims to do more good than harm to individuals as well as to society. Any human activity involving radioactive materials or radiation-generating equipment may be separated, for the purposes of radiological protection, into a 'practice' or an 'intervention'. A '**practice**' is defined as any human activity that introduces additional sources of exposures or exposure pathways or extends exposure to additional people or modifies the network of exposure pathways from existing sources, so as to increase the exposure or the likelihood of exposure of people [1]. An **intervention**, on the other hand, is defined as any human action that is intended to reduce or avert exposure or the likelihood of exposure to sources which are not part of a controlled practice. The objectives of radiation protection are to limit exposures from 'practices' so that more good than harm ensues.

Exposures can be divided into: (i) **occupational exposures**, (ii) **medical exposures** and (iii) **public exposures**. The basic principles of radiation protection apply to all of these exposures. However, a slightly different mode of justification is required for medical exposures.

## 3.4 Radiological Protection in Practice

The implementation of radiological protection principles is carried out under the '**system of radiological protection**' as recommended by the ICRP in its publication 26 [2] and carried forward in publication 60 [1]. The three fundamental recommendations under the system are: justification of a practice, optimisation of protection, and individual dose and risk limits.

### 3.4.1 Justification of a Practice

The principle of justification of a practice states

"No practice involving exposures to radiation should be adopted unless it produces sufficient benefit to the exposed individuals or to society to offset the radiation detriment it causes."

The very concept of sufficient benefit introduces the aspect of cost-benefit analysis which needs to be considered in order to make a decision regarding the introduction of a practice involving radiation. In this cost-benefit analysis, the net benefit, $B$, of a practice is given as

$$B = V - (P + X + Y) \tag{3.8}$$

where $V$ is the gross benefit which includes tangible and intangible social, economic, commercial and other benefits. $P$ is the basic production cost which includes costs of non-radiological detriments and the costs of protecting against these hazards. $X$ is the cost of radiation protection, and $Y$ is the cost of radiation detriment.

These terms are not always easy to quantify, particularly when such intangible items as human satisfaction etc. in $V$ are involved. However, broadly speaking, the benefit includes all the positive aspects accruing to the society and not just those received by particular groups or individuals. The costs are considered to comprise the sum total of all the negative aspects of a practice, including monetary costs and any damage to human health or to the environment. Since the distribution of the benefits and costs are unlikely to be uniform across the population, a process of broadly balancing these effects should be undertaken. The introduction of a practice is only justified if $B$ is positive and is increasingly justified at higher positive values of $B$.

### 3.4.2 Optimisation of Radiation Protection

The use of a particular source within a practice is only allowed after adequate provisions have been made to ensure that the exposure is As Low As Reasonably Practicable (ALARP), economic and social factors being taken into account. It should be noted that ALARP is an adaptation in the UK from the internationally accepted principle of As Low As Reasonably Achievable (ALARA). This procedure is constrained by restrictions on doses to individuals (**dose**

**constraints**), or risks to individuals, in the case of potential exposures (**risk constraints**) [1]. This recommendation can be separated into optimisation on a collective level and optimisation on an individual level. Both of these are dealt with below.

The first aim of this quantitative analysis is to assess how far exposures may be reduced before further reduction would not justify the incremental cost required to accomplish it. This assessment is made by a differential cost-benefit analysis intended to maximise the net benefit, $B$. Here the independent variable is the collective effective dose, $S$. When $B$ is differentiated with respect to $S$, one obtains

$$\frac{dB}{dS} = \frac{dV}{dS} - \left( \frac{dP}{dS} + \frac{dX}{dS} + \frac{dY}{dS} \right)$$

At the maximum value of $B$, $dB/dS$ would be zero and hence

$$\frac{dV}{dS} - \left( \frac{dP}{dS} + \frac{dX}{dS} + \frac{dY}{dS} \right) = 0 \tag{3.9}$$

As $V$ and $P$ are invariant with respect to $S$ for a given practice

$$\frac{dX}{dS} = -\frac{dY}{dS} \tag{3.10}$$

Equation (3.10) is the optimisation condition. This effectively means that when the rate of variation of the cost of radiation protection with $S$ becomes equal and opposite to the rate of variation of the cost of radiation detriment with $S$, an optimum situation has been reached. Figure 3.4 shows the variations of X and $Y$ with $S$ and the point where these two lines intersect gives the optimised collective dose value, $S_0$

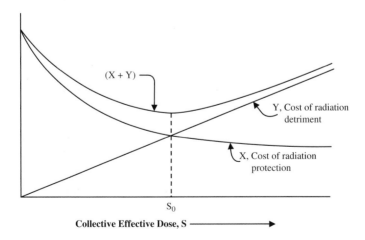

**Fig. 3.4** *Optimisation of radiation protection*

At the level of individual doses, the choice of dose constraints from a source is an important part of the optimisation process. For many types of occupation, it is possible to reach conclusions about the level of individual doses likely to be incurred in well-managed operations. This information can be used to establish a dose constraint for the type of operation.

### 3.4.3 Individual Dose and Risk Limits

The third recommendation from the ICRP states that the exposure of individuals resulting from nuclear practices should be subject to dose limits, or risk limits in the case of potential exposures.

#### 3.4.3.1 Statutory Dose Limits

On the basis of recommendations of the ICRP in 1990 [1], the Council of the European Union produced the Council Directive 96/29/Euratom [5] in May 1996. Conforming to this directive is obligatory to all member states. Subsequently in 1996 the International Atomic Energy Agency (IAEA) produced a safety standards document, now generally known as the International Basic Safety Standards (BSS) [6], incorporating the ICRP-60 recommendations [1] and the EC Directive [5]. In the UK, the responsibility for initiating national legislation related to radiological protection lies with the Health and Safety Commission (HSC) (see Section 6.3.1 for details of the relationship between the HSC and the Health and Safety Executive (HSE)). A new set of regulations, known as the **Ionising Radiations**

**Table 3.3** *Statutory dose limits (IRR99)*

| Dosimetric quantity | Dose limits | | |
|---|---|---|---|
| | **Workers aged 18 years or over (\*)** | **Trainees under 18 years** | **Other persons** |
| Effective dose | 20 mSv.y$^{-1}$(\*\*) | 6 mSv.y$^{-1}$ | 1 mSv.y$^{-1}$(\*\*\*) |
| Equivalent dose in | | | |
| lens of the eye | 150 mSv.y$^{-1}$ | 50 mSv.y$^{-1}$ | 15 mSv.y$^{-1}$ |
| skin | 500 mSv.y$^{-1}$(\*\*\*\*) | 150 mSv.y$^{-1}$ | 50 mSv.y$^{-1}$ |
| hands, feet and ankles | 500 mSv.y$^{-1}$ | 150 mSv.y$^{-1}$ | 50 mSv.y$^{-1}$ |

* Workers here refer to both men and women, aged 18 and 65. However, there is a special restriction for women of reproductive capacity; the equivalent dose from external radiation averaged throughout the abdomen shall be limited to 13 mSv in any consecutive three-month period. Also when a woman becomes pregnant, the equivalent dose to the foetus should be limited to 1 mSv for the remainder of the pregnancy.
** Although an effective dose of 20 mSv.y$^{-1}$ averaged over five years (100 mSv in 5 y) is recommended, there is a provision that in any single year the effective dose shall not exceed 50 mSv.
*** Although the annual dose for other persons is specified as 1 mSv, it is also stated that under special circumstances a higher effective dose could be allowed provided that the average over five years does not exceed 1 mSv.y$^{-1}$.
**** Skin dose is averaged over an area of 1 cm$^2$ regardless of the area exposed.

**Regulations 1999** (IRR1999) [7], came into force at the beginning of 2000. These dose limits are legally enforceable and are shown in Table 3.3.

For compliance with the regulatory dose limits, the effective dose or, more precisely, the total effective dose is calculated as [6]

$$E_T = H_p(d) + \sum_j e(g)_{j,ing} I_{j,ing} + \sum_j e(g)_{j,inh} I_{j,inh} \qquad (3.11)$$

where $H_p(d)$ is the personal dose equivalent from exposure to a penetrating directional radiation; $e(g)_{j,ing}$ and $e(g)_{j,inh}$ are dose coefficients (i.e. dose per unit intake) from ingestion and inhalation of radionuclide, $j$, respectively; $I_{j,ing}$ and $I_{j,inh}$ are the intake of $I$ radionuclides of type $j$ by ingestion and inhalation, respectively.

## 3.5  UK MoD Requirements

Although the Euratom treaty does not apply to military activities and the Ministry of Defence (MoD) embracing the navy, army and air force with all their headquarters, units, establishments, the fleet, ships etc. is exempt, legislation emanating from Euratom can form the basis of EU directives relating to health and safety at work. These directives are then incorporated into the national legislation and in the UK they are brought under the framework of the Health and Safety at Work etc. Act, 1974 (HSWA74) and these are directly applicable to the MoD.

There is no general Crown exemption for the MoD from the HSWA74. The MoD is bound by the general duties imposed by the Act and by regulations made under it except where specific exemptions are given. The Crown is, however, exempt from certain enforcement provisions: in particular, **Improvement and Prohibition Notices** are not legally enforceable on Crown premises although the HSE have instituted a procedure for issuing **Inspector's Notices** in lieu.

Until May 1996, the MoD was exempt from the inspection and monitoring of safety standards by HSE inspectors. However, under a 'general agreement' in May 1996 between the MoD and the HSE, the restrictions on the HSE inspectors had been partially removed. But there are still some restrictions in place so that national security is not compromised in order to maintain rigid safety standards.

## 3.6  MoD Standards

In view of the Basic Safety Standards Directive from the EU (see Section 6.2.6) and the consequent IRR99, the MoD has introduced stringent individual dose and risk limits which are contained in the MoD's *Safety Principles and Safety Criteria* document [8]. The criteria for individual doses from the normal operation of a plant are given in Table 3.4. The MoD uses the Basic

Safety Level (BSL) and the BSO (Basic Safety Objective) to define dose criteria (see Section 7.6.1 for definitions of BSL and BSO).

**Table 3.4** *Individual dose criteria for normal operation of a nuclear facility*

|  | BSO | Constraint | BSL |
|---|---|---|---|
| **Radiation worker** |  |  |  |
| Effective dose | 2 mSv.y$^{-1}$ | 15 mSv.y$^{-1}$ | 20 mSv.y$^{-1}$ |
| Average dose from a single site | 1 mSv.y$^{-1}$ | – | 10 mSv.y$^{-1}$ |
| **Non-radiation worker** |  |  |  |
| Effective dose | 0.5 mSv.y$^{-1}$ | – | 5 mSv.y$^{-1}$ |
| **General public** |  |  |  |
| Effective dose to a critical member | 0.02 mSv.y$^{-1}$ | 0.3 mSv.y$^{-1}$ | 1 mSv.y$^{-1}$ |

The MoD also requires that the total dose level in five consecutive years shall not exceed 100 mSv with the constraint that if the dose level in five consecutive years reaches 75 mSv, an investigation shall be initiated. From 1995, when an individual is likely to receive or exceed 6 mSv of dose per year, that person should be designated as a classified radiation worker. If the person exceeds a dose level of 6 mSv in any calendar year, the work practice will be investigated. This constitutes the initial level of investigation. A second investigation level has been set at 15 mSv whole body dose. This local investigation is instituted to establish whether all steps are being taken to keep the radiation level as low as is reasonably practicable. If the cumulative dose to an individual over a period of five years reaches 75 mSv, the third level of investigation will be carried out.

In addition, the MoD lists planning targets and planning limits for the collective dose from normal operation of a plant (see Table 3.5).

**Table 3.5** *Planning targets and planning limits for collective dose for radiation workers*

|  | Planning target | Planning limit |
|---|---|---|
| Total collective dose per plant from critical operation | 20 man-mSv.y$^{-1}$ | 200 man-mSv.y$^{-1}$ |
| Total collective dose per plant during refit | 500 man-mSv | 5 man-mSv |
| Total collective dose per plant during maintenance (excluding refit) | 20 man-mSv.y$^{-1}$ | 200 man-mSv.y$^{-1}$ |
| Total collective dose per plant during DDLP | 100 man-mSv | 500 man-mSv |

# Revision Questions

1. Explain the significance of the term 'Linear No Threshold (LNT)' effect of radiation in matters of radiation protection.

2. Define the following quantities with units, where applicable:
   - (i)   absorbed dose
   - (ii)   radiation weighting factor
   - (iii)   equivalent dose
   - (iv)   tissue weighting factor
   - (v)   effective dose
   - (vi)   committed effective dose
   - (vii)   collective effective dose

3. What is the 'system of radiological protection'? Describe the main elements of this recommendation.

4. What are the legal dose limits applicable to a radiation worker and a member of the public in the EU?

5. What are the BSL and BSO values for effective doses to a radiation worker, a non-radiation worker and a general member of the public, as applied by the UK MoD?

## REFERENCES

[1] International Commission on Radiological Protection, Recommendations of the International Commission on Radiological Protection, ICRP Publication 60, *Annals of the ICRP*, 1990, Vol. 21, No. 1–3.

[2] International Commission on Radiological Protection, Recommendations of the ICRP, ICRP Publication 26, Vol. 1 No. 3., 1977.

[3] UNSCEAR, Sources, Effects and Risks of Ionizing Radiation, Annex F, Radiation Carcinogenesis in Man, United Nations Scientific Committee on the Effects of Atomic Radiation, United Nations, New York, USA, 1988.

[4] Gaines M., *New Scientist*, Inside Science, 18 March 2000.

[5] EU Council Directive 96/29/Euratom of 13 May 1996, laying down Basic Safety Standards for the Protection of Health of Workers and the General Public Against the Dangers Arising from Ionising Radiation, *Official Journal of the European Union* L 159, 29 June 1996.

[6] International Atomic Energy Agency, International Basic Safety Standards for Protection against Ionising Radiation and for the Safety of Radiation Sources, jointly sponsored by FAO, IAEA, ILO, OECD/NEA, PAHO, WHO. Safety Series No. 115, IAEA, Vienna, Austria, 1996.
[7] UK Health and Safety Executive, The Ionising Radiations Regulations 1999, Statutory Instruments No. 3232.
[8] UK Ministry of Defence, Safety Principles and Safety Criteria for the Naval Nuclear Propulsion Programme, Issue 2, August 1996.

# 4

# STATISTICAL METHODS

## 4.1 Basic Statistical Quantities

### 4.1.1 Mean, Median and Mode

The arithmetic mean of a variable is the average value of that variable. The variable may be the anything such as the height of individuals, concentration of pollutants in air, the activity of radioactive sample or the levels of exposure of individuals etc. If a variable, $x$ takes on a number of discrete values, $x_i$ where $i = 1, 2, 3 ,... n$, then the arithmetic **mean,** $\bar{x}$ is given by

$$\bar{x} = \frac{\sum_{i=1}^{n} x_i}{n} \tag{4.1}$$

where $n$ is the total number of discrete values of the variable.

Here, each value of $x$ has been assumed to occur only once. There may be occasions when a variable takes on a specific value a number of times. In the measurement of the activity of a radioactive sample over a period of time, there may be a number of the times when the measured activity could be the same. This multiple occurrence of the same value of a variable is called the frequency, $f$ and the distribution of values of the variable is called the **frequency distribution**.

The arithmetic mean, $\bar{x}$ of a frequency distribution is calculated as

$$\bar{x} = \frac{\sum_{i=1}^{n} f_i x_i}{N} \tag{4.2}$$

where

$$N = \sum_{i=1}^{n} f_i$$

If the numerical values of a variable are large or unwieldy, then arithmetic simplification can be made by subtracting an arbitrary value from each of these values and then adding that arbitrary value to the mean of the differences to obtain the mean of the original values. The equation for the mean is thus:

$$\bar{x} = \frac{\sum_{i=1}^{n} f_i(x_i - m)}{N} + m \qquad (4.3)$$

where $m$ is a chosen arbitrary number.

The **median** of a distribution, on the other hand, is the central value of a variable when the values are arranged in ascending or descending order (or the mean of two centre values for an even number of values).

The **mode** of a distribution is the value of a variable which occurs most frequently. In a frequency distribution, there may be more than one mode.

---

## Example 4.1

Calculate the mean, median and mode of the following frequency distribution of a variable (counts per minute of a radioactive sample).

| $i$ | 1 | 2 | 3 | 4 | 5 | 6 |
|---|---|---|---|---|---|---|
| $x_i$ | 23 | 25 | 26 | 29 | 31 | 35 |
| $f_i$ | 1 | 1 | 2 | 1 | 3 | 2 |

### Solution

The mean $\bar{x} = \dfrac{1\times23+1\times25+2\times26+1\times29+3\times x31+2\times35}{10} = 29.2$

The median $= \dfrac{29+31}{2} = 30.0$

The mode $= 31$

---

The variable, $x_i$ may assume a large number of different values. For example, the heights of adult males can assume literally infinite numbers of discrete values (within the upper and lower limits) depending on how precisely the measurements are made and recorded. The collection of all of these possible values of a variable constitutes a population. However, a population does not always need to be infinite. For example, the number of people in a town or in a county, the number of patients in the National Health Service, the number of students in schools are all examples of populations. The number in each case may be large but it is numerically finite. However, the measurement of each and every member of the population is neither feasible nor desirable. So a selected number of items from the population may be chosen and that is known as **sampling**. The total number of values of a variable in a sample is normally denoted by the quantity $n$, whereas the total number of values in a population is denoted by the quantity, $N$. So $N$ may be taken to be the sum of all the '$n$'s. Whereas the sample mean is denoted by $\bar{x}$, the population mean is normally denoted by the parameter, $\mu$ (mu). When a sample chosen is without any preference or bias, then that sample is called the unbiased or random sample. The techniques of drawing samples and the properties of those samples in representing the characteristic of the population are described in Section 4.3.

## 4.1.2 Variance and Standard Deviation

A measure of the spread or dispersion of a variable, $x_i$ about the mean, $\bar{x}$ is given by a quantity called the variance, $\sigma^2$ and is defined as,

$$\sigma^2 = \frac{1}{N} \sum_{i=1}^{n} f_i (x_i - \bar{x})^2 \tag{4.4}$$

$$= \frac{1}{N} \sum_{i=1}^{n} f_i x_i^2 - \frac{2\bar{x}}{N} \sum_{i=1}^{n} f_i x_i + \frac{\bar{x}^2}{N} \sum_{i=1}^{n} f_i$$

$$= \frac{1}{N} \sum_{i=1}^{n} f_i x_i^2 - 2\bar{x}^2 + \bar{x}^2$$

$$= \frac{1}{N} \sum_{i=1}^{n} f_i x_i^2 - \bar{x}^2 \tag{4.5}$$

If the numerical values of $x_i$ are large, then the variance calculation may be very tedious. However, it may be simplified by subtracting an arbitrary quantity $m$ from each of the $x_i$ values and then taking account of $m$. The equation for variance is then given by

$$\sigma^2 = \frac{\sum_{i=1}^{n} f_i d_i^2}{N} - \left( \frac{\sum_{i=1}^{n} f_i d_i}{N} \right)^2 \tag{4.6}$$

where $d_i = (x_i - m)$.

The unit of variance is the square of the unit of $x$. For the derivation of this equation, see Annex 4A at the end of this chapter.

The standard deviation is defined as the square root of the variance, $\sigma^2$. So the standard deviation, $\sigma$ is given by

$$\sigma = \sqrt{\sigma^2} = \sqrt{\frac{1}{N} \sum_{i=1}^{n} f_i (x_i - \bar{x})^2} \tag{4.7}$$

The unit of standard deviation is the same as that of $x$. It should also be noted that as $\sigma$ is taken as the positive square root of $\sigma^2$, $\sigma \geq 0$. A large standard deviation (or variance) implies that there is a large spread of data values around the mean.

When the standard deviation is expressed as a percentage of the mean of the variable, a quantity, called the **coefficient of variation**, $v$ (nu) is obtained. This is given by

$$v = \frac{100\sigma}{\bar{x}} \tag{4.8}$$

The coefficient of variation measures the dispersion of the variable in terms of the mean. A low value of the coefficient of variation indicates close clustering of values whereas a large value indicates a wide scatter.

Quite often a limited number of samples from a large population are used to characterise the whole population. This introduces an element of error which is called the standard error. The standard error of the means of a number of samples is calculated as the standard deviation, $\sigma_s$ of the means of samples divided by the square root of the number of observations, $n$. This is given by

$$\text{standard error of the mean} = \frac{\sigma_s}{\sqrt{n}} \tag{4.9}$$

where $\sigma_s$ is the standard deviation of the means of $n$ observations.

A limited sample size will have an effect on the estimate of the mean which is utilised in the calculation of $\sigma_s$. So a correction, called the Bessel's correction, is applied when the standard error of the mean is given by

$$\text{standard error of the mean} = \frac{\sigma_s}{\sqrt{n-1}} \tag{4.10}$$

where $(n-1)$ is the number of degrees of freedom.

## 4.1.3 Percentile and Quartile

Sometimes it is important to know the position of an observation in relation to all the observations. Percentile is one such quantity. A percentile is a data value that is greater than or equal to a given percentage of data values. For example, if $x$ is the $p$th percentile, then $p\%$ of the values in the data set are less than or equal to $x$, and so $(1-p)\%$ of the values are greater than or equal to $x$. The calculation of a percentile value is carried out by arranging the data set in ascending order and then identifying the data value which corresponds to the required percentage of the number of data items. Percentiles are also called quantiles.

Some important percentiles are quartiles of the data – 25th, 50th and 75th percentiles. The 50th percentile is the sample median. Figure 4.1 shows the quartile values of an arbitrary data set.

## Example 4.2

Calculate the 90th and 95th percentiles from the following data set of 11 observations of radioactivity reading.

8, 10, 14, 8, 14, 8, 20, 16, 10, 12 and 17 cpm

### Solution

The data set when arranged in ascending order becomes

8, 8, 8, 10, 10, 12, 14, 14, 16, 17, 20 cpm

Calculation of the 90th percentile:
90 percent of 11 data points is $0.9 \times 11 = 9.9$.
Now, 9.9 data points is rounded up higher to 10 data points.
So, the 90th percentile value is the 10th data point which is 17 cpm.
Calculation of the 95th percentile:
95 percent of 11 data points is $0.95 \times 11 = 10.45$. As 10.45 is between 10 and 11, the 95th percentile value is the average of 10th and 11th data points.
So, 95th percentile = $(17 + 20) / 2 = 18.5$ cpm.

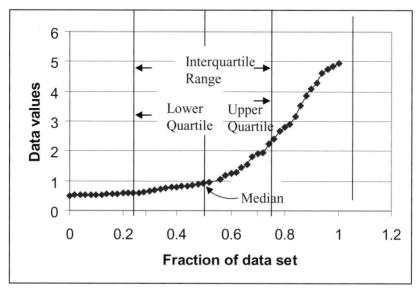

**Fig. 4.1** *Example of quartile values*

## 4.2 Probability Distributions

### 4.2.1 The Binomial Distribution

The binomial distribution is applicable when the variable in question can only take two possible values. One of these values may be considered as a success and the other would then be a failure. When a coin is tossed, one can obtain only a head or a tail. If obtaining a head is considered a success, then not obtaining a head e.g. obtaining a tail would be a failure. When a die is thrown, getting a desired number, say 2, may be considered a success, whereas not getting a 2 e.g. getting any other number would then be a failure. Other binomial variables could be the state of a switch – either on or not on (off); birth of a child – either a boy or not a boy (a girl) and so on.

Let $p$ be the probability of success e.g. obtaining a head in a toss of a coin and $q$ be the probability of failure e.g. getting a tail, then obviously

$$p + q = 1 \tag{4.11}$$

If there are $n$ independent trials or observations, then the probability of receiving $0,1,2,3...r,...n$ successes is given by the successive terms of the binomial series $(q + p)^n$

$$(q+p)^n = q^n + {}^nC_1 q^{n-1} p + {}^nC_2 q^{n-2} p^2 + ... + {}^nC_r q^{n-r} p^r + ... + p^n \tag{4.12}$$

where the binomial coefficient, ${}^nC_r$ is

$${}^nC_r = \frac{n!}{r!(n-r)!} \tag{4.13}$$

The general expression for the probability of $r$ successes and hence $(n-r)$ failures is given by

$$P(r) = {}^nC_r q^{n-r} p^r \tag{4.14}$$

It can be shown mathematically that the mean or expectation value and the variance of the binomial distribution are

$$\text{mean} = n\,p \tag{4.15}$$

$$\text{variance} = n\,p\,q \tag{4.16}$$

where $n$ is the number of trials or observations.

## Example 4.3

In a family of 5 children, what is the probability of having 3 boys and 2 girls? (Given that the probability of a male birth is 0.51 and a female birth is 0.49).

*Solution*

Let $b$ be the probability of birth of a boy and $g$ be the probability of birth of a girl. The binomial distribution is

$$(g+b)^5 = g^5 + 5g^4b + 10g^3b^2 + 10g^2b^3 + 5gb^4 + b^5$$

The probability of having 3 boys and 2 girls is $10\,g^2\,b^3 = 10 \times (0.49)^2 \times (0.51)^3 = 0.3185 = 32\%$.

### 4.2.2 Poisson's Distribution

This distribution, named after the mathematician Siméon-Denis Poisson (1781–1840), is a special case of binomial distribution when the binomial probability of success, $p$, is set equal to $m/n$, where $m$ is a constant quantity and $n$ is a variable which may increase indefinitely. In other words, the probability of success, $p$, is made a very small quantity by increasing $n$ indefinitely. An example of Poisson's distribution may be given from the radioactive decay process. Radioactive decay is fundamentally a random process and so the number of decays taking place at any time interval, $t$, would be different at different $t$ intervals. If the time interval, $t$, is then divided into $n$ equal sub-intervals (each of $t/n$) where $n$ is sufficiently large, the probability of detecting decay events in a sub-interval, $t/n$, is going to be very small and hence $p \ll 1$. The probability of detecting $r$ decays in the time interval, $t/n$, is then given by the corresponding term of the binomial distribution, $(q + p)^n$.

$$P(r) = {}^nC_r\,q^{n-r}\,p^r \tag{4.17}$$

The mean and variance of Poisson's distribution can be deduced from the binomial distribution by putting $p = m/n$, where $m$ is a constant quantity and letting $n$ increase indefinitely. So the mean becomes,

$$\text{mean} = \lim_{n \to \infty} np = m \tag{4.18}$$

$$\text{variance} = \lim_{n \to \infty} npq = \lim_{n \to \infty} mq = m \tag{4.19}$$

as, $q \to 1$ when $p \to 0$.

It can be shown mathematically that equation (4.17) under the stated limiting condition becomes

$$\lim_{n\to\infty} P(r) = \lim_{n\to\infty} {}^nC_r q^{n-r} p^r = \frac{m^r e^{-m}}{r!} \tag{4.20}$$

This is Poisson's distribution. It shows that the probability of detecting $r$ events becomes independent of both $n$ and $p$ individually, it depends only on their product. From equation (4.20), one can also deduce that,

$$P(r+1) = \frac{m^{r+1} e^{-m}}{(r+1)!} = \frac{m}{(r+1)} \frac{m^r e^{-m}}{r!} = \frac{m}{(r+1)} P(r) \tag{4.21}$$

Figures 4.2 and 4.3 show examples of Poisson's distribution with $m = 3$ and $m = 10$. This shows that as $m$ becomes larger and larger, the Poisson distribution becomes similar to the normal distribution which is described in the next section.

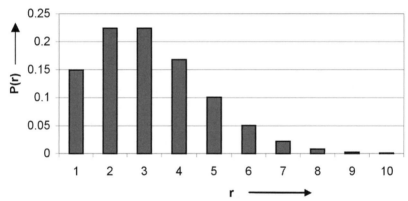

**Fig. 4.2** *Poisson's distribution with m=3*

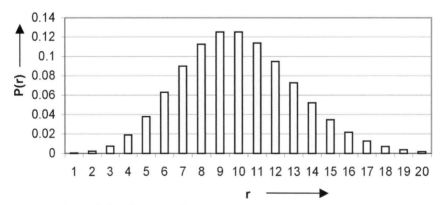

**Fig. 4.3** *Poisson's distribution with m=10*

### 4.2.3 Normal Distribution

The normal distribution plays a pivotal role throughout science, particularly in the physical and biological sciences. It is also referred to as the Gaussian distribution, after Friedrich Karl Gauss (1777–1855), who derived this distribution.

In this distribution the variable, $x$, is a continuous one, as against discrete values in binomial and Poisson distributions. The value of $y$, given as the ordinate, for a certain value of $x$ is given by

$$y = \frac{1}{\sigma \sqrt{2\pi}} \exp\left(-\frac{(x-\bar{x})^2}{2\sigma^2}\right) \qquad (4.22)$$

where $x$ is the independent variable ranging from $-\infty$ to $+\infty$, $\bar{x}$ is the arithmetic mean of the variable, and $\sigma$ is the standard deviation of $x$.

Mathematically it can be shown that the infinite integral of the exponential term in equation (4.22) is equal to $\sigma \sqrt{2\pi}$. In other words,

$$\int_{-\infty}^{\infty} \exp\left(-\frac{(x-\bar{x})^2}{2\sigma^2}\right) dx = \sigma \sqrt{2\pi} \qquad (4.23)$$

$$\text{so,} \quad \frac{1}{\sigma \sqrt{2\pi}} \int_{-\infty}^{\infty} \exp\left(-\frac{(x-\bar{x})^2}{2\sigma^2}\right) dx = 1 \qquad (4.24)$$

This shows that the LHS of equation (4.22) without the integration is, in fact, the probability value (also known as the probability density) whose integral over the whole range (from $-\infty$ to $+\infty$) is the total probability of 1.

This normal distribution is characterised by two parameters – the arithmetic mean and the standard deviation (or variance). The arithmetic mean of the whole distribution is normally denoted by the population mean, $\mu$, and so $\bar{x}$ will be replaced by $\mu$ from now on. Three normal distributions with two different arithmetic means ($\mu = 0$ and 1) and ($\sigma = 1$) are shown in Figure 4.4. The standard deviation, $\sigma$, determines the spread of the normal distribution.

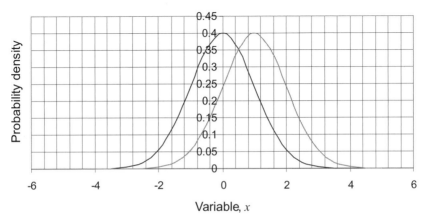

**Fig. 4.4** *Normal distributions with $\mu = 0$ and $\sigma = 1$*

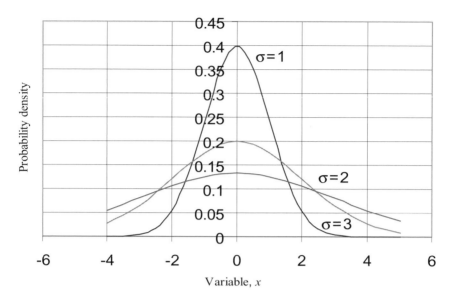

**Fig. 4.5** *Normal distributions with standard deviations of 1, 2 and 3 and μ = 0*

Figure 4.5 shows three normal distributions with different standard deviations ($\sigma$ = 1, 2 and 3) and an arithmetic mean of 0.

The properties of the normal distribution can be summarised as:

- A normal distribution is symmetrical about the arithmetic mean. This implies that the area to the right of the mean is exactly equal to the area to the left of the mean, each being equal to 0.5. This specifies that the mean is at the middle of the distribution.

- The mean, median and mode of a normal distribution are coincident and that is at the maximum height of the distribution.

- As the total area under the normal curve is 1, the wider the spread of the distribution i.e. the larger the standard deviation, the smaller is the height of the probability distribution.

- The fraction of the normal distribution which lies between limits on either side of the mean is:
  68.3 per cent between ($\mu + \sigma$) and ($\mu - \sigma$)
  95.5 per cent between ($\mu + 2\sigma$) and ($\mu - 2\sigma$)
  99.7 per cent between ($\mu + 3\sigma$) and ($\mu - 3\sigma$)
  99.99 per cent between ($\mu + 4\sigma$) and ($\mu - 4\sigma$)

Thus it can be stated that in a normal distribution, there is 68 per cent probability that a variable picked up at random will lie within ($\mu \pm \sigma$) or 95 per cent probability that it would lie within ($\mu \pm 2\sigma$).

### 4.2.3.1 The Probability Calculation

The probability of finding a value of $x$ within a certain range can be found by calculating the area of the normal curve within that range. The probability of finding $x$ anywhere within the range is obviously 1, as the area under the normal curve is 1.

As mentioned above, a normal distribution is identified by parameters $\mu$ and $\sigma$ i.e. it can be described as $f(x,\mu,\sigma)$. The variable, $x$, can be standardised by transforming it to $z$ by using the equation

$$z = \frac{x-\mu}{\sigma} \qquad (4.25)$$

This $z$ value, the deviation of $x$ from its mean in units of $\sigma$, is called the standardised normal variable. The transformed standard normal distribution is then identified by $f(z,0,1)$. The areas under this normal distribution for various values of $z$ are given in Table 4.1. It may be noted that the values given are the areas to the right of the mean (where $\mu$ is equal to 0) for various values of $z$. The areas to the left of the mean are same, as the distribution is symmetrical about the mean. For example, the area for $z = 0$ to $z = 1$ from the table is 0.3413. So the area up to $z = -1$ is also 0.3413. Thus the total area is $0.6826 \approx 0.683 \approx$ 68.3 %. This means that the probability of finding $z$ between $\pm\,\sigma$ is 68.3%, as stated above. Similarly, the area from $z = 0$ to $z = 2$ is 0.47725 and area from $z = 0$ to $z = -2$ is also 0.47725 (from Table 4.1). So the total area is $0.9545 \approx 0.955 \approx$ 95.5%. So the probability of finding $z$ between $\pm\,2\,\sigma$ is 95.5%.

---

## Example 4.4

What is the probability of measuring an activity with a count rate between 75 cpm and 90 cpm in a contaminated room, when the mean is 85 cpm and the standard deviation is 5 cpm? Assume that the activity distribution follows normal distribution.

*Solution*

First we find the probability of measuring an activity between 75 cpm and 85 cpm. The $z$ value for 75 cpm is

$$z = \frac{75-85}{5} = -2.0$$

The minus sign indicates that the value of the variable is below the mean. From Table 4.1, the area between $z = -2$ and the mean is 0.47725.

Then we need to find the probability of measuring an activity between 85 cpm and 90 cpm. Following the same procedure, $z$ becomes $+1$. The area from Table 4.1 for $z = 1.0$ is 0.3413.

So the total probability is $0.4772 + 0.3413 = 0.8185 = 81.85$ %.

---

**Table 4.1** *Cumulative normal distribution as a function of z for z ≥ 0*

| z | .00 | .01 | .02 | .03 | .04 | .05 | .06 | .07 | .08 | .09 |
|---|---|---|---|---|---|---|---|---|---|---|
| .0 | .0000 | .0040 | .0080 | .0120 | .0160 | .0199 | .0239 | .0279 | .0319 | .0359 |
| .1 | .0398 | .0438 | .0478 | .0517 | .0557 | .0596 | .0636 | .0675 | .0714 | .0753 |
| .2 | .0793 | .0832 | .0871 | .0910 | .09488 | .0987 | .1026 | .1064 | .1103 | .1141 |
| .3 | .1179 | .1217 | .1255 | .1293 | .1331 | .1368 | .1406 | .1443 | .1480 | .1517 |
| .4 | .1554 | .1591 | .1628 | .1664 | .1700 | .1736 | .1772 | .1808 | .1844 | .1879 |
| .5 | .1915 | .1950 | .1985 | .2019 | .2054 | .2088 | .2123 | .2157 | .2190 | .2224 |
| .6 | .2257 | .2291 | .2324 | .2357 | .2389 | .2422 | .2454 | .2486 | .2517 | .2549 |
| .7 | .2580 | .2611 | .2642 | .2673 | .2703 | .2734 | .2764 | .2794 | .2823 | .2852 |
| .8 | .2881 | .2910 | .2939 | .2967 | .2995 | .3023 | .3051 | .3078 | .3106 | .3133 |
| .9 | .3159 | .3186 | .3212 | .3238 | .3264 | .3289 | .3315 | .3340 | .3365 | .3389 |
| 1.0 | .3413 | .3438 | .3461 | .3485 | .3508 | .3531 | .3554 | .3577 | .3599 | .3621 |
| 1.1 | .3643 | .3665 | .3686 | .3708 | .3729 | .3749 | .3770 | .3790 | .3810 | .3830 |
| 1.2 | .3849 | .3869 | .3888 | .3907 | .3925 | .3944 | .3962 | .3980 | .3997 | .40147 |
| 1.3 | .40320 | .40490 | .40658 | .40824 | .40988 | .41149 | .41309 | .41466 | .41621 | .41774 |
| 1.4 | .41924 | .42073 | .42220 | .42364 | .42507 | .42647 | .42785 | .42922 | .43056 | .43189 |
| 1.5 | .43319 | .43448 | .43574 | .43699 | .43822 | .43943 | .44062 | .44179 | .44295 | .44408 |
| 1.6 | .44520 | .44630 | .44738 | .44845 | .44950 | .45053 | .45154 | .45254 | .45352 | .45449 |
| 1.7 | .45543 | .45637 | .45728 | .45818 | .45907 | .45994 | .46080 | .46164 | .46246 | .46327 |
| 1.8 | .46407 | .46485 | .46562 | .46638 | .46712 | .46784 | .46856 | .46926 | .46995 | .47062 |
| 1.9 | .47128 | .47193 | .47257 | .47320 | .47381 | .47441 | .47500 | .47558 | .47615 | .47670 |
| 2.0 | .47725 | .47778 | .47831 | .47882 | .47932 | .47982 | .48030 | .48077 | .48124 | .48169 |
| 2.1 | .48214 | .48257 | .48300 | .48341 | .48382 | .48422 | .48461 | .48500 | .48537 | .48574 |
| 2.2 | .48610 | .48645 | .48679 | .48713 | .48745 | .48778 | .48809 | .48840 | .48870 | .48899 |
| 2.3 | .48928 | .48956 | .48983 | .49010 | .49036 | .49061 | .49086 | .49111 | .49134 | .49158 |
| 2.4 | .49180 | .49202 | .49224 | .49245 | .49266 | .49286 | .49305 | .49324 | .49343 | .49361 |
| 2.5 | .49370 | .49396 | .49413 | .49430 | .49446 | .49461 | .49477 | .49492 | .49506 | .49520 |
| 2.6 | .49534 | .49547 | .49560 | .49573 | .49586 | .49598 | .49609 | .49621 | .49632 | .49643 |
| 2.7 | .49653 | .49664 | .49674 | .49683 | .49693 | .49702 | .49711 | .49720 | .49728 | .49736 |
| 2.8 | .49744 | .49752 | .49760 | .49767 | .49774 | .49781 | .49788 | .49795 | .49801 | .49807 |
| 2.9 | .49813 | .49819 | .49825 | .49830 | .49836 | .49841 | .49846 | .49851 | .49856 | .49861 |
| 3.0 | .49865 | .49869 | .49874 | .49878 | .49882 | .49886 | .49889 | .49893 | .49896 | .49900 |
| 3.1 | $.49^203$ | $.49^206$ | $.49^209$ | $.49^213$ | $.49^216$ | $.49^218$ | $.49^221$ | $.49^224$ | $.49^226$ | $.49^229$ |
| 3.2 | $.49^231$ | $.49^234$ | $.49^236$ | $.49^238$ | $.49^240$ | $.49^242$ | $.49^244$ | $.49^246$ | $.49^248$ | $.49^250$ |
| 3.3 | $.49^253$ | $.49^253$ | $.49^255$ | $.49^257$ | $.49^258$ | $.49^260$ | $.49^261$ | $.49^262$ | $.49^264$ | $.49^265$ |
| 3.4 | $.49^266$ | $.49^267$ | $.49^269$ | $.49^270$ | $.49^271$ | $.49^272$ | $.49^273$ | $.49^274$ | $.49^275$ | $.49^276$ |

## 4.2.4 The Log-normal Distribution

Quite often in environmental calculations, a distribution is encountered which shows asymmetric behaviour. There are quite a few asymmetric distributions such as the gamma distribution, beta distribution, Rayleigh distribution, Cauchy distribution, Weibull distribution and so on, but the log-normal distribution is particularly important. In a log-normal distribution, when the probability density function is plotted against the value of the variable, it does not show the normal symmetrical behaviour at all, but it becomes normally bell shaped when the natural logarithm of the variable is used. The log-normal distribution is bounded by zero on the left and has a flatter tail than a normal distribution, as shown in Figure 4.6. So, to carry out any statistical test, the variable is transformed by taking the natural logarithm.

In a normal distribution, the probability density function is given by equation (4.22) where the variable is $x$. In a log-normal distribution, the probability density function for the variable, $y$ where $y = \ln x$ is given by

$$g(y) = \frac{1}{\sigma\sqrt{2\pi}}\exp(-\frac{(y-\bar{x})^2}{2\sigma^2}) \tag{4.26}$$

It should be noted that the infinite integral ($-\infty$ to $+\infty$) of the RHS of this equation with respect to the variable, $y = \ln x$ must be equal to 1. Now, to change the variable $y$ to variable $x$; one considers

$$\frac{d(\ln x)}{dx} = \frac{1}{x}$$

hence

$$d(\ln x) = \frac{dx}{x}$$

So, in terms of the variable $x$, the equation (4.26) becomes,

$$g(y) = \frac{1}{x\sigma\sqrt{2\pi}}\exp(-\frac{(\ln x-\bar{x})^2}{2\sigma^2}) \tag{4.27}$$

where $x > 0$.

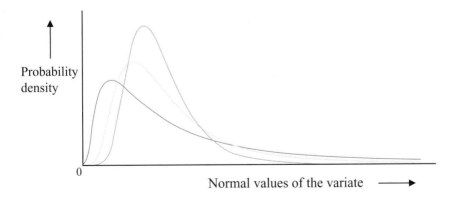

**Fig. 4.6** *Three different log-normal distributions*

### 4.2.5 Central Limit Theorem

It had been assumed for a long time that the normal distribution is the natural distribution of the physical or biological world. But with the advent of statistical tests of significance, this has been proved not to be the case. There are various other types of distribution, as mentioned earlier. However, there is a central limit theorem which gives a special character to the normal distribution.

The central limit theorem states that the distribution of the means of independent samples of size, say, $n$ from any distribution, or even up to $n$ different distributions, with finite mean and variance approaches a normal distribution as $n$ approaches a large value. Although a large value ($\rightarrow \infty$) for the number of observations in a sample is desirable, the central limit theorem is valid even for a relatively small number of observations in a sample, as long as the sample does not come from extremes. This theorem plays a crucial role in statistical calculations. As the estimated means of the samples form the normal distribution, the statistical calculations relating to the population mean, the variance as well as the test of the significance applicable to the normal distribution can be applied.

## 4.3 Sampling Design

The proper design of a sample such that the characteristic of the sampled data can represent the target population is a very important consideration in the statistical study. There are four basic factors in the sampling plan [3] and these are:

- Objectives of the study
- Cost-effectiveness of various sampling designs
- Anticipated patterns of the variable
- Practical considerations such as site accessibility for sampling; convenience, reliability, and availability of sampling equipment.

A sampling plan can be drawn up within these parameters. The primary aim is to have a sufficient but limited number of data points in the sample that will represent the true characteristics of the whole population. It is obvious that larger the sample size, the better will be the representation of the population, provided that a proper sampling plan and technique have been followed. But that will entail excessive costs which may not be sustainable in the commercial world. Various types of samples that are used in practical applications are described in the following section [1].

### 4.3.1 Types of Sampling

There are basically four methods of sampling: (i) haphazard sampling, (ii) subjective sampling, (iii) probability sampling, and (iv) search sampling. The choice depends on a number of factors such as the intended use of the sample (e.g. scoping study or detailed survey), the desired accuracy, prior knowledge of the population from which samples are drawn, the ability to perform statistical analysis on samples, etc. Each of these methods is described briefly.

### 4.3.1.1 Haphazard Sampling

This is a technique in which samples are drawn haphazardly, without any systematic plan or convention. This type of sampling may be appropriate if the parameter under consideration is uniformly distributed throughout the survey area. For example, if one wishes to measure surface contamination and the surface is known to be uniformly contaminated, then haphazard sampling could be carried out. However, one must be extremely careful in pre-judging the distribution of the parameter.

### 4.3.1.2 Subjective Sampling

This type of sampling is applicable when a judgement about the selection of areas, volumes, times, etc. from which sampling is to be carried out needs to be made. Again subjective sampling may require prior knowledge of the population distribution. Expert knowledge is a pre-requisite for this type of sampling. For example, if one is required to draw samples of airborne activity following an inadvertent release of activity from a chimney stack of a nuclear installation, then relevant parameters such as stack height, wind direction, wind speed, release duration, etc. need to be known or elicited by experts to obtain a representative sample.

### 4.3.1.3 Probability Sampling

This type of sampling is carried out on a probability basis. A number of probability sampling techniques are available: (a) simple random sampling, (b) stratified random sampling and (c) systematic sampling. A simple random sampling is one when samples are taken at random, without any preference from the whole population. This method may give a good estimate of the mean if there is no strong clustering, cyclic variation or pattern in the parameter under consideration. Figure 4.7(a) shows simple random sampling. Stratified sampling occurs when the heterogeneous population can be broken down into somewhat homogeneous strata and samples are taken from these strata. This method ensures that all sections of the population are represented. Figure 4.7(b) shows stratified sampling. Systematic sampling is carried out when a trend or pattern is suspected in the population. A scheme for systematic sampling is shown in Figure 4.7(c).

### 4.3.1.4 Search Sampling

Search sampling is when a specified value of a parameter is being searched. For example, if 'hot spots' or elevated contamination points are suspected, then search sampling is an ideal technique. Prior information about the location of 'hot spots' may be found from historical data, records of incidents and accidents etc.

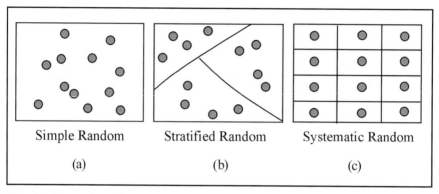

Simple Random     Stratified Random     Systematic Random

(a)             (b)             (c)

**Fig. 4.7** *Types of sampling*

## 4.3.2 Properties of Samples

Once the samples have been collected, the first task is to estimate certain characteristic properties of the samples, which should reflect the characteristics of the population from which the samples have been drawn, then the precision of the estimates can be judged. The method of estimating the properties of the samples is heavily dependent on the techniques utilised in the sampling process itself. For example, the mathematical technique applied for simple random sampling, which is very widely utilised in statistical evaluation, cannot be applied, without modifications, to stratified random sampling. An estimation of the properties of the samples based on simple random sampling will be presented here.

When the sample size is small, then the probabilities of success of 0, 1, 2,..., $n$ in a simple random sample of $n$ trials, where $n$ is finite, are the terms of the binomial expansion of $(q + p)^n$, as shown earlier in equation (4.12). The mean is $n.p$ and the standard deviation (s.d.) is $(npq)^{1/2}$. The s.d. is usually called the Standard Error (S.E.) of the number of successes in a sample size of $n$. The deviation from the mean $(n.p)$ is regarded as an 'error' [2]. The proportion of successes in a sample is obtained by dividing the number of successes by $n$. So,

$$\text{S.E. of the number of successes} = \left(npq\right)^{1/2} \tag{4.28}$$

$$\text{S.E. of the proportion of success} = \left(pq/n\right)^{1/2} \tag{4.29}$$

The precision of the proportion of success is inversely proportional to the 'S.E. of the proportion of success' and hence the precision is proportional to the square root of $n$. In other words, if the precision of the S.E is to be increased by, say, two-fold, the sample size, $n$, is to be increased by four-fold. This is quite important in practical work where the precision needs to be improved.

In a large sample size where $n$ becomes larger and larger, the distribution tends to change from the binomial to normal distribution and in the extreme case when $n$ increases indefinitely, the distribution becomes effectively normal. The probability of choosing a random variable of a normal distribution lying outside the $2\sigma$ value on either side of the mean is 0.0456 or 4.5% and the probability of it lying outside $\pm3\sigma$ of the mean is only 0.0027 or 0.27% (see Section 4.2.3). In other words, it can be said that if a sample value (or a number of sample values) shows a value outside the $\pm3\sigma$ value, then the difference (0.27%) is considered to be highly significant. On the other hand, if it is outside $\pm2\sigma$ value (or $\approx5\%$), it is considered significant. The exact value when the difference is considered significant or highly significant is not well defined. The implication of this test of significance is that it raises questions. When the difference is significant, it may be that either the sampling technique, in this case a simple random sampling, has not been followed properly or the characteristic of the sample distribution is not fundamentally normal. For example, in a contaminated area where the activity is assumed to be normally distributed, one or more high values outside $\pm2\sigma$, which may be considered significant, may furnish evidence that the distribution may not inherently be normal or there may be localised 'hot spots'.

Besides the normal distribution of a continuous variable, there are other distributions in mathematics such as the half-normal distribution, log-normal distribution, exponential distribution, gamma distribution, Cauchy distribution, Weibull distribution and so on, which describe different characteristics and which are applicable in various circumstances. A detailed description of these distributions and their potential applications may be found in [3]. One particular distribution which is often encountered in atmospheric pollution calculations is the log-normal distribution, which is described in Section 4.2.4.

## 4.4 Test of Hypotheses

A statistical hypothesis refers to a condition (or a set of conditions) which is to be tested mathematically using the data which have been collected in samples. This hypothesis forms the basis for a decision with regard to the quality of the data and the inferences to be made from the sample data. The primary statistical hypothesis to be tested on collected data may include a null hypothesis ($H_0$), which is a baseline condition that is presumed to be true in the absence of strong evidence to the contrary, and an alternative hypothesis ($H_A$), which requires burden of proof [4]. In other words, the baseline condition is to be upheld unless the alternative hypothesis is thought to be true due to the preponderance of evidence. In both null hypothesis and the alternative hypothesis, a population parameter is compared to either a fixed value (for a one-sample test) or another population parameter (for a two-sample test). The population parameter may be the mean, median or mode, measures of dispersion such as the variance, standard deviation, inter-quartile range, etc. In one population, the fixed value may be the regulatory limit, such as the threshold value or the clearance level. For two populations, the null and alternative hypothesis may be based on the comparison of a true value of one population to the corresponding true value of the other population, where both of the parameters are variables. An example of this two-sample problem is when a contaminated land site is compared to another land site or background (uncontaminated) condition. The hypothesis will be stated in terms of the difference between the two parameters.

For site remediation as well as for the decontamination process, one of the pre-requisites is to ascertain that the population parameter of the facility under consideration shows levels of activity higher than that of the threshold value or of the background levels. If both the facility and the background level data are normally distributed, then a simple t-test may be used. If data from both of these populations are log-normally distributed, then a t-test on the logarithms of the data values needs to be carried out. If the probability distribution of the data points does not show normal or log-normal pattern or the distribution is unknown, then a non-parametric (distribution free) test can be carried out. The Wilcoxon Rank Sum (WRS) test and the Sign test are example of non-parametric tests. In fact, the WRS test for a skewed distribution is generally preferred to the two sample t-test [4].

## 4.4.1 Illustration of the Test of Hypothesis

Let us consider a practical example to clarify the methodology for the test of a hypothesis (Student's t-test) when samples have been collected on the basis of simple random sampling and the distribution is assumed to be normal. A parameter such as the mean, median or mode, or variance, or some other quantity is used to test the hypothesis. Quite often it is the mean that is used. A one-sample test against a fixed value, which may be the regulatory value, is to be conducted. The hypothesis to be tested is:

$$\text{null hypothesis: } H_0 : \ \mu \leq A$$
$$\text{alternative hypothesis: } H_A : \ \mu > A$$

where $\mu$ is the population mean and $A$ is the given value such as the threshold of activity in cpm. If the mean of the population exceeds $A$, the data user may wish to take action.

At this stage numerical probability limits are set, which would trigger a false rejection or false acceptance as a result of the uncertainty in the data. A false rejection error occurs when the null hypothesis is rejected when it should not have been rejected; in other words when the hypothesis is true. A false acceptance error occurs when the null hypothesis is accepted although it is false. Students t-test is conducted to test the hypothesis as the sample mean and standard deviation are very sensitive to outliers.

Let us now consider 10 simple random data points (or composite data points of $m$ measured activities of each): 41.5, 52.3, 51.6, 54.1, 48.2, 56.6, 45.6, 45.2, 48.6 and 54.6 cpm. These data points will be used to test the hypothesis: $H_0$: $\mu \leq 47.6$ cpm as against $H_A$: $\mu > 47.6$ cpm. (This value of 47.6 cpm has been chosen arbitrarily, as if it were the specified regulatory set limit). The false rejection error limit, $\alpha$, has been specified as 5% at 47.6 cpm and false acceptance error limit, $\beta$, has been specified as 20% .

The mean of the samples $= \Sigma$ (sample values)/10 $= 49.83$ cpm

The SD $= 4.571$ cpm

Using Table 4.2, the critical values of the t-distribution with (10−1) i.e. 9 degrees of freedom and $t_{(1-0.05)}$ is 1.833. Now,

$$t = \frac{\bar{X} - A}{\sigma/\sqrt{n}} = \frac{49.83 - 47.6}{4.571/\sqrt{10}} = 1.543 \qquad (4.30)$$

As 1.543 is less than 1.833, there is not enough evidence to reject the null hypothesis. In other words, the true mean is likely to be less than 47.6 cpm. It should be noted that the false acceptance error rate should also be verified by calculating the sample size. If the calculated sample size is less than or equal to the given data points, then the false acceptance error rate has been satisfied. (The methodology for calculating the sample size is somewhat involved and a standard statistical book [2] should be consulted).

**Table 4.2** *Critical values of Student's t-distribution*

| Degrees of Freedom | 1 - α | | | | | | | | |
|---|---|---|---|---|---|---|---|---|---|
| | 0.70 | 0.75 | 0.80 | 0.85 | 0.90 | 0.95 | 0.975 | 0.99 | 0.995 |
| 1 | 0.727 | 1.000 | 1.376 | 1.963 | 3.078 | 6.314 | 12.706 | 31.821 | 63.657 |
| 2 | 0.617 | 0.816 | 1.061 | 1.386 | 1.886 | 2.920 | 4.303 | 6.965 | 9.925 |
| 3 | 0.584 | 0.765 | 0.978 | 1.250 | 1.638 | 2.353 | 3.182 | 4.541 | 5.841 |
| 4 | 0.569 | 0.741 | 0.941 | 1.190 | 1.533 | 2.132 | 2.776 | 3.747 | 4.604 |
| 5 | 0.559 | 0.727 | 0.920 | 1.156 | 1.476 | 2.015 | 2.571 | 3.365 | 4.032 |
| 6 | 0.553 | 0.718 | 0.906 | 1.134 | 1.440 | 1.943 | 2.447 | 3.143 | 3.707 |
| 7 | 0.549 | 0.711 | 0.896 | 1.119 | 1.415 | 1.895 | 2.365 | 2.998 | 3.499 |
| 8 | 0.546 | 0.706 | 0.889 | 1.108 | 1.397 | 1.860 | 2.306 | 2.896 | 3.355 |
| 9 | 0.543 | 0.703 | 0.883 | 1.100 | 1.383 | 1.833 | 2.262 | 2.821 | 3.250 |
| 10 | 0.542 | 0.700 | 0.879 | 1.093 | 1.372 | 1.812 | 2.228 | 2.764 | 3.169 |
| 11 | 0.540 | 0.697 | 0.876 | 1.088 | 1.363 | 1.796 | 2.201 | 2.718 | 3.106 |
| 12 | 0.539 | 0.695 | 0.873 | 1.083 | 1.356 | 1.782 | 2.179 | 2.681 | 3.055 |
| 13 | 0.538 | 0.694 | 0.870 | 1.079 | 1.350 | 1.771 | 2.160 | 2.650 | 3.012 |
| 14 | 0.537 | 0.692 | 0.868 | 1.076 | 1.345 | 1.761 | 2.145 | 2.624 | 2.977 |
| 15 | 0.536 | 0.691 | 0.866 | 1.074 | 1.340 | 1.753 | 2.131 | 2.602 | 2.947 |
| 16 | 0.535 | 0.690 | 0.865 | 1.071 | 1.337 | 1.746 | 2.120 | 2.583 | 2.921 |
| 17 | 0.534 | 0.689 | 0.863 | 1.069 | 1.333 | 1.740 | 2.110 | 2.567 | 2.898 |
| 18 | 0.534 | 0.688 | 0.862 | 1.067 | 1.330 | 1.734 | 2.101 | 2.552 | 2.878 |
| 19 | 0.533 | 0.688 | 0.861 | 1.066 | 1.328 | 1.729 | 2.093 | 2.539 | 2.861 |
| 20 | 0.533 | 0.687 | 0.860 | 1.064 | 1.325 | 1.725 | 2.086 | 2.528 | 2.845 |
| 21 | 0.532 | 0.686 | 0.859 | 1.063 | 1.323 | 1.721 | 2.080 | 2.518 | 2.831 |
| 22 | 0.532 | 0.686 | 0.858 | 1.061 | 1.321 | 1.717 | 2.074 | 2.508 | 2.819 |
| 23 | 0.532 | 0.685 | 0.858 | 1.060 | 1.319 | 1.714 | 2.069 | 2.500 | 2.807 |
| 24 | 0.531 | 0.685 | 0.857 | 1.059 | 1.318 | 1.711 | 2.064 | 2.492 | 2.797 |
| 25 | 0.531 | 0.684 | 0.856 | 1.058 | 1.316 | 1.708 | 2.060 | 2.485 | 2.787 |
| 26 | 0.531 | 0.684 | 0.856 | 1.058 | 1.315 | 1.706 | 2.056 | 2.479 | 2.779 |
| 27 | 0.531 | 0.684 | 0.855 | 1.057 | 1.314 | 1.703 | 2.052 | 2.473 | 2.771 |
| 28 | 0.530 | 0.683 | 0.855 | 1.056 | 1.313 | 1.701 | 2.048 | 2.467 | 2.763 |
| 29 | 0.530 | 0.683 | 0.854 | 1.055 | 1.311 | 1.699 | 2.045 | 2.462 | 2.756 |
| 30 | 0.530 | 0.683 | 0.854 | 1.055 | 1.310 | 1.697 | 2.042 | 2.457 | 2.750 |
| 40 | 0.529 | 0.681 | 0.851 | 1.050 | 1.303 | 1.684 | 2.021 | 2.423 | 2.704 |
| 60 | 0.527 | 0.679 | 0.848 | 1.046 | 1.296 | 1.671 | 2.000 | 2.390 | 2.660 |
| 120 | 0.526 | 0.677 | 0.845 | 1.041 | 1.289 | 1.658 | 1.980 | 2.358 | 2.617 |
| ∞ | 0.524 | 0.674 | 0.842 | 1.036 | 1.282 | 1.645 | 1.960 | 2.326 | 2.576 |

# Revision Questions

1. Define the mean, median and mode of a statistical distribution and explain the significance of each of these terms.

2. What is the statistical significance of the SD of a variable? Write down the mathematical expression for the SD, $\sigma$ and state its unit in relation to that of the variable. Can $\sigma$ be negative?

3. Define the following terms:
   - (i) standard error of the mean
   - (ii) coefficient of variation
   - (iii) percentile
   - (iv) quartile

4. What are the parameters of a binomial distribution? Give the mathematical expression for the probability of '$r$' successes in '$n$' trials. Write down the expressions for mean and variance of a binomial distribution.

5. What conditions are applied to binomial distribution to reach Poisson's distribution? Write down the expression for the mean and variance of Poisson's distribution.

6. Write down the expression for the normal distribution and define the parameters involved. Why is it also described as the probability density?

7. Briefly describe the properties of a normal distribution. How are mean, median and mode related in a normal distribution? What is the probability that a variable will be within $\pm 2\sigma$ of the mean, if the underlying distribution is a normal distribution?

8. What is the standardised normal distribution? How is standardised normal variable related to the variable of the normal distribution?

9. What is the central limit theorem? Describe its significance in the evaluation of the statistical properties of a variable.

10. What is the purpose of sampling? Describe briefly the sampling objectives and the main types of sampling.

11. What is the test of hypothesis? Describe briefly the significance of the null hypothesis vis-à-vis the alternative hypothesis.

## REFERENCES

[1]  Richard O. Gilbert, *Statistical Methods for Environmental Pollution Monitoring*, John Wiley & Sons, 1987, ISBN 0 471 28878 0.
[2]  C E Weatherburn, *A First Course in Mathematical Statistics*, Cambridge University Press, 1968.
[3]  Hahn & Shapiro, *Statistical Models in Engineering*, John Wiley & Sons, 1994, ISBN 0 471 04065 7.
[4]  United States Environmental Protection Agency, Guidance for Data Quality Assessment, Practical Methods for Data Analysis, EPA QA/G-9, QA00 Version, Final July 2000.

---

# Annex 4A

$$\sigma^2 = \frac{\sum f_i (x_i - \bar{x})^2}{\sum f_i}$$

As $(x_i - \bar{x}) = (x_i - m) + (m - \bar{x})$

so, $(x_i - \bar{x})^2 = \left[ (x_i - m)^2 + 2(x_i - m)(m - \bar{x}) + (m - \bar{x})^2 \right]$

Now putting $(x_i - m) = d_i$

$$\sigma^2 = \frac{\sum f_i d_i^2}{\sum f_i} + 2(m - \bar{x}) \frac{\sum f_i (x_i - m)}{\sum f_i} + (m - \bar{x})^2 \frac{\sum f_i}{\sum f_i}$$

$$= \frac{\sum f_i d_i^2}{\sum f_i} + 2(m - \bar{x})(\bar{x} - m) + (m - \bar{x})^2$$

$$= \frac{\sum f_i d_i^2}{\sum f_i} - (\bar{x} - m)^2$$

$$= \frac{\sum f_i d_i^2}{\sum f_i} - \left[ \frac{\sum f_i (x_i - m)}{\sum f_i} \right]^2$$

$$= \frac{\sum f_i d_i^2}{\sum f_i} - \left[ \frac{\sum f_i d_i}{\sum f_i} \right]^2$$

# 5

# DECOMMISSIONING OF NUCLEAR FACILITIES

## 5.1 Introduction

Decommissioning is the final phase in the life-cycle of a nuclear facility. It comes after siting, design, construction, commissioning and operations. Decommissioning refers to the administrative and technical actions taken to allow the removal of some or all of the regulatory controls from a nuclear facility at the end of its useful life [1]. The process of decommissioning incorporates a number of activities which can broadly be categorised as

- facility cleanout
- decontamination
- dismantling
- demolition and site clearance
- delicensing and release of the site from regulatory control

All of these activities involving the removal of radioactive and non-radioactive materials and waste from the site need to be carried out in a safe, systematic and environmentally acceptable way so that the site can eventually be returned to non-radioactive use. The non-radioactive use is the ultimate end-point for the decommissioning process. However, the site can only be returned to non-radioactive use (without any restriction) if all traces of radioactivity are removed such that there is no further danger from any ionising radiation. There are a number of possible end-points and associated conditions that regulators may attach and these are described in detail in Chapter 6.

Although decommissioning involves much lower risks than those of the operational phase, there are, nonetheless, specific problems associated with the process of decommissioning. In the operational phases plans, procedures and schedules flow smoothly. However, in the decommissioning phase, there is no certainty that plans will proceed smoothly. There is a large element of unpredictability which arises due to the uncertainty regarding the nature, extent and composition of the radioactivity. This is even more so if the plant under consideration had been shut down due to an abnormal incident or accident. Consequently decommissioning work needs to be carried out in a cautious and careful way such that any untoward situation can be accommodated within the programme schedule. The whole process needs to be carried out such that radiation doses and risks to the workers and the public are maintained As Low As Reasonably Achievable (ALARA). In the UK, ALARA has been slightly modified to As Low As Reasonably Practicable (ALARP). Full details of the ALARA/ALARP principle can be found in Section 7.6.

## 5.2 Facilities to be Decommissioned

Any nuclear facility will eventually be required to be decommissioned following the cessation of operation. The nuclear facilities include uranium/thorium mines, nuclear material conversion facilities, enrichment plants, fuel fabrication plants, nuclear reactors, fuel reprocessing plants, wastes storage and disposal facilities. A schematic diagram outlining the fuel cycle associated with the nuclear power programme is shown in Figure 5.1.

In addition, there are numerous other facilities associated with medical and industrial uses of radioactive materials and radiation-generating equipment which would also require decommissioning. The full extent of the decommissioning problem is described in Section 5.3.

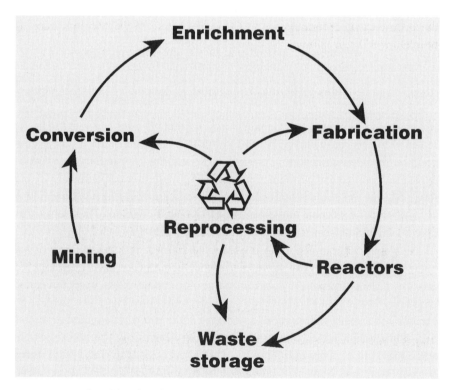

**Fig. 5.1** *Broad outline of nuclear fuel cycle*

## 5.3 Scale of the Decommissioning Problem

As of April 2006, there were 580 nuclear power reactors worldwide, of which 443 were operating, 110 reactors had been shutdown and 27 were under construction [2]. In addition, 169 power reactors worldwide had either been cancelled or suspended due to changes in government policy, financial problems during the construction phases of reactors, public opposition etc. Most of these reactors were in Western Europe and Japan. However, as a result of climatic changes in the world and the ensuing global warming, there is now a marked shift in public attitude towards the nuclear option which may result in a resurgence of the nuclear power industry. A number of new plants numbering over 40 worldwide are already planned, mostly in Asia and South America.

The existing 443 reactors now operating are not uniformly spaced in time. An overall picture of the age distribution of operating nuclear reactors can be seen in Figure 5.2. A closer analysis shows that more than 75% of the operating plants have been operating for more than half of their design lives. Although life extensions are possible, it is most likely that many of them will shortly be coming to the end of their operating lives and will require decommissioning. In addition, there are a large number of other nuclear facilities associated with the power reactors such as fuel enrichment and fuel fabrication plants, reprocessing plants, fuel storage and disposal facilities etc., which will be candidates for nuclear decommissioning. Thus the extent of the decommissioning problem worldwide is huge.

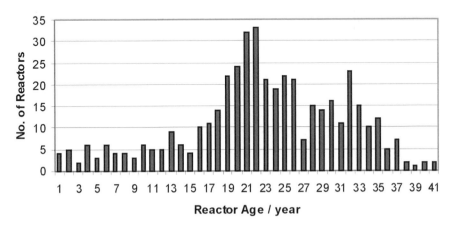

**Fig. 5.2** *Number of reactors in operation by age*

In the UK alone, decommissioning of civil nuclear facilities and restoration of civil nuclear sites are anticipated to cost about £50 billion (discounted) (see Section 8.7 for the discounting concept) over the next 40–50 years. To carry out this huge task of managing the decommissioning and clean up of civil nuclear facilities, the UK government set up an organisation called the Nuclear Decommissioning Authority (NDA) in 2005. The full details of this organisation, its task and responsibilities are given in Section 5.7.

## 5.4 Reasons for Decommissioning

There are a wide variety of reasons for taking a nuclear facility out of service and to decide to decommission it. These include [3]

- End of operating life: a facility is designed by the owner/operator/ licensee to operate for a certain length of time and is licensed by the regulatory body to do so. Unless the licence period is extended by the regulatory body, the facility needs to be taken out of service and decommissioned.

- Uneconomical operation: the operation of the facility may be uneconomical due to rising operating costs or falling revenues or both.

- Technical obsolescence: the facility may become technically obsolete and hence requires to be retired.

- Safety considerations: a facility may be shut down if the regulatory body requires substantial safety improvement which becomes too expensive for the owner/operator to implement.

- Change in government policy: a facility may be required to be shut down if its operation is contrary to the government's stated national policy.

- Accident situation: a facility may be shut down following an accident or major incident.

## 5.5 Overall Decommissioning Strategies at the End of the Useful Life

Nearer the end of the useful life of a facility, a decision has to be made by the owner/operator of the facility as to the best way to take it out of service and reduce its risk. In doing so, the owner/operator must weigh the various options and find an optimal solution which is acceptable to the various stakeholders including the regulatory bodies. The overall options that may be considered are:

- No action: This effectively means that the decommissioning operation would be kept in abeyance until further decision regarding the facility is taken. This may also sometimes be referred to as the 'wait and see' option. It should be noted that this option is the one least likely to be acceptable to the regulatory body or to the national government. Even if it is allowed, the owner/operator should nonetheless be required to carry out some measure of clean-up operation.

- Safe-store: This is the safe storage of the facility under a care and maintenance programme. This safe-storage provision can be applied for a short period of time or for an extended period. If it is for short period, the major active components are removed and the reactor is defuelled. In the case of long-term safe-storage, there may not be any need for the removal of highly active components, although the reactor is to be defuelled.

- Entomb: This involves encasing and maintaining the active parts of the facility in a safe and structurally sound material enclosure for a long period of time until the radioactivity decays to manageable levels. Both the safe-store and entomb methods are known as deferred dismantling.

- Decommissioning: This implies removing all radioactive and non-radioactive contaminants to levels which would permit the release of the site for either partially restricted use or unrestricted use. It may include the provision of safe-storage or entombing as ways to achieve decommissioning.

In making a decision as to which of these choices is to be taken, a number of factors needs to be taken into account and these are: present and long-term safety of the facility; type of facility under consideration; operating history of the facility; mode and cause of cessation of operation of the facility (normal or accident conditions); location of the facility, particularly with regard to population centres; availability and retention of key staff; national laws; political imperatives and, of course, financial considerations. All of these factors collectively, with differing levels of emphasis, will lead to the identification of the optimum or most suitable option. The detailed description of the option study is given in Chapter 12.

## 5.6 Decommissioning Operations

As mentioned earlier, the term decommissioning includes all the operations subsequent to the cessation of operations covering decontamination, dismantling and removal of all contaminations from the site to such an extent that the site would pose 'no further danger to human beings from ionising radiation'. The process of Decommissioning and Dismantling (D&D) is often used, particularly in the USA, as synonymous to decommissioning. However, in the context of this book, decommissioning is used to include all activities from the cessation of operation to D&D operations, management of wastes arising thereof, remediation of the site right up to the stage of delicensing the site. The site may then be used for either a non-nuclear activity (requiring no licence) or a nuclear activity with a new licence, if applicable.

In the UK, when a nuclear facility is taken out of service, the responsibility for the decommissioning of the plant remains with the site licence holder under the Nuclear Installations Act 1965 (NIA65, as amended) (see Chapter 6) under the site licence condition [4]. This implies that the safety provisions which were applicable to the operating plant will also apply to the decommissioning operations. In some European countries, however, the responsibility for carrying out decommissioning operations is transferred to a different organisation(s) and separate licences may be issued. Appendix 5 gives the institutional and regulatory aspects in some of the EU member states where a clear demarcation is placed between the operational phase and the decommissioning phase.

Following the cessation of operation, the decommissioning phase starts. But the transition is not always very well-defined. Some activities leading to decommissioning may be carried out after the shutdown of the facility under the operating licence provisions. Such activities may include: management of operational waste, removal of nuclear fuel and other stored radioactive materials, specification of the operational waste inventory and preliminary decontamination. Spent fuel may have been stored in fuel stores on the site of a nuclear power plant in some countries where a central storage facility or reprocessing facility is not available. Management of such waste may be carried out under an operating site licence. All of these activities are put together as the Post-Operational Clean Out (POCO) operation in the UK, which is similar to

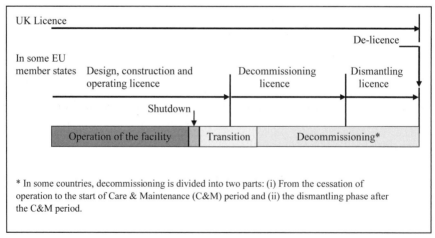

**Fig. 5.3** *Activities during the life-cycle of a nuclear facility*

the transition phase specified in the IAEA Technical Reports Series 420 [5]. If major refurbishment or modification to the plant needs to be carried out after the shutdown in preparation for the decommissioning operation, then that work may be transferred to the decommissioning phase. The exact demarcation of activities with regard to the end of the operational phase and the start of decommissioning phase should be clearly defined as it may have significant financial implications. Figure 5.3 shows the licensing regimes and various phases in the life-cycle of a nuclear facility in the EU.

## 5.7 Nuclear Decommissioning Authority

In July 2002, the British government announced in a White Paper, *Managing the Nuclear Legacy: a strategy for action* [6], its intention to set up a new body which would be made responsible for managing a programme of decommissioning of civil nuclear facilities and for cleaning up of civil nuclear sites in the UK. To implement this decision, the government initiated an Act of Parliament under the Energy Act 2004 [7] which announced the setting up of the NDA. After the Act received the Royal Assent in 2004, the NDA was set up as a Non-Departmental Public Body (NDPB) functionally responsible to the Secretary of State for the Department for Business, Enterprise and Regulatory Reform (DBERR) for activities in England and Wales, and jointly responsible to the Secretary of State and the Scottish Ministers for activities in Scotland. It started operations on 1 April 2005.

The primary function of the NDA is to decommission and clean-up of all civil nuclear legacy sites in the UK safely, securely and cost-effectively. These sites would include Research and Development (R&D) facilities managed and run by the United Kingdom Atomic Energy Authority (UKAEA); Magnox and old Advanced Gas-cooled Reactor (AGR) power stations under the management of the then British Nuclear Fuels Limited (BNFL); facilities used for storing, treating, transporting and disposing of radioactive and mixed hazardous materials. Altogether 20 civil public sector sites were identified in the UK and the ownership of these sites and associated assets was transferred to the NDA on the vesting day, 1 April 2005. The secondary functions of the NDA include: carrying out or sponsoring research into matters related to decommissioning of nuclear installations, helping to maintain a skilled workforce by educating and training persons in these matters and maintaining liaison with the public. To perform its function efficiently, the NDA has been structurally divided into four regional offices to cover the whole of England, Wales and Scotland and these regional offices are: Region 1 in the south with its office at Abingdon; Region 2 in the north-west with its office at Springfields; Region 3 to cover Sellafield with its office at Westlakes; and Region 4 for the whole of Scotland with its office at Dounreay. The head office is at Westlakes Science and Technology Park, Cumbria. It is anticipated that the total budget to complete the task assigned to NDA would require about £50 billion (discounted) (£80 billion undiscounted) with an initial annual expenditure of approximately £2 billion. The details of the management of the decommissioning operations in the UK covering the operating model and supply chain provisions are given in Appendix 4.

The role of the NDA had been subsequently extended by a government decision, announced on 25 October 2006, as a result of the Committee on Radioactive Waste Management (CoRWM, see Section 16.6) report in the summer of 2006, that the NDA will be responsible for the geological disposal of high-level wastes and that the UK Nirex will be merged with the NDA. As a consequence of this decision, the NDA becomes effectively responsible for the management of all categories of waste. In effect, the NDA becomes responsible for the complete fuel cycle.

# 5.8 Decommissioning of Nuclear Facilities

There is a wide variety of nuclear facilities that are available today which may require decommissioning. However, nuclear facilities may be separated into the civil and military facilities. Civil nuclear facilities include: nuclear power plants, reprocessing plants, fuel fabrication and enrichment plants, as well as medical, industrial and research facilities. The military nuclear facilities include non-stationary facilities such as nuclear powered submarines and nuclear powered aircraft carriers as well as nuclear weapon production facilities. There are some intrinsic differences in the implementation strategy for the decommissioning processes of the civil and military (nuclear submarines) reactors.

Decommissioning of a nuclear facility may be divided into a number of implementation stages, which may be separated by periods of relative dormancy. The International Atomic Energy Agency (IAEA) in Safety Series No. 105 [8] has recommended delineating decommissioning operations of civil nuclear facilities into three separate stages, as follows.

## 5.8.1 Decommissioning of Civil Nuclear Power Plants

**Stage 1 (Monitored Closure)**
This stage starts when the reactor is shut down and control of the facility (reactor or otherwise) is removed from the operators. It involves de-fuelling the reactor, removal of process materials and other non-fixed items of plant. The irradiated fuels are transferred for reprocessing or for long-term storage. This stage may take a number of years, from a minimum of two years up to ten years, depending on the management strategy and the prevalent condition of the plant. At the end of this stage, 95–99% of the total activity should have been removed from the facility.

Quite a significant part of this stage of work can be considered to be POCO. But beyond the POCO some work may be carried out such as the radiological survey to identify radiological hazards, removal of non-fixed contaminated equipment and plants etc.

## Stage 2 (Partial or Conditional Release)

This stage involves decontaminating and dismantling all active and non-active plants and buildings outside the biological shield of the reactor and sealing the reactor within the biological shield. The major part of the total decommissioning operation is undertaken at this stage. The first part of this stage can also be considered as the Care and Maintenance Preparation (C&MP) stage. It may take 20–30 years. When this C&MP is complete, the rest of the site outside the sealed reactor may go through the delicensing process and be released for re-use. The sealed reactor within the biological shield enters the period of Care and Maintenance (C&M) period when radioactivity inside the reactor decays to manageable levels. During this period the site remains in a quiescent state during which no significant dismantling work is undertaken. This quiescent period may last for 50 years or more.

## Stage 3 (Unconditional Release)

Following the C&M period, the stage is reached when complete dismantling of the biological shield, the reactor and all associated structures can be undertaken. At this point, the Final Site Clearance (FSC) process is undertaken when all radioactive material and other hazardous materials are removed from the site to release the site from regulatory control. This stage may take about ten years. It should be noted that before the site or facility can be released for restricted or unrestricted release, the licensee must demonstrate to the regulator that there is 'no longer any danger from ionising radiation'.

Although three stages have been defined, in practice the exact delineation of the stages may vary significantly. It should be noted that this specification of three stages only gives overall guidance and includes some flexibility in the decommissioning operation. The final strategy is dependent on a number of factors such as: operational imperatives, workforce requirement, safety and environmental impact, socio-political considerations, financial provisions etc. Chapter 10 describes the details of the decommissioning plan and the various factors which come into play in defining the adopted strategy. Table 5.1 gives the extent of the decommissioning of nuclear power plants in the UK and their lifetime plans under the management of the NDA [9].

Of all these sites, Sellafield is the largest and most challenging decommissioning site. It is not only because it has a myriad of nuclear chemical facilities but also it carried out the back-end of the research and development of the nuclear activities generating wastes of all descriptions, both civil and military, within the country right from the beginning of the nuclear age. Figure 5.4 shows an aerial view of the Sellafield site with its wide diversity of operations.

**Table 5.1** *Decommissioning of nuclear facilities in the UK under NDA ownership*

| Location | SLC* | Site details |
|---|---|---|
| 1. Berkeley Power Station, Gloucestershire | Magnox Electric South | This 17.4 hectares (ha) site contains the power plant (two Magnox reactors: Berkley 1 and Berkley 2) and associated laboratories and Magnox Electric's head office. The plant operated from 1962 to 1989, when it was taken out of service. It has largely been decommissioned and it is due to enter care and maintenance in 2009 |
| 2. Bradwell Power Station, Essex | Magnox Electric South | This 30 ha site contains a power plant which ceased electricity generation in 2002 (1962–2002). It is now being defuelled. Its Environmental Impact Assessment for Decommissioning (EIAD) has been approved by the HSE/NII and the decommissioning work has already started |
| 3. Calder Hall Power Station, Sellafield, Cumbria | Sellafield Ltd | This 30 ha site contains Calder Hall power station which was shutdown in 2003 (1956–2003). It was the first commercial nuclear power plant in the world. It is now undergoing defuelling and its EIAD is being prepared for HSE/NII approval to enable decommissioning to begin |
| 4. Capenhurst, Ellesmere Port, Cheshire | Sellafield Ltd | The 40 ha site contains an uranium enrichment plant and associated facilities. The plant ceased operation in 1982 (1953–1982). Most of the plant equipment has already been removed from the site and decommissioning is expected to be completed by 2009. However, the site will be used to store the UK stockpile of uranium materials until 2120 |
| 5. Chapelcross Power Station, Dumfrieshire, Scotland | Magnox Electric North | Chapelcross Power Station covering an area of 92 ha was the first nuclear power station in Scotland. It ceased operation in June 2004 (1959–2004) and is now being defuelled. An EIAD will be produced for the HSE/ Nuclear Installations Inspectorate (NII) approval for decommissioning to proceed |
| 6. Culham, Joint European Torus (JET) Plant, Oxfordshire | UKAEA (self-licensed) | This 73 ha site contains the world's largest fusion research programme (JET machine) which is financed under the European Fusion Development Agreement (EFDA). It is in operation (1960– ), but the operation is expected to cease in 2007. Decommissioning will then start. It should be completed by 2022 |

**Table 5.1** *Continued*

| Location | SLC* | Site details |
|---|---|---|
| 7. Dounreay, Caithness, Scotland | UKAEA | This 55 ha site was established in the early 1950s as a fast breeder reactor research site and fuel treatment facility. There were three reactors, the last of which ceased operation in 1994. The site is now being decommissioned. The plan is to have long-term storage for radioactive material and wastes on site in a passively safe condition by 2036 |
| 8. Dungeness A Power Station, Kent | Magnox Electric South | This 91 ha site contains the power station which ceased generating electricity at the end of 2006 (1995–2006). An EIAD was approved in 2006. The reactor will be defuelled, this may take up to three years |
| 9. Harwell Research Establishment, Didcot, Oxfordshire | UKAEA | This 110 ha site was established in 1946 as Britain's first atomic energy research establishment. It accommodated five research reactors of various types, a number of other facilities and nuclear waste treatment and storage facilities. These facilities were operated from 1946 to 1990. Legacy wastes are being retrieved and repackaged for longer term storage. It is expected that decommissioning will be complete by 2025 |
| 10. Hinkley Point A Power Station, Somerset | Magnox Electric South | This 26 ha site contains a power station which ceased operation in 2000 (1965–2000). Defuelling was completed in November 2004. Decommissioning work then started |
| 11. Hunterston A Power Station, West Kilbride, Ayrshire, Scotland | Magnox Electric North | This 65 ha site contains Hunterston A power station which ceased operation in 1989 (1964–1989). It had already been defuelled and decommissioning work is now in progress |
| 12. LLW Repository Drigg, Cumbria | Sellafield Ltd | This (LLW) Low Level Waste repository has been in operation as a national LLW disposal facility since 1959. Since 1995, all LLWs have been compacted and placed in containers and disposed of in engineered concrete vaults |
| 13. Oldbury Power Station, Thornbury, Gloucestershire | Magnox Electric North | This 71 ha site accommodates Oldbury power station which will cease operation at the end of 2008 (1967–2008). It will then be defuelled, which may take about three years. An EIAD will then be prepared for decommissioning work to commence |

**Table 5.1** *Continued*

| Location | SLC* | Site details |
|----------|------|--------------|
| 14. Sellafield site, Cumbria | Sellafield Ltd | The largest nuclear site in the UK over an area of 232 ha in Cumbria. It contains a variety of nuclear chemical facilities including MOX fuel fabrication, nuclear material and nuclear waste facilities. It started operating in 1947 and decommissioning of some parts started in the 1980s |
| 15. Sizewell A Power Station, Suffolk | Magnox Electric South | This 10 ha site contains Sizewell A power plant which started operation in 1966. It ceased generating electricity at the end of 2006 and the EIAD was approved in 2006. It is expected that the defuelling operation and removal of fuel to Sellafield for treatment will take about three years |
| 16. Springfields,Preston, Lancashire | Springfields Fuels | This 60 ha site accommodates nuclear fuel manufacturing facilities for nuclear power stations in the UK |
| 17. Trawsfynydd Power Station, North Wales | Magnox Electric North | This 65 ha site contains Trawsfynydd power station which ceased operation in 1991 (1965–1991). Fuels have been removed and decommissioning work is well underway |
| 18. Windscale, Cumbria | UKAEA | This 14 ha site, located on the Sellafield site, comprises three reactors two of them were shutdown in 1957 following an accident and the third one in 1981. It was mainly a research facility. Decommissioning work began in the 1980s and will continue until 2015. At that stage, all nuclear facilities will be in passive safe condition |
| 19. Winfrith, Dorset | UKAEA | This 88 ha site contained eight research reactors and one prototype commercial reactor (Steam Generating Heavy Water Reactor (SGHWR)). Five of the research reactors have been removed from the site and the remaining three are in various stages of decommissioning. Part of the site has been delicensed and the whole work is expected to be complete by 2018 |
| 20. Wylfa Power Station, Angelsey, North Wales | Magnox Electric North | This 50 ha site contains the Wylfa power station which was the last Magnox reactor (1971–). It is expected to be operational until 2010. Decommissioning work will then begin, after the approval of the EIAD |

\* SLC stands for Site Licence Company. These are the present Site Licence Companies. A restructuring and re-allocation of site licences is underway.

**Fig. 5.4** *Sellafield site (Courtesy of Sellafield Limited)*

### 5.8.2 Decommissioning of Civil Nuclear Facilities (other than reactors)

Decommissioning of nuclear facilities other than reactors can be separated into three stages as given below. It should be noted that these stages came more from operational imperatives, rather than from well-defined, internationally accepted stages.

- Initial decommissioning: This work is carried out following the POCO phase at the termination of operation. There is no defined timescale or even demarcation of work for this stage. It depends on the type of facility to be decommissioned, the present state of the facility and the operational imperatives.

- Dismantling: This stage involves taking all non-active or lightly contaminated parts, components or systems of a plant apart. Actual timing depends on a number of factors. Plutonium plant needs to be dismantled soon after operation for safety reasons. Generally, a period of 50 years from the cessation of operation may be earmarked for this stage. Following this stage, the highly active areas of the building may be kept under a C&M regime.

- Demolition: After an extended period under a C&M regime, the buildings may be demolished, the rubbish cleared and site is returned for unrestricted or partially restricted use. The exact timescale for this demolition activity to be undertaken is dependent on the type of plant in question, the nature and extent of the contamination and, of course, the management strategy.

## 5.9 Decommissioning of Nuclear Submarines

The decommissioning of a nuclear submarine differs in its implementation strategy from that of a civil nuclear power reactor. The present decommissioning strategy for a nuclear submarine can be divided into five basic stages.

- Stage 1 Naval decommissioning: When a submarine ceases to be operational, it undergoes a naval decommissioning ceremony and all on-board weapons are removed.

- Stage 2 Defuel, De-equip and Lay-up Preparation (DDLP): The DDLP involves several basic operations to prepare the submarine for long-term storage afloat. The primary circuit is decontaminated using the MODIX process. A Decontamination Factor (DF) of 3 to 8 is achieved at this stage. All the fuel is removed and at present transported to Sellafield and the radioactive resins are removed to storage tanks ashore, as in a normal refit. The primary circuit and associated pipe-works are drained and sealed. The submarine is then militarily declassified with equipment removed for use elsewhere in the fleet.        The next process involves careful preparation to provide watertight integrity and an effective containment boundary for a long period of storage afloat. The Reactor Compartment (RC) is isolated to provide the initial shielding and containment. All the hull valves are shut and the hull openings are blanked, hatches welded shut and the main ballast tanks and flood-free spaces are blanked and dehumidified. Cathodic protection anodes are fitted to reduce corrosion. The submarine is then moored in a tideless basin within the dockyard for further decommissioning.

- Stage 3 Afloat Storage: The submarine may be stored in the basin for a period of 30 years or more. During this time, it is important to continually assess the submarine's ability to float and the integrity of the containment and shielding. Therefore, a Planned Maintenance Schedule (PMS) needs to be in operation. Weekly external visual inspections are carried out to determine the state of the submarine and any potential problems. Radiological and structural surveys are conducted annually and any remedial or preservation work is carried out. After five years, the hull and casing above waterline are painted. The submarine is dry docked every ten years for a more detailed inspection below the waterline. The main area of concern is the external structure and the ballast tanks. Any necessary preservation work is carried out during dry docking.

At present only one submarine, *Dreadnought*, had undergone two 10-year dry dockings in 1991 and in 2001. The pressure hull was in a good-to-fair condition and the ballast tanks were in good condition. Only a few of the blanks that were welded and bolted to the external pressure hull needed attention.

A more extensive maintenance period is planned for the third dry docking, i.e. after 30 years. It is expected that the pressure hull, made of thick, high-grade steel, will hardly require any repair work and will provide an effective containment barrier. It is, however, possible that the ballast tanks may need to be strengthened.

- Stage 4 Dismantling and Disposal of Radioactive Material: The next major stage is the dismantling and disposal of the radioactive material prior to the final disposal of the rest of the submarine. This stage presents the greatest difficulty. The radioactive components are mainly within the RC and, therefore, it is sensible to remove the whole of the RC from the submarine. The whole RC will be cut out and then taken to a handling facility where it can be cut out and the wastes separated before eventual disposal. The basic operation for RC removal is quite simple. All piping and other intrusions which penetrate the RC bulkheads are cut out and blanked off to maintain the containment integrity. The submarine is dry docked and the RC is cut out several feet forward and aft of the RC bulkheads to allow for the structure to be strengthened and adapted for ease of handling. It is expected that the pressure hull and the RC bulkheads will meet the structural integrity requirement.

- Stage 5 Disposal of the submarine: With the RC removed, effectively all the radioactive material has been taken away from the submarine. It can then be disposed off by cutting out and recycling 3500 tonnes of high-grade steel, copper alloys and other scrap metals.

The strategy of following these five stages of decommissioning of nuclear submarines is being pursued in the UK. However, there had been severe criticisms of this strategy, both by the public as well as by the national policy makers – partly because there is no well-defined time frame for the stages, particularly for stages 4 and 5. This strategy can at best be defined as 'wait and see'.

## 5.9.1 Recent Developments in Nuclear Submarine Decommissioning

Following the erstwhile Radioactive Waste Management Advisory Committee (RWMAC) report in 1997 criticising the MoD for not having a long-term strategy for the disposal of decommissioned nuclear submarines and the MoD realising the likely shortage of space for 'afloat storage' of submarines beyond 2012, a study was commissioned by the Warship Support Agency (WSA) at Bath. (It should be noted here that the RWMAC has since been disbanded and the CoRWM was set up in 2003. For further details on the CoRWM, see Sections 6.5 and 16.6). The study produced a report called 'The ISOLUS Investigation' [10]. The main conclusions of this study are:

(i) Land storage of the separated RC of the submarine is the favoured option and that there will be cost savings from its early implementation. It should be noted that only interim storage of separated RC was considered in the report, not the permanent disposal. This is due to the fact there is no disposal route for Intermediate Level Waste (ILW) or High Level Waste (HLW) in the UK.

(ii) Afloat storage (stage 3 and beyond) should be considered only as a 'stop-gap' measure. The shortage of storage space at the dockyard as additional submarines are decommissioned precludes this option from being a viable long-term option.

## Revision Questions

1. Show schematically the various components which constitute a nuclear fuel cycle and briefly describe the functions of these components.

2. What are the prime reasons for decommissioning a nuclear facility?

3. What is meant by the term 'decommissioning of a nuclear facility'? List the activities that are involved in decommissioning and describe them briefly.

4. Regulation of the decommissioning operation in the UK is somewhat different from that in many EU member states. Describe the differences. Show diagrammatically the decommissioning stages in the life-cycle of a nuclear facility in the UK and in another EU member state.

5. What role does NDA plays in the decommissioning and clean-up of legacy nuclear sites in the UK? Briefly describe the primary functions of the NDA.

6. What are the various stages of decommissioning of a civil nuclear power plant? Briefly describe each one of these stages with an indicative time frame.

7. Briefly describe the stages of decommissioning of civil nuclear facilities (other than reactors).

8. Describe the stages of decommissioning of mobile nuclear reactors, such as nuclear submarines, and indicate why they are different from those of nuclear power plants.

## REFERENCES

[1] International Atomic Energy Agency, Decommissioning of Nuclear Power Plants and Research Reactors, Safety Guide, Safety Standards Series No.WS-G-2.1. IAEA, Vienna, Austria, 1999.

[2] International Atomic Energy Agency, Nuclear Power Reactors in the World, IAEA-RDS-2/26. IAEA, Vienna, Austria, April 2006, http://www.iaea.org/programmes/a2/index.html.

[3] International Atomic Energy Agency, Planning and Management for the Decommissioning of Research Reactors and Other Small Nuclear Facilities. Technical Reports Series No. 351, IAEA, Vienna, Austria, 1993.

[4] UK Health and Safety Executive, Guidance for Inspectors on Decommissioning on Nuclear Licensed Sites, Nuclear Safety Directorate, 2001.

[5] International Atomic Energy Agency, Transition from Operation to Decommissioning of Nuclear Installations, Technical Reports Series No. 420, IAEA, Vienna, Austria, 2004.

[6] UK Government, Cm 5552, A White Paper on 'Managing the Nuclear Legacy: A Strategy for Action', July 2002.

[7] UK Government, Energy Act 2004, The Stationary Office, 2004.

[8] International Atomic Energy Agency, The Regulatory Process for the Decommissioning of Nuclear Facilities, Safety Series No. 105, IAEA, Vienna, Austria, 1990.

[9] UK Nuclear Decommissioning Authority, 2006, http://www.nda.gov.uk.

[10] Warship Support Agency, The ISOLUS Investigation, A Review of Options for Interim Storage of Reactor Compartments and Associated Hull and Structure of UK Nuclear Submarines following their Withdrawal from Service and pending Final Disposal, 26 May 1999.

# 6

# REGULATORY ASPECTS IN DECOMMISSIONING

## 6.1 Introduction

Decommissioning is a multi-faceted process starting with the cessation of operation of the plant, continuing with various phases of decommissioning operations, managing radioactive and non-radioactive wastes and finally culminating with the delicensing of the plant and/or site for unrestricted or restricted use. Consequently, the institutional and regulatory framework controlling decommissioning is quite involved. Although the hazards associated with decommissioning are inherently lower than those of the operating phase, the involvement of various regulatory bodies overseeing safety, security and environmental issues (setting aside financial and competitive issues) makes decommissioning a very management intensive process.

In order to understand the workings and the inter-relationships of the various regulatory bodies covering the multitude of activities associated with decommissioning, there is a need to understand the overall legislative framework. This chapter first describes the international bodies contributing to international and national regulations for the protection of workers, the public and the environment from the effects of ionising radiations. It then describes the supra-national organisations such as the EU and finally the national regulatory regimes in the UK. The hierarchy of these organisations is shown in Figure 6.1. Needless to say, the national regulatory regimes vary from country to country and so, to limit such country-specific details, only the regulations specific to decommissioning in the UK are included here. However, overarching issues relating to institutional and legal framework and standards in some of the European countries such as France, Germany and Italy are given in Appendix 5.

It should be noted again, that there is no separate decommissioning regulatory framework or specific decommissioning licence requirement in the UK. Decommissioning is carried out under the same operational licence with the same safety standards (see Section 7.5). In some European countries, decommissioning is carried out as a distinct activity separate from operational activity and hence requires a separate licence (see Appendix 5). This chapter covers only regulatory issues and brief descriptions of major regulations; the implementation of such regulations and safety aspects will be described in detail in Chapter 7.

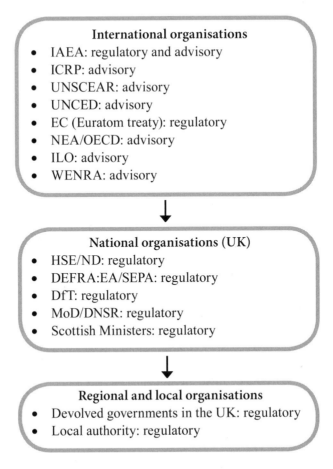

**Fig. 6.1** *Hierarchy of organisations regulating decommissioning activities*

## 6.2 International Organisations

As decommissioning work involves exposures to ionising radiations of the workers as well as of the public, regulatory standards overseeing such activities are no different from those of the operational phase of a nuclear facility in the UK. A number of international organisations and agencies are active in the fields of nuclear safety, radiological protection, chemical and industrial safety associated with decommissioning work. They carry out studies and research on various aspects of radiological protection. The outcome of their work is then put forward in reports and publications. The recommendations from some of these organisations are mandatory for the member states while others are purely advisory. The roles and status of some of the important organisations are given below.

## 6.2.1 International Atomic Energy Agency

The International Atomic Energy Agency (IAEA) was founded in 1957 by the United Nations General Assembly as an autonomous inter-governmental UN agency. The main objectives of this organisation are: assisting research, encouraging exchange of expertise among member states in the peaceful use of nuclear energy, establishing international standards for the protection of health of workers and the public from the hazards of ionising radiation, establishing and administering safeguards against non-peaceful use of nuclear energy. Its other function is to recommend practices for the minimisation of danger to life and property from abnormal incidents that may occur during the generation of nuclear energy. It produces treaties, conventions and agreements, some of which are purely voluntary whereas others are legally binding on the signatory states. Since 1986, five conventions have been ratified in the areas of nuclear safety, nuclear transport and waste safety and these are

- Convention on Nuclear Safety [1]. This aims to maintain a high level of safety in land-based nuclear power plants by setting international benchmarks to which Member States would subscribe. It is a purely voluntary undertaking and is not designed to enforce obligations on Member States. It came into force on 24 October 1996. Recently the IAEA produced, in association with other international organisations, namely the FAO, ILO, OECD/NEA, WHO and PAHO, a document entitled 'International Basic Safety Standards for Protection against Ionising Radiation and the Safety of Radiation Sources' [2] which may form the basis of safety standards.
- Convention on the Physical Protection of Nuclear Material [3]. This convention obliges contracting states to ensure physical protection of nuclear material within their territory or on board their ships or aircraft during international transport of nuclear materials. It came into force on 8 February 1987. It had subsequently been amended as the Convention on the Physical Protection of Nuclear Material and Nuclear Facilities [4]. A diplomatic conference was held in July 2005 under the auspices of the IAEA to amend the convention and strengthen its provisions in peaceful domestic use, storage and transport of radioactive material.
- Convention on Early Notification of a Nuclear Accident [5]. This convention establishes a notification system for nuclear accidents that have the potential for trans-boundary release of activity that could have radiological significance for another country. It came into force on 27 October 1986. An addendum to this convention was issued on 25 September 2000 [6].

- Convention on Assistance in the Case of a Nuclear Accident (Radiological Emergency) [7]. This convention sets out an international framework for cooperation to facilitate prompt assistance and support in the event of nuclear accidents or radiological emergencies. It came into force on 26 February 1987. An addendum to this convention was issued on 25 September 2000 [8]. The materials from both of the above mentioned conventions have been amalgamated into a single document entitled 'Preparedness and Response for a Nuclear or Radiological Emergency' [9].

- Joint Convention on the Safety of Spent Fuel Management and on the Safety of Radioactive Waste Management [10]. This is a legally binding international treaty in these areas (see Section 17.3 for further details).

## 6.2.2 International Commission on Radiological Protection

This international body was originally set up in 1928 as the International X-ray and Radium Protection Committee. In 1950 it was reorganised and renamed as the International Commission on Radiological Protection (ICRP). It is an independent, registered charitable body, neither a UN body nor an intergovernmental organisation. It was established with the purpose of advancing, for the benefit of the public, the science of radiological protection by providing recommendations and guidance on all aspects of radiological protection. It is composed of a main commission and four standing committees. The main commission comprises 12 members and a chairman, assisted by a scientific secretary. Members of the main commission are elected for a period of four years. The standing committees are tasked, by the main commission, to look at various aspects of radiological protection. Each standing committee is supported by a number of task groups. Over the years the ICRP has established an unparalleled reputation such that its recommendations are regarded by all national and international organisations and regulatory bodies as the most advanced and authoritative in the field. It has, however, no statutory role or responsibilities.

The ICRP in its 1990 recommendations, ICRP Publication 60 [11], as well as in its previous ICRP Publication 26 [12], put forward a system of radiological protection which essentially contains three basic principles for the protection of people from the harmful effects of ionising radiation. These radiological protection principles are generic in nature, but contain all the basic elements required for the protection of human beings from the harmful effects of ionising radiation. They are: (i) justification of a practice; (ii) optimisation of protection; and (iii) dose and risk limits. These principles and operational limits were described in Section 3.4.

### 6.2.3  UN Scientific Committee on the Effects of Atomic Radiation

The United Nations Scientific Committee on the Effects of Atomic Radiation (UNSCEAR) was set up by the UN General Assembly in 1955 with the prime objective of reviewing the effects of radiation on a global basis. It meets annually and reports its findings on effects of radiation on a five-yearly basis. Its findings may form the basis for national or international standards, but the findings on their own have no statutory role.

### 6.2.4  United Nations Conference on Environment and Development

The Globally Harmonised System of Classification and Labelling of Chemicals (GHS) [13] was set up under an international mandate adopted in the 1992 United Nations Conference on Environment and Development (UNCED), often called the 'Earth Summit'. The GHS is neither a regulation nor a standard. It is a system of harmonised classification and labelling of chemicals in order to improve safety from chemical hazards. The GHS document (referred to as 'The Purple Book') establishes an agreed hazard classification and labelling of chemicals, which the regulatory authorities of all the countries would use to develop national programmes to specify hazardous properties of chemicals and prepare safety data sheets as appropriate. The full official text of the system is available on their website [13].

### 6.2.5  European Atomic Energy Community

The European Atomic Energy Community (Euratom) was created by a treaty to assist in the development of civil nuclear industry in Europe. This treaty was signed in Rome on 25 March 1957, on the same date as the Treaty of Rome establishing the European Economic Community was signed. This treaty was one of the founding treaties of the European Union (EU). It has the power to set up and enforce protection standards on all aspects of nuclear energy on all member states. However, the emphasis of this treaty is on research and development for a healthy nuclear industry.

### 6.2.6  European Union Legislation

Under the provisions of the Euratom treaty, the European Commission (EC) within the EU acquired the status of a supra-national regulatory authority in three areas of nuclear activity: radiation protection, supply of nuclear fissile materials and nuclear safeguards. A number of EU legislative instruments are applied. The Council Directives are produced in order to establish uniformity in safety standards across the whole of the EU and are binding on all member states. However, the Member States are given some flexibility as to the mode and timing of introduction of the required measures into national legislation or administrative procedures. A particular Council Directive that was promulgated by the EU under the Euratom Treaty is the 'basic safety standard' in radiological protection [14], which came to be known as the European Basic Safety Standard (BSS). Regulations, on the other hand, are directly applicable in law in all Member States. Other legislative instruments include recommendations, decisions, resolutions and opinions of the EU. Council decisions relate to specific cases and may be addressed to Member States, organisations or individuals. It should also be noted that the directives and regulations promulgated by the EU institutions have primacy over the law of the Member States.

Decommissioning is one of the activities where the EC requires submission by the governments of the Member States, under Article 37 of the Euratom treaty, identifying potential impacts on Member States of the decommissioning activities. In the UK, the Department for Environment, Food and Rural Affairs (DEFRA) for England and Wales and the Scottish Executive for Scotland are the Authorising Bodies to make submissions which are prepared by the relevant environment agencies in consultation with the HSE. Further details on Article 37 can be found in Section 17.3.3.

## 6.2.7 Nuclear Energy Agency of the Organisation for Economic Cooperation and Development

The Nuclear Energy Agency (NEA) is a specialised agency within the Organisation for Economic Cooperation and Development (OECD), an intergovernmental organisation of EU member states and other industrialised countries. The OECD was set up in 1948 as the Organisation for European Economic Co-operation (OEEC) for the reconstruction of Europe after the second world war. Its membership was extended to non-European states, and in 1961 it was reformed into the OECD. It is a forum where peer pressure can act as powerful incentive to improve policy and non-binding instruments which may lead to binding treaties. There are currently 28 full members of the OECD, mostly from the EU. The major non-EU industrialised members are: USA, Canada, Switzerland, Japan, Australia and New Zealand. There are an additional 24 non-members who participate as regular observers or participate in OECD committees.

The NEA was established in 1956, within the framework of the OECD. Its objectives are the furtherance of peaceful use of nuclear energy, protection of workers and the public from the hazards of ionising radiation and the preservation of the environment. Another objective of the NEA is the promotion of third-party liability and insurance with regard to nuclear damage. Its publications and recommendations are advisory in nature.

## 6.3  UK National Regulations

### 6.3.1  Health and Safety at Work etc. Act 1974

The main legislation governing the health, safety and welfare of the occupational workers and the public at the nuclear installations is the Health and Safety at Work etc. Act 1974 (HSWA74) [15]. This is a major piece of legislation which brings together and rationalises all the fragmentary pieces of legislation concerning the health and safety of workers and the public. It came into force on 31 July 1974. It places a fundamental duty on employers to ensure, so far as is reasonably practicable (SFAIRP), the health, safety and welfare of their employees. It has four parts. Part I deals with 'health, safety and welfare in connection with work and control of dangerous substances and certain emission into the atmosphere', Part II deals with 'the employment medical advisory service', Part III deals with 'building regulations, and amendment of building (Scotland) Act 1959' and Part IV deals with 'miscellaneous and general' issues. This is an enabling Act, which means that further regulations may be enacted (by Parliament) under this Act.

The Health and Safety Commission (HSC) was established under this Act in 1974. It is appointed by, and reports to, the Secretary of State, Department for Work and Pensions (DWP), though it may report to other secretaries of state on other specific matters. For example, on nuclear matters in England and Wales, it advises the Secretary of State, Department for Business, Enterprise and Regulatory Reform (DBERR) and, in Scotland, the Secretary of State for Scotland. For the health and safety laws and standards, the HSC relies on the advice of the HSE. The HSC is also advised by the Nuclear Safety Advisory Committee (NuSAC).

The HSC maintains an overview of the work of the HSE and has the power to delegate some of its responsibilities. However, the HSC cannot give directions to the HSE about the enforcement of any particular aspect of the HSWA74. The HSE is a corporate body consisting of three persons statutorily appointed to enforce health and safety laws in the UK. The HSE is the licensing authority for nuclear installations in the UK. The Nuclear Directorate (ND) is a free-standing directorate within the HSE. Its objective is to control safety and security of radioactive material and radioactive waste at nuclear facilities and maintain health and safety of the workers and the general public as a whole.

The ND consists of six divisions
- Division 1: Civil Nuclear Power Generation
- Division 2: Nuclear Chemical and Research Site Regulation (including UK Safeguards office)
- Division 3: Defence Nuclear Facilities Regulation (DNFR)
- Division 4: Nuclear Research, Strategy and Business Systems
- Division 5: Office for Civil Nuclear Security (OCNS)
- Division 6: Nuclear Reactor Generic Design Assessment

The Nuclear Installations Inspectorate (NII) is a part of the ND, within these divisions, to which day-to-day exercise of the HSE's inspection and licensing function of civil nuclear facilities is delegated. The OCNS (Division 5) is responsible for the security of the UK civil nuclear industry (see Section 6.3.15 below). The DNFR (Division 3) maintains close liaison with the defence facilities for the maintenance of health and safety of workers and the public (see Section 6.6).

## 6.3.2 Nuclear Installations Act 1965 (as amended)

The Nuclear Installations Act 1965 (NIA65) (as amended) is a very important piece of legislation which the nuclear industry in the UK must comply with. It is the amended version of the Nuclear Installations (Licensing and Insurance) Act 1959, under which the NII was established in April 1960. The NII had been reorganised as part of the HSE/ND under the HSWA74. In 1971, the jurisdiction of the NIA65 was extended by the promulgation of Nuclear Installations Regulation 1971 (NIR71) to cover sites operated by the then British Nuclear Fuels Limited (BNFL) and the GE Healthcare (erstwhile Amersham plc).

The NIA65 has three main objectives [16]

- It requires licensing of sites which are to be used for the operation of nuclear reactors (except where reactors form part of a transport vehicle) and certain other classes of nuclear installations which may be prescribed.
- It provides for the control of security associated with the enrichment of U and extraction of Pu or U from irradiated fuel.
- It sets out the liability of the licensees towards third parties.

The first of these requirements, the licensing and inspection of sites, is under the ambit of the HSE/ND while the other two are the responsibility of the Secretary of State for DBERR for sites in England and Wales, and of the Scottish Ministers for Scotland.

Under this Act, no site may be used for the purposes of installing or operating a nuclear installation unless a licence has been granted by the HSE. A site licence is granted to a corporate body covering a defined site for specified purposes. It is granted for an indefinite term and one licence can cover the lifetime of a facility from design, siting, construction, commissioning, operation and modification through to eventual completion of decommissioning. A licence is not transferable, but a replacement licence may be granted by the HSE to another corporate body if that body can demonstrate that it is fit to hold a licence. Circumstances under which re-licences may be issued include changes to the site boundary and changes to the types of activity.

At the time of granting the site licence, the HSE attaches conditions under the terms of Section 4 of the Act, which are known as the Licence Conditions (LCs), in the interest of safety or with regard to handling, treatment and disposal of radioactive waste. The HSE may subsequently attach further conditions, vary or revoke conditions. There are 36 licence conditions which are attached to all nuclear site licences. Some of the important ones which relate to decommissioning are given below.

### Licence Condition 4: Restriction on Nuclear Matter on the Site

Under this LC, the licensee is required to control the introduction and storage of radioactive matter on a licensed site so as to ensure nuclear safety. Nuclear matter here covers nuclear fuel, radioactive material and radioactive waste.

### Licence Condition 6: Documents, Records, Authorities and Certificates

Under this LC, the licensee is required to preserve records, consents, certificates etc. for 30 years or such other periods as the HSE may direct. This is important as the records concerning design, construction and modification to plants; operational records covering incidents and accidents etc. are maintained and made available to the Decommissioning Safety Case (DSC).

### Licence Condition 10: Training

This LC requires the licensee to make and implement adequate arrangements for suitable training for all those on site who may have responsibility for any operations which may affect safety.

### Licence Condition 11: Emergency Arrangements

Under this condition, the licensee is to ensure that adequate emergency arrangements are in place to respond effectively to any incident ranging from a minor on-site event to large incidents or emergencies which can result in a significant release of radioactivity to the environment (see Section 6.3.7).

### Licence Condition 12: Duly Authorised and Other Suitably Qualified and Experienced Persons

This LC requires the licensee to make and implement adequate arrangements to ensure that only Suitably Qualified and Experienced Persons (SQEPs) perform any duties which may affect the safety of operations on the site or any duties assigned.

### Licence Condition 14: Safety Documentation

Under this LC, the licensee is required to assess and produce safety cases consisting of documentation to justify safety during design, construction, manufacture, commissioning, operation and decommissioning of the installation.

### Licence Condition 15: Periodic Review

This LC requires that the licensee carries out review of the safety case at regular intervals. The objective of this review is to compare the safety case against modern standards to see if there are reasonably practicable improvements that can be made.

### Licence Condition 17: Quality Assurance

This LC requires that the licensee establishes adequate Quality Assurance (QA) arrangements covering managerial and procedural arrangements to ensure safety of operation.

### Licence Condition 18: Radiological Protection

This LC complements the Ionising Radiations Regulations 1999 (IRR99) (see Section 6.3.4) to protect workers from the hazards of radiation. The licensee is required to make and implement adequate arrangements to assess the average effective dose to classified radiation workers and notify the HSE if the dose level exceeds a specified level.

### Licence Condition 23: Operating Rules (ORs)

This LC requires that the licensee produces an adequate safety case for any operation to demonstrate that operation can be carried out safely and to identify any limits and conditions in the interest of safety. These limits and conditions are referred to as operating rules.

### Licence Condition 25: Operational Records

The purpose of this LC is to ensure that adequate records are kept regarding operations, inspection and maintenance of any safety related matter. The records must include the amount and location of all radioactive material, including nuclear fuel and radioactive waste used, stored or accumulated on site.

### Licence Condition 27: Safety Mechanisms, Devices and Circuits

This LC states that a plant is not operated, inspected, maintained or tested unless suitable and sufficient safety mechanisms, devices and circuits are properly connected and in good working order.

### Licence Condition 31: Shutdown of Specified Operations

Under this LC, the licensee may be directed by the HSE to shut down any plant, operation or process on the site. Once shut down, it will not be started up without the consent of the HSE.

### Licence Condition 32: Accumulation of Radioactive Waste

This LC requires that the licensee has adequate arrangements to ensure that the production and accumulation of radioactive waste on site is minimised. The radioactive waste must be stored under suitable conditions and records must be kept.

### Licence Condition 33: Disposal of Radioactive Waste

This LC condition gives discretionary powers to the HSE to direct the licensee that the radioactive waste should be disposed of in a specified manner. This is similar to the powers available to the environmental agencies under the Radioactive Substances Act 1993 (see Section 6.3.5). This power is exercised by the HSE in conjunction with the appropriate environmental agency.

## Licence Condition 34: Leakage and Escape of Radioactive Material and Radioactive Waste

This LC requires that the licensee ensures, as far as is reasonably practicable, that radioactive material and radioactive waste on site are adequately controlled and contained so as to prevent leaks or escapes, and that in the event of any fault or accident resulting in a leak or escape, they are detected, recorded and reported to the HSE.

## Licence Condition 35: Decommissioning

This LC requires the licensee to have adequate arrangements for safe decommissioning of the facility. The HSE has the power under this condition to direct the licensee to commence decommissioning of any plant or facility to prevent it being left in a dangerous condition or to ensure that decommissioning takes place in accordance with any national strategy. Once an arrangement has been approved by the HSE, no alteration or amendment can be made unless the HSE has approved such alteration or amendment. It also gives the HSE power to halt any decommissioning activity if it has concerns about safety.

## Licence Condition 36: Control of Organisational Change

This LC requires the licensee to take adequate arrangements to control any changes to its organisational structure or resources which could affect safety. These arrangements include assessment of safety implications of the proposed changes. The HSE has the power to direct the licensee to submit its safety case and prevent any change from taking place until the HSE is satisfied that the safety implications are understood and that there will be no lowering of safety standards.

The licence conditions provide the basis for regulation by the ND which exercises the delegated licensing responsibility of the HSE. In carrying out this responsibility, the ND exercises a number of controls from the NIA65. These are

- The ND/NII assesses the adequacy of the safety case produced by the licensee. A **safety case** is the totality of documented information and arguments produced by the licensee to substantiate the safety of the plant, operation or modification. This is done prior to issuing the licence and attaching conditions. The chief inspector of the NII is empowered to issue a licence.
- Issues 'Consent' to a particular action of the licensee, prior to the commencement of the activity.
- 'Approve' particular arrangements or documents. Once 'Approved' by the NII, it cannot be changed without an 'Agreement'.
- Give 'Directions' whereby the licensee is directed to take an action which NII considers necessary.
- NII inspectors may use their enforcement powers to issue 'Prohibition' and 'Improvement Notices' to the licensee in case of

violation of safety. Breaches of the licence conditions may lead to prosecution.

### 6.3.3 The Environment Act 1995

The Environment Act 1995 [17] is the regulatory framework in the UK for the protection of human beings and the environment from discharge of pollutants in solid, liquid and gaseous forms. The Environment Agency (EA) in England and Wales and the Scottish Environment Protection Agency (SEPA) in Scotland were set up under this Act. In Northern Ireland, the same role is played by the Industrial Pollution and Radiochemical Inspectorate (IPRI). These agencies are the 'Authorising Bodies' for environmental protection.

### 6.3.4 Ionising Radiations Regulations 1999

This document [18] contains specific requirements for radiological protection of employees and the public in England, Scotland and Wales. In Northern Ireland, Ionising Radiations Regulations (Northern Ireland) 2000 apply. The HSE enforces these regulations under the provisions of HSWA74. These regulations, incorporating the EU Council Directive (96/29/Euratom) [14], are the statutory regulations for radiological protection in the UK. The EU Council Directive itself was derived from the 1990 recommendations of the ICRP [11]. These regulations came into force on 1 January 2000.

This is the fundamental piece of legislation for the protection of workers and the public from the hazards of ionising radiation. It imposes responsibilities on any employer undertaking any work with ionising radiation such that all necessary steps must be taken to limit the exposures of the employees and other persons so far as is reasonably practicable. The statutory dose limits (also called the legal limits) applicable to the workers and the members of the public are defined (see Section 3.4.3.1) and these limits must not be exceeded under normal conditions. Any violation of these legal limits may lead to prosecution. The clause 'so far as is reasonably practicable' is applicable only when exposures are below the legal limits and in that sense it is part of the ALARP principle (see Section 7.6). It is incumbent on employers to take all necessary steps to limit exposures by engineering controls and design features, safety measures and warning devices as well as by means of personal protective equipment.

### 6.3.5 Radioactive Substances Act 1993

The 1960 Radioactive Substances Act had been significantly amended by the Environmental Protection Act 1990 (EPA90) [19] with regard to discharge authorisation, withdrawal of UKAEA exemptions from requirements of 1960 Act etc. These amendments were consolidated into the Radioactive Substances

Act 1993 (RSA93) [20]. This provides for registration of the use of radioactive materials as well as authorisation for accumulation or disposal of radioactive wastes (solid, liquid or gaseous) from licensed nuclear sites.

Under the RSA93, radioactive substances are categorised as either radioactive material or radioactive waste. The precise definitions of radioactive materials and radioactive wastes can be found in Section 16.2. If a substance is radioactive waste, then removal from any site constitutes disposal and that needs to be undertaken in accordance with the conditions of an authorisation. If the wastes are below the defined clearance levels, then the disposal is without conditions and there would be no further restrictions or controls. If the substance is a radioactive material, then it is subject to the requirements of registration for keeping and use.

In England and Wales, control under RSA93 is exercised by the chief inspector of the EA. In Scotland, control under the Act is exercised in all cases by the SEPA. These agencies regulate discharges of pollutants to controlled waters, disposal and management of waste, release to the environment from major industrial processes and the management of radioactive substances. These agencies have statutory powers under the RSA93 for regulating and enforcing certain provisions of the Act and operational responsibilities for acting on behalf of or in support of the DEFRA.

As licensed nuclear sites are regulated under the NIA65 (as amended) regulations for the safe management of radioactive materials and wastes, these sites are exempt from those provisions of RSA93 which relate to the registration for keeping and using radioactive material and to the authorisation for the accumulation of radioactive waste. This exemption does not, however, apply to an operator of a facility on a licensed site who is not the licensee. The operator is required to be a registered user of radioactive material under this Act. This Act provides for the disposal and dispersal of radioactive wastes (solid, liquid or gaseous) from licensed as well as non-licensed nuclear sites by way of authorisation.

## 6.3.6 Nuclear Reactors (Environmental Impact Assessment for Decommissioning) Regulations 1999; and (Amendment) Regulations 2006

The Environmental Impact Assessment for Decommissioning Regulations 1999 (EIADR99) [21] came into force on 19 November 1999 as a result of the EU Council Directive, 97/11/EC [22] requiring the licensee to produce an Environmental Impact Assessment (EIA) on a project such as dismantling or decommissioning of a nuclear facility that may affect the environment. An amendment to this piece of legislation was introduced in 2006 [21a]. The EIA means a process that identifies, describes and assesses the direct and indirect effects of the proposed project on

- human beings, flora and fauna
- soil, water, air, climate and landscape
- material assets and cultural heritage
- the interaction between these factors

In order to comply with these regulations, the licensee is required to produce an Environmental Statement (ES) which should include

- a description of project comprising information of the site, design and extent of the project, an estimate of the type and quantity of expected residues and emissions
- an outline of the main alternatives considered by the licensee and the selection of the alternative with reasoned opinion
- a description of the aspects of the environment that is likely to be significantly affected by the project including, in particular, population, flora, fauna, soil, water, air, climatic factors, landscape etc.
- a description of the effects on the environment covering direct, indirect, secondary, cumulative, short, medium and long-term positive and negative effects on the project
- a description of the measures envisaged to avoid, reduce and remedy significant adverse effects

The ES is submitted to the HSE by the licensee for consent to carry out the project. The HSE then checks the adequacy of information provided by the licensee, consults with the relevant bodies such as the local planning authority, local highway authority, the EA in England and Wales or the SEPA in Scotland and the public before giving its opinion on the proposal. The HSE may attach conditions to any consent in the interest of limiting the impact of the project on the environment. Further details on the ES and its implementation methodology can be found in Chapter 12.

## 6.3.7 Radiation Emergency Preparedness and Public Information Regulations 2001

The Radiation Emergency Preparedness and Public Information Regulations 2001 (REPPIR2001), replacing the Public Information for Radiation Emergencies Regulations 1992 (PIRER92), provide for the dissemination of information to members of the public who may live nearby and who may be affected by a reasonable radiation emergency at a site. Although EU Council Directive [14] includes articles on emergency preparedness for radiation emergencies, the HSE board decided that these articles would not be incorporated in IRR99 [18]. Instead a new set of regulations, to be called REPPIR [23], was produced which subsumed PIRER92.

### 6.3.8 Management of Health and Safety at Work Regulations 1999

This is an extension of the HSWA74 and specifically addresses the issues relating to the reduction of risks by information dissemination and training [24]. Under these regulations, a relevant risk assessment is required and actions are taken to reduce risks.

### 6.3.9 Ionising Radiations (Outside Workers) Regulations 1993

The Ionising Radiations (Outside Workers) Regulations 1993 (OWR93) address the control of doses to the contract employees who may move from site to site. These are particularly applicable to contractors of a decommissioning project. These regulations are now being subsumed in the IRR99 regulations.

### 6.3.10 High-Activity Sealed Radioactive Sources and Orphan Sources Regulations, 2005

The purpose of these regulations, commonly referred to as the HASS regulations [25], is to prevent exposure of workers and the public to ionising radiation arising from inadequate control of high-activity sealed radioactive sources and orphan sources. These regulations which came into force on 20 October 2005 in the UK were derived from the EU Council Directive 2003/122/Euratom [26], commonly known as the HASS Directive. In Scotland, the HASS regulations are accompanied by the HASS (Scotland) Directions 2005.

Whilst many aspects of the HASS Directive are already implemented by the RSA93 and the IRR99, the HASS regulations were required to have further controls on

- the movement and records associated with all high activity sources
- the level of security at each site registered to hold high activity sealed sources

The HASS regulations give new powers to the regulators of the RSA93 (EA and SEPA) to ensure that appropriate controls are in place to prevent unauthorised access to, or loss or theft, of a HASS material before issuing a registration or authorisation, and to consult with the police or other appropriate persons where necessary.

### 6.3.11 Control of Substances Hazardous to Health Regulations 2002 (as amended)

The Control of Substances Hazardous to Health (COSHH) Regulations 2002 (as amended) [27] are applicable as the use of chemicals or other hazardous substances at work can put workers' health at risk. It is incumbent on the employers to control exposure to hazardous substances to prevent ill health of the workers. Hazardous substances under this law include

- substances used directly in work activities such as adhesives, paints, cleaning agents etc.
- substances generated during work activities, namely fumes from welding, soldering etc.
- naturally occurring substances such as grain dust
- biological agents such as bacteria and other micro-organisms

The list of hazardous substances is given under the Chemicals (Hazard Information and Packaging for Supply) Regulations 2002 (CHIP). It should be noted that this regulation does not apply to: (i) asbestos and lead as they have their own regulations; and (ii) radioactive substances which are regulated under the RSA93.

### 6.3.12  Control of Asbestos at Work Regulations 2002 and the Control of Asbestos Regulations 2006

These regulations were promulgated under the HSWA74 and the HSE was made the enforcing authority. The Control of Asbestos at Work Regulations 2002 [28] places requirements on the employers, whereas the Control of Asbestos Regulations 2006 [29] applies to self-employed person(s). Asbestos here means various forms of fibrous silicates.

### 6.3.13  Hazardous Waste Regulations 2005

In addition to radioactive wastes, there may be many other types of non-radioactive hazardous chemicals which may arise in the course of the decommissioning process. These hazardous wastes are regulated by the Hazardous Waste (England and Wales) Regulations 2005 [30]. These regulations came into force on 16 July 2005. The list of chemicals which are considered to be hazardous under these regulations are given in the List of Wastes (England) Regulations 2005 [31].

### 6.3.14  The Provision and Use of Work Equipment Regulations 1998

These regulations [32] have been promulgated under the provisions of HSWA74 and they delegate specific responsibilities to employers on behalf of their employees or on the self-employed persons themselves with regard to the use of work equipment. The work equipment means any machinery, appliance, apparatus, tool or installation for use at work, whether or not it is used exclusively at work. The term 'use' here includes: starting, stopping, programming, setting, transporting, repairing, modifying, servicing and cleaning. Proper inspection and testing should be carried out on work equipment to make it suitable for use. An associated piece of legislation produced by the HSE is the Operational Circular on the Lifting Operations and Lifting Equipment Regulations 1998 [33]. These regulations (also known as LOLER) require assessments of risks from falling loads which could cause injury to people to be carried out. Other characteristics of the load such as its flammability, toxicity, corrosiveness or radioactivity may form an important element of this lifting plan, but should not be a major element as other regulations such as those mentioned in Sections 6.3.10, 6.3.11 and 6.3.13 may be more applicable. This LOLER replaces nearly all earlier regulations concerning lifting equipment and machineries such as cranes, fork lift truck etc. as well as those concerned with lifting accessories.

Another piece of legislation brought in under the provisions of HSWA74 is the Personal Protective Equipment at Work Regulations 1992 [34]. The Personal Protective Equipment (PPE) means all equipment which is intended to be worn or held by a person at work and which protects the person from hazards to health and safety. Section 7.8.7 describes the PPE that are commonly used in the industry.

## 6.3.15  Anti-terrorism, Crime and Security Act 2001

This is a comprehensive piece of legislation covering terrorism, crime and security in nuclear as well as in non-nuclear fields [35]. There are 14 parts dealing with aspects such as: terrorist property, asylum and immigration, race and religion as well as weapons of mass destruction, security of the nuclear industry and other issues. A subordinate piece of legislation is the Nuclear Industries Security Regulations (NISR) 2003 [36] which deals with storage and transportation of nuclear material in a safe and secure way.

The security arrangements for the protection of nuclear materials within the civil nuclear industry are carried out by the OCNS of the HSE/ND. The primary aim of this security system is to oversee and approve nuclear operators' security arrangements and to enforce compliance. It also undertakes vetting of nuclear industry personnel who may have access to sensitive material or information. This vetting process involves employment checks and, if necessary, further scrutiny.

The access to civil nuclear licensed sites as well as to MoD sites is controlled by the security clearance process conducted by the MoD Defence Vetting Agency (DVA). Individuals requiring access must be security cleared – the level of clearance depends on the security status of the site and the individual's access requirement On a very basic level, employment checks comprising Baseline Personnel Security Standard (BPSS) (formerly Basic Check) and Enhanced Baseline Standard (EBS) (formerly Enhanced Basic Check) are carried out. For national security checks, progressively stringent checks comprising Counter-Terrorist Check (CTC), Security Check (SC) and Developed Vetting (DV) are carried out.

## 6.4 Regulatory Controls

Once the site licence has been granted by the HSE/NII under the NIA65 (as amended), the licensee remains responsible for the safety of the site until the site is granted to another organisation or the HSE declares that there has 'ceased to be any danger from ionising radiations from anything on the site or, as the case may be, on that part thereof'. When the HSE is satisfied on this 'no danger' criterion, it can de-licence the site or part thereof, thereby bringing to the 'end of the period of responsibility of the licensee under the NIA 65 (as amended)'.

The HSE criterion of 'no danger' is difficult to comprehend and even more so to implement. If it is rigorously applied, it will imply zero risk which is almost impossible for the licensee to demonstrate. Consequently no site can be delicensed under this criterion. Realising the conceptual and implementation difficulties, HSE produced a paper on 'HSE Criteria for De-licensing Nuclear Sites' [37] in March 2004 where it is argued that 'no danger' cannot literally be taken as 'no risk' or 'completely safe'. It suggests that following rigorous decontamination and clean up, there may remain a small but finite radiological hazard, whose further detection and reduction would necessitate a disproportionate effort and cost. Under such circumstances, the site may be delicensed. HSE would require the licensee to show that any residual hazard from radioactivity over and above natural background radioactivity will not pose a significant ongoing risk to any person, regardless of the future use of the site. Quantitatively, the risk of death of 1 in a million per year ($10^{-6}$ $y^{-1}$) in a site would be broadly acceptable to the HSE [38, 39].

If, on completion of the decontamination and clean up processes, there remains some residual risks which may not be considered insignificant, then that site would not be delicensed and may be placed under 'requiring continuing institutional control'. Even when a site is delicensed on the basis of 'no danger' criterion, there may still be some restrictions placed on its future use and that site would be called a 'brown field' site. A brown field site may be used for industrial purposes, but not for housing or agricultural purposes. A site that had previously been used for nuclear purposes will remain a brown field site. A green field site, on the other hand, is one which attracts no such restrictions.

Under the EPA 90 [19], non-radioactive land can be remediated to a risk-based end-point which would determine what the land can be used for. There is no equivalent 'no danger' criterion.

No person or organisation may dispose of radioactive waste except in accordance with an authorisation under the RSA93, unless the waste is below the clearance level. The operator of a facility is expected to show that the Best Practicable Environmental Option (BPEO) has been selected in the disposal of radioactive waste. The operator must also demonstrate that a process which represents the Best Practicable Means (BPM) for minimising the creation of waste and for limiting the radioactivity in releases such that the ALARP principle is adhered to, has been chosen. Full details of the EIA and BPEO can be found in Chapter 12.

In addition, there are a number of regulations controlling environmental pollution and land remediation practices. The most important ones are the Pollution Prevention and Control (England and Wales) Regulations 2000 [40] and the Contaminated Land (England) Regulations 2000 [41].

## 6.5  UK Government Policy

The government decides on matters of overall policy on decommissioning and radioactive waste management. The regulators then ensure that the policy matters are implemented within the framework of national and international regulations. In formulating the policy, Government is advised by a number of organisations. A new committee, called the Committee on Radioactive Waste Management (CoRWM), was set up in 2003 by the government replacing the erstwhile Radioactive Waste Management Advisory Committee (RWMAC). This new committee is to report the Secretary of State for the DEFRA. It is an independent advisory body consisting of a chairperson and 12 experienced members drawn from various disciplines. Its task is to review the options for managing high-level and intermediate-level solid radioactive wastes in the UK and to recommend options to provide long-term solutions which provide protection to people and the environment. The committee will also provide solutions for some LLW which is deemed unsuitable for disposal at the LLW Repository at Drigg. In addition, the government is being assisted on specific issues of radioactive waste management by the Radioactive Waste Policy Group (RWPG). This is a group made up of UK government officials, devolved admin-istration representatives and regulatory body representatives. The Advisory Committee on the Safety of Nuclear Installations (ACSNI) advises the HSC on issues of nuclear safety. The Radiation Protection Division (RPD) of the Health Protection Agency (HPA) provides technical services on matters of protection from ionising radiation. The RPD used to be known as the National Radiological Protection Board (NRPB) but on 1 April 2005 it became a part of the HPA.

Government policy on decommissioning of nuclear installations or facil-ities is set out in the Government White Paper on 'Review of Radioactive Waste Management Policy – Final Conclusions' [42]. The major elements of the policy are

- The process of decommissioning should be undertaken as soon as is reasonably practicable, taking account of all relevant factors.
- It is the responsibility of the operator to draw up a decommis-sioning strategy for submission to the regulators and seek their approval.

- Regulatory approval by HSE/NII for decommissioning will continue to be given on a case-by-case basis. Consequently, some flexibility as to the timing of various stages of decommissioning may be approved. However, it would be unwise for the operators to foreclose technically or economically any options with regard to timing of various stages of decommissioning.
- Hazards associated with decommissioning should be systematically and progressively reduced.
- Segregated funds to finance decommissioning should be established by privatised nuclear companies.

In order to clean up the legacy of nuclear activities, the government has set up the Nuclear Decommissioning Authority (NDA) [43, 44], to oversee the decommissioning and radioactive waste management arising from the R&D programmes of the last 50 years or so as well as decommissioning of fuel reprocessing and Magnox power stations, now being operated by Sellafield Limited (SL), Magnox Electric South and Magnox Electric North.

## 6.6 MoD Regulatory Regime

The MoD is a major player in the nuclear industry in the UK as it is in charge of both the Nuclear Weapons Programme (NWP) and the Naval Nuclear Propulsion Programme (NNPP) overseeing the nuclear submarine fleet and its shore facilities. Unlike civil nuclear plants which are land-based and fixed, the MoD nuclear plants such as nuclear submarines are non-fixed and mobile. These nuclear submarines may visit foreign ports and foreign nuclear submarines and nuclear powered warships may visit UK ports. All of these activities create situations which do not occur in civil nuclear plants and consequently they need to be tackled differently. There are significant differences in regulatory matters between the MoD and civil nuclear organisations.

It should also be noted that whereas civil nuclear establishments are operated for commercial reasons, defence nuclear establishments, particularly the nuclear submarine fleet and its supporting shore facilities and other army, navy and air force nuclear facilities, are only operated for defence purposes. Consequently there may be some unavoidable compromises to be made on operational grounds, which are contrary to civil nuclear safety standards. For example, in a nuclear submarine there are severe functional constraints with regards to space and weight which may necessitate higher operational doses to submariners. Submarines require a stronger and high quality metallic hull to operate in a very challenging environment: high hydrostatic pressure of deep oceans, dynamic forces of waves, manoeuvring and combat requirements. All of these demands may require the safety emphasis to be on physical safety, rather than on radiological protection. However, the general principle is that, so far as is reasonably practicable, the standards of safety in defence establishments should be at least as good as those required by statute for civil nuclear plants.

The operational activities of the MoD under the NWP and the NNPP have diminished recently following the end of the cold war and the signing of the Non-Proliferation Treaty (NPT). However, activities associated with the removal of nuclear warheads to make them inactive and decommissioning activities following removal of nuclear submarines from service have increased. These activities are carried out by a host of nuclear organisations under a complex regulatory regime, national as well as international. These matters are discussed in this section.

## 6.6.1 Legal Requirements

All sites in the UK, whether civil or military, are subject to the requirements of the provisions of the HSWA74. As the IRR99 were promulgated under this enabling Act, these ionising radiation regulations are de facto applicable to MoD establishments, regardless of whether or not sites are exempt from licensing requirements. However, the MoD has come to an agreement with the HSE that in meeting these regulations national security and defence imperatives should be taken into account. The MoD facilities covering army, navy and air force facilities are also exempt from the provisions of the Euratom treaty as long as national security and defence imperatives can be claimed. However, standards emanating from Euratom can form the basis of the EU Council Directive which may then be incorporated into the national legislation. In fact, the Euratom safety standards [14] have been brought, with minor amendments, into the UK legislation under the HSWA74 and this is directly applicable to the MoD.

The regulations which are directly applicable to the MoD establishments are

- Ionising Radiations Regulations 1999 (IRR99). There are, however, some exemptions in matters of national security and defence imperatives
- Radioactive Materials (Road Transport) Regulations 1996
- Ionising Radiations (Outside Workers) Regulations 1993 (OWR93)
- Radiation Emergency Preparedness and Public Information Regulations 2001 (REPPIR2001).

## 6.6.2 Institutional Control in the MoD: Licensing and Authorisation

The institutional control of the regulation of defence nuclear establishments is quite complex as it encompasses a variety of situations such as nuclear reactors within transport vehicles and fixed facilities under the army, navy and air forces. In addition, there is a wide variety of organisations controlled and administered by the MoD, legally and/or financially, requiring a complex regulatory structure. Fundamentally, the Secretary of State for Defence is responsible for the management of safety and security of all of these establishments and is accountable to Parliament.

The nuclear sites and mobile Nuclear Reactor Plants (NRPs) within transport vehicles under direct control or operated directly by the MoD are exempt from the licensing requirements of the NIA65. These sites include the naval bases at Devonport, Faslane, RNAD Coulport and NRTE Vulcan. The privatised dockyards at Devonport and Rosyth as well as sites operated under the Government Owned, Contractor Operated (GOCO) arrangements such as the AWE Aldermaston and Burghfield are not exempt and hence they do require site licences under the NIA65 to operate. Whether a facility is exempt from licensing requirement or not, safety responsibility lies with the Secretary of State for Defence. In order to execute this responsibility, he instituted an MoD regulatory regime, called the Defence Nuclear Safety Regulator (DNSR), whose primary role is to regulate defence nuclear activities which are exempt from statutory legislation and jointly regulate with ND which are not exempt. The DNSR subsumes the erstwhile the Naval Nuclear Regulatory Panel (NNRP) and the Nuclear Weapons Regulator (NWR) in April 2006 by a decision of the Navy Board. The chairman of the NNRP became the director of the DNSR.

The DNSR applies regulatory control over MoD sites in a way similar to that of the ND over civil nuclear establishment by non-prescriptive elicitation of safety principles and safety objectives. It is up to the organisation to demonstrate that those principles and objectives can be complied with by instituting adequate safety provisions in the safety documentation. The DNSR issues 'authorisation' along with Authorisation Conditions (ACs) to an individual or organisation who would be in effective control of the site and to an organisation who would be the NRP authorisee. The site authorisee or licensee are responsible for all activities carried out on the site, including those on a nuclear submarine, whereas an NRP authorisee is responsible for all activities within the NRP when it is not in an authorised or licensed site. The NRP authorisee is also the Approving Authority (AA) for the control and approval of the design of the NRP.

### 6.6.3  Regulatory Process

The MoD-controlled sites under the DNSR and the contractorised sites regulated by the ND may exist side by side and in many cases activities and potential hazards do overlap. In order to improve safety standards of those sites regulated by two separate regulatory bodies and sometimes with overlapping regulations, there is a need for a joined-up approach by these two regulatory bodies. In May 1996, a 'General Agreement' between the MoD and the HSE was signed where clear demarcation of responsibilities was stated taking into consideration the defence imperatives of the MoD-controlled sites. In particular, the existing restrictions on HSE inspectors were partially removed [45]. The DNSR and the ND agreed that through consultation a system may be made to operate to ensure a consistency of regulatory approach and control. This agreement sets out the principles that apply to the MoD's observance of the health and safety of service personnel, MoD civilian employees and others affected by MoD activities. Under this agreement, the Secretary of State for Defence states that where MoD has been granted exemptions from specific regulations of the HSWA74, it is the MoD policy that health and safety standards and arrangements will be, so far as is reasonably practicable, at least as good as those required by statute.

The detailed description of regulatory requirements of the NNPP can be found in the MoD's Safety Principles and Safety Criteria (SPSC) document [46]. The safety principles, derived from the IAEA guidance notes, define the standards to be achieved in any NNPP activity, taking account of the mobile character of the nuclear reactor in a defence environment. The safety criteria define the risk targets against which the NNPP activities would be judged. The ACs for authorisees are similar to the ND license conditions. Under the joint regulation regime, a mixed team of inspectors from DNSR, ND, EA and SEPA is likely to carry out inspection of authorised sites at least every 30 months. The DNSR is planning to go one step further by aligning the principles and practices of the SPSC with those of the HSE/ND Safety Assessment Principles (SAPs). In doing so, the safety standards of the NNPP and of the civilian facilities would be made coherent and easier to implement. The DNSR is also proposing to adopt SAPs for the assessment of the NWP safety provisions.

It may, however, be noted that the Crown is exempt from certain enforcement provisions of the HSE. The Crown cannot be prosecuted for breaches of the law, including failure to comply with the improvement and prohibition notices that may apply to the civil nuclear installations. In lieu of this, HSE has made arrangements for censuring Crown bodies in respect of offences which would have led to prosecution if they occurred in the private sector and has instituted a procedure for issuing Crown Notices. There are two types of Crown Notices: Crown Improvement and Crown Prohibition notices. Failure to comply with the requirements of a Crown Notice may lead to Crown Censure. If in the opinion of the HSE, the MoD commits an offence relating to health and safety, the HSE would initiate Crown Censure in lieu of prosecution. It may be noted that the Censure is against the MoD, not against the individual named on the summons. The HSE, however, retains the authority to prosecute MoD individuals if the HSE feel they have been negligent of their duty of care or have by their consent, connivance or neglect allowed a health and safety offence to be committed.

# Revision Questions

1. What are the major international or multi-national organisations which are involved in the nuclear industry in advisory or regulatory roles? Briefly describe the functions of these organisations.

2. What are the 'International Basis Safety Standards' and the 'European Basic Safety Standards'? Briefly describe the origins of these standards and how they are implemented.

3. Briefly describe the legislation entitled 'Health and Safety at Work etc. Act 1974' in the UK. Describe also how this piece of legislation covers all nuclear and non-nuclear safety matters in the UK.

4. What is the significance of the Nuclear Installations Act (NIA) 1965 in the UK? Briefly describe the main elements of this Act.

5. Write short notes on the following pieces of legislation:
    (i) IRR99
    (ii) RSA93
    (iii) EIADR99
    (iv) REPPIR2001
    (v) COSHH2002 Regulations (as amended)
    (vi) Control of Asbestos Regulations 2002

6. What role does the OCNS play in the security of nuclear matters in the UK? Briefly describe its functions.

7. How are the defence nuclear activities in the UK regulated? What is the name of the MoD regulatory body and how does it operate?

## REFERENCES

[1] International Atomic Energy Agency, Convention on Nuclear Safety, INFCIRC/449, 5 July 1994.

[2] International Atomic Energy Agency, International Basic Safety Standards for Protection Against Ionising Radiation and for the Safety of Radiation Sources, jointly sponsored by FAO, IAEA, ILO, OECD/NEA, PAHO and WHO, IAEA Safety Series No. 115, Vienna, Austria, 1996.

[3] International Atomic Energy Agency, Convention on the Physical Protection of Nuclear Material, INFCIRC/274/Rev. 1 May 1980.

[4] International Atomic Energy Agency, Convention on the Physical Protection of Nuclear Material and Nuclear Facilities, INFCIRC/225/Rev. 4 (corrected), 1998.

[5] International Atomic Energy Agency, Convention on Early Notification of a Nuclear Accident, INFCIRC/335, 18 November 1986.

[6] International Atomic Energy Agency, An Addendum to Convention on Early Notification of a Nuclear Accident, INFCIRC/335/Add. 10, 25 September 2000.

[7] International Atomic Energy Agency, Convention on Assistance in the case of a Nuclear Accident (Radiological Emergency), INFCIRC/336, 18 November 1986.

[8] International Atomic Energy Agency, An Addendum to Convention on Assistance in the case of a Nuclear Accident or Radiological Emergency, INFCIRC/336/Add. 11, 25 September 2000.

[9] International Atomic Energy Agency, Preparedness and Response for a Nuclear or Radiological Emergency, jointly sponsored by FAO, IAEA, ILO, OECD/NEA, PAHO, OCHA, WHO, IAEA Safety Standards Series No. GS-R-2, Vienna, Austria, 2002.

[10] International Atomic Energy Agency, The Joint Convention on the Safety of Spent Fuel Management and on the Safety of Radioactive Waste Management, INFCIRC/546, IAEA, Vienna, Austria, 24 December 1997.

[11] International Commission on Radiological Protection, 1990 Recommendations of the International Commission on Radiological Protection, ICRP Publication 60, Annals of the ICRP, 1991, Vol. 21, No. 1–3.

[12] International Commission on Radiological Protection, Recommendations of the ICRP, ICRP Publication 26, Vol. 1, No. 3.

[13] United Nations Economic Commission for Europe, The Globally Harmonised System of Classification and Labelling of Chemicals (GHS), http://www.unece.org/trans/danger/publi/ghs/ghs_rev00/00files_e.html.

[14] European Union, Council Directive 96/29/Euratom of 13 May 1996, Basic Safety Standards for the Protection of the Health of Workers and the General Public Against the Dangers from Ionising Radiation, Official Journal of the European Union L159, 29 June 1996.

[15] UK Government, Health and Safety at Work etc. Act 1974. http://www.healthandsafety.co.uk/haswa.htm

[16] UK Health and Safety Executive, The Regulation of Nuclear Installations including Notes for Licence Applicants, HSE, 1994.

[17] UK Government, Environment Act 1995, http://www.opsi.gov.uk/acts/acts1995/Ukpga_19950025_en_1.htm.

[18] UK Health and Safety Executive, The Ionising Radiations Regulations 1999, Statutory Instrument No. 3232, HSE, 1999.

[19] UK Government, Environmental Protection Act 1990, http://www.opsi.gov.uk/acts/acts1990/Ukpga_19900043_en_1.htm.

[20] UK Government, Radioactive Substances Act 1993, The Stationary Office.

[21] UK Health and Safety Commission, Nuclear Reactors (Environmental Impact Assessment for Decommissioning) Regulations 1999, Statutory Instruments No. 2892.

[21a] Health and Safety Commission, Nuclear Reactors (Environmental Impact Assessment for Decommissioning) (Amendment) Regulations 2006, Statutory Instrument 2006 No. 657.

[22] European Commission, Council Directive 97/11/EC of 3 March 1997 amending Directive 85/337/EEC Assessment of the Effects of Certain Public and Private Projects on the Environment, 1997.

[23] UK Health and Safety Executive, The Radiation (Emergency Preparedness and Public Information) Regulations, Statutory Instrument 2001 No. 2975, The Stationary Office.

[24] UK Health and Safety Executive, The Management of Health and Safety at Work Regulations 1999, Statutory Instrument 1999 No. 3242, The Stationary Office.

[25] UK Government, The High-activity Sealed Radioactive Sources and Orphan Sources Regulations 2005, Statutory Instrument 2005 No. 2686.

[26] European Union, Council Directive 2003/122/Euratom on the Control of High-activity Sealed Radioactive Sources and Orphan Sources, *Official Journal of the European Union* L 346, 31 December 2003.

[27] UK Government, Control of Substances Hazardous to Health Regulations 2002 (COSHH) (as amended), www.hse.gov.uk/pubns/indg136.pdf.

[28] UK Government, Control of Asbestos at Work Regulations 2002, Statutory Instrument 2002 No. 2675.

[29] UK Government, The Control of Asbestos Regulations 2006, Statutory Instrument 2006 No. 2739.

[30] UK Government, The Hazardous Waste (England and Wales) Regulations 2005, Statutory Instrument 2005 No. 894, http://www.opsi.gov.uk/si/si2005/20050894.htm.

[31] UK Government, The List of Wastes (England) Regulations 2005, Statutory Instrument 2005 No. 895, http://www.opsi.gov.uk/si/si2005/uksi_20050895_en.pdf.

[32] UK Health and Safety Executive, Statutory Instrument 1998 No. 2306, The Provision and Use of Work Equipment Regulations 1998.

[33] UK Health and Safety Executive, The Lifting Operations and Lifting Equipment Regulations 1998, Operational Circular, OC 234/11.

[34] UK Health and Safety Executive, Statutory Instrument 1992 No. 2966, The Personal Protective Equipment at Work Regulations 1992.

[35] UK Government, Anti-terrorism, Crime and Security Act 2001.

[36] UK Government, The Nuclear Industries Security Regulations 2003.

[37] UK Health and Safety Commission, HSE Criteria for De-licensing Nuclear Sites, HSE/04/119.

[38] UK Health and Safety Commission, The Tolerability of Risk from Nuclear Power Stations, HSE 1988.

[39] Health and Safety Commission, Reducing Risks, Protecting People, HSE 2001.

[40] UK Government, The Pollution Prevention and Control (England and Wales) Regulations 2000, Statutory Instrument 2000 No. 1973.

[41] UK Government, Contaminated Land (England) Regulations 2000 & Statutory Guidance: Regulatory Impact Assessment (Final).

[42] UK Government, Review of Radioactive Waste Management Policy – Final Conclusions, Cm 2919, July 1995.

[43] UK Government White Paper, Managing the Nuclear Legacy – a Strategy for Action, Cm 5552, July 2002.

[44] UK Government, Energy Act 2004.

[45] UK Health and Safety: General Agreement between the Ministry of Defence and the Health and Safety Executive, DCI GEN 275-279, September 1996.

[46] UK Ministry of Defence, Regulation of the Naval Nuclear Propulsion Programme, Joint Services Publication 518, Issue 2, April 2004.

# 7

# SAFETY ASPECTS IN DECOMMISSIONING

## 7.1 Introduction

The safety in a nuclear facility/installation during its operation, outage and maintenance and eventual final shutdown as well as during decommissioning is of paramount importance. National and international regulations require that safety should be pervasive throughout the whole of the life-cycle of a plant: from design, construction, operation to eventual shutdown and decommissioning. In some countries, such as in the UK and Spain, decommissioning is conducted using the same framework of regulations as were applied during the operational period. In other countries, such as Belgium, the Netherlands, Germany and many others in the EU, decommissioning is regarded as a distinct and separate activity from the operational stage and hence safety provisions and safety requirements may be somewhat different.

It should be noted that decommissioning presents much lower radiological risks than the operational phase, as most of the high-level radioactive materials such as the irradiated nuclear fuel or stored fuel in a nuclear power plant would have been removed before the start of the decommissioning work. This phase of work is called the POCO phase in the UK, whereas the IAEA calls it the transition phase (see Section 5.6). However, there may be other types of risks in the decommissioning work arising from industrial activities such as decontamination and dismantling operations, removal of carcinogenic and chemo-toxic waste etc. An additional point to remember is that whereas in the operational phases plans, procedures and schedules flow smoothly; in the decommissioning phase, there is no certainty that plans will move smoothly. In fact, there is a large element of uncertainty arising from lack of predictability or lack of knowledge about the distribution and content of hazardous materials.

The word 'safety' here encompasses all aspects associated with safety covering principles, criteria, standards, guides, implementation methodology and so forth. It applies to the workforce, the public and the environment. As decommissioning of a nuclear facility involves nuclear as well as non-nuclear industrial activities, all of these aspects need to be addresses in a logical and sequential manner.

## 7.2 Safety Objectives

The fundamental safety objectives in a nuclear facility can be described under the following two broad headings [1].

### 7.2.1 General Nuclear Safety Objective

The general nuclear safety objective is to protect individuals, society and the environment from harm by establishing and maintaining in nuclear facilities effective defences against radiological hazards.

The radiological hazards have been specifically mentioned as only nuclear facilities are being considered here. However, other forms of hazards such as chemical, radio-chemical, industrial etc. arising during the process of decommissioning are also addressed here. It should also be noted that the hazards from nuclear operations arise not only within the site but also from planned/authorised discharges of radioactive material to the environment.

### 7.2.2 Technical Safety Objective

The technical safety objective is to take all reasonably practicable measures to prevent accidents in nuclear facilities and to mitigate their consequences should they occur; to ensure with a high level of confidence that, for all possible accidents taken into account in the design of the installation, including those of very low probability, any radiological consequences would be minor and below prescribed limits; and to ensure that the likelihood of accidents with serious radiological consequences is extremely low.

## 7.3 Strategy for Achieving Objectives

The above-mentioned objectives are very much inter-linked and hence cannot be totally separated from each other. However, for the purposes of clarity of presentation, they are described sequentially below.

The first objective can be met by fulfilling the following requirements:

(i)  No person shall receive doses of radiation in excess of the statutory dose limits.

(ii)  The exposure of any person to radiation shall be kept As Low As Reasonably Achievable (ALARA) (In the UK, ALARA has been replaced by ALARP).

(iii)  The collective effective dose to operators and to the public shall be kept ALARA.

The second objective stated above can be met by:

(iv)  All reasonable practical steps shall be taken to prevent accidents.

(v) All reasonable practical steps shall be taken to minimise the consequences, radiological or otherwise, of any accident.

The first three items relate to radiological protection applicable to nuclear operations during operations and decommissioning. They can be implemented by incorporating safety measures, operating procedures, training of workforce and management control. The fourth item concerning the provision of accident prevention can be achieved by ensuring integrity of the plant and by providing for the detection and control of abnormal conditions. The fifth item is related to mitigating the consequences of nuclear accidents.

Before going into the details of radiological protection, it would be appropriate to describe the design philosophy and operational practice of the facility as a means of ensuring technical safety. This effectively means that the design and engineering substantiation process of the safety aspect will be dealt with first.

## 7.4 Technical Safety

### 7.4.1 Defence in Depth

Defence in depth is an important principle in safety, radiological or non-radiological. Under this principle, successive layers (structures, systems, components, procedures or a combination of all of these) of overlapping safety provisions are put in place such that if one layer fails, other layers become effective in preventing or mitigating the failure. Hence it is also referred to as multi-barrier protection. One important requirement is that these layers must be independent of each other. Otherwise, if one layer is dependent on another layer, then the failure of one layer may lead to the failure of the dependent second layer, thereby negating the whole concept of multiple layers of protection. When these layers are independent, then the overall failure probability becomes multiplicative. For example, if the failure probability of layer A is 1 in 100 ($10^{-2}$) and that of B is 1 in 1000 ($10^{-3}$), then the joint failure probability of both A and B is 1 in 100,000 ($10^{-5}$).

These layers may be constructed from technical considerations as well as from human actions. However, technical layers of defence are usually preferred to reliance on human actions. The hierarchy of safety sequence in a system or operation can be given as

 (i) design for minimum hazard

 (ii) reduction of hazards through safety systems (protective system, interlocks etc.)

 (iii) safety monitoring system (installed monitors, alarms etc.)

 (iv) working procedures and practices

(v)   training of workforce

(vi)   management review to identify residual hazards

Technical specification of safety systems involves **redundancy**, where two or more copies of the same protective layer are available in parallel; **diversity**, where alternative modes of protection are available and **independence**, where common mode of failure or common cause failure cannot jeopardise the diverse systems.

This defence in depth principle can be applied in diverse circumstances. It can be applied to the management provision for an operation, to the containment of radiation sources, to the disposal of radioactive wastes, to the physical security of sources, etc.

## 7.4.2 Design Basis Accident Analysis

The safety of a nuclear facility or installation from design to decommissioning is ensured by proper safety assessments based on the Design Basis Accident (DBA) and Probabilistic Safety Assessment (PSA) methodologies. These two approaches complement each other in the analysis of accident and abnormal conditions.

The purpose of DBA analysis is to demonstrate the fault tolerance of the facility, the effectiveness of the safety systems and to set limits for safe operation. This analysis should show that there are sufficient safety measures available (multi-barrier protection) either to prevent the design basis faults developing into accidents or to ensure that, given the accident, the consequences are mitigated to the extent that they are not significant and the facility can be brought to a safe state.

The DBA analysis is undertaken on existing facilities, new facilities or facilities undergoing modifications. Although decommissioning may be perceived by uninitiated individuals as no more than mere dismantling and demolition, there are quite a few operations in existing facilities undergoing decommissioning that may require DBA analysis. This analysis should cover design basis initiating faults against which the facility is designed/modified to cope. These initiating faults are brought out in the HAZard and Operability (HAZOP)/ fault schedule. If an existing facility or the facility undergoing modifications fails to satisfy the modern DBA criteria, a robust ALARA argument may be required.

An initiating fault is one which, if not prevented from developing further, could result in radiological consequences to the workforce or to the public. Thus an initiating fault would require the safety system to function or an operator action to be taken to prevent or mitigate the consequence of this fault. All initiating events which have fault frequencies of

$> 10^{-7} \, y^{-1}$ for external (man-made) faults

$> 10^{-4} \, y^{-1}$ for external (natural) faults

$> 10^{-5} \, y^{-1}$ for internal faults

and could give unmitigated doses of

> 20 mSv.y$^{-1}$ to the operator and/or

> 1 mSv.y$^{-1}$ to the public

are considered to be Design Basis Initiating Faults (DBIFs) and they need to be addressed.

The methodology of DBA analysis comprises the following steps:

    (i)   examine all possible initiating faults and identify the DBIFs

    (ii)  examine the DBIFs and quantify the frequencies of these event sequences

    (iii) estimate the unmitigated consequences of these event sequences

    (iv) estimate the mitigated consequences of various event sequences and categorise equipment and systems according to their safety requirements

Following the identification of faults or event sequences, radiological consequences (in the form of HAZard ANalyses (HAZANs)) to the workforce and to the public need to be carried out. Initially this analysis is to be conducted without claiming any protective measures and this would constitute the unmitigated consequence assessment. If this unmitigated consequence is within the bounds of the upper limits of radiological consequences specified by the site licence holder or owner/operator, then further DBA assessment may not be necessary. But the application of the ALARA/ALARP principle, described in Section 7.6, still applies. On the other hand, if the unmitigated consequence exceeds the specified dose limits, further assessment claiming protective measures need to be carried out to bring the consequences within the acceptable limits. Those protective measures (monitors, alarms, operating procedures, safety rules etc.) are to be classified as safety critical. An ALARA/ALARP assessment may also be necessary.

## 7.4.3 Probabilistic Safety Assessment

PSA is a powerful tool not only in assessing individual and collective health risks from a nuclear practice but also for managing risks, identifying plant weaknesses, improving operating procedures and safety mechanisms, defining countermeasures as well as applying cost-benefit analysis in a nuclear practice. It requires a structured approach and an amalgamation of a number of diverse disciplines [2]. PSA is customarily categorised into three levels.

**Level 1**: This level is concerned with the identification and quantification of probabilities or frequencies of events that may lead to plant damage states following beyond design basis accidents/faults. To identify initiating event frequencies, reliability database for components, human reliabilities etc. are used in the event tree and fault tree formalisms.

**Level 2**: This level is concerned with the progression of an accident scenario within the plant leading to the failure of the activity containment provisions, and transport of activity from the plant to the environment.

**Level 3**: This level deals with all the processes, following environmental release of activity, such as the atmospheric dispersion, meteorological conditions and other related parameters and takes into consideration the population distribution, dosimetric effects, economic activities etc. The end-points could be health effects to individuals and to public including mortality or morbidity; non-health effects such as the loss of agricultural production due to land contamination, decontamination costs, costs of food banning, countermeasure strategies such as sheltering, evacuation and relocation.

Different levels of PSA are carried out to satisfy different purposes. Quite a large number of national and international codes of varying degrees of complexity and details encompassing one or more levels of PSA are available. Levels 1 and 2 giving failure frequencies leading to accident scenarios and the corresponding source terms are essential input to level 3. The major codes for level 3 are: (i) COSYMA, developed jointly by the then NRPB (now reorganised as the RPD) of the UK and Kfk of Germany in the early 1980s for use in Europe; (ii) MACCS, developed by the American Nuclear Regulatory Commission for use in the USA; (iii) CONDOR, developed jointly by the UK AEAT (now SERCO), Nuclear Electric (now British Energy) and NRPB (now RPD) in the 1990s. Level 3 normally deals with collective or societal health risks, economic risks, countermeasure strategies and so forth. Having assessed the accident consequence for the societal risk assessment, comparison with the societal risk target should be carried out in order to establish whether or not the plant meets the criteria. It should, however, be noted that as many of the parameters used in the assessment are statistically variant in nature, a parametric sensitivity and uncertainty analysis is desirable in order to show the degree of variability of the end-points on these parametric values.

## 7.4.4 Applicability of DBA and PSA

The DBA is intended to demonstrate the tolerance to faults, whereas the PSA provides residual risks from accident conditions on probabilistic grounds. The DBA considers all those faults which are likely to take place and against which design and engineering substantiation must be provided. On the other hand, the PSA considers all those accident scenarios which are beyond the design basis (and consequently highly unlikely) and assesses health risks to the workforce and to the public as well as other risks such as environmental risks, economic risks etc.

The methodologies of these two approaches are quite different. The DBA analysis is a deterministic analysis that is carried out on a conservative basis. It predicts the course of events following certain design basis faults and assesses their consequences to demonstrate that the plant is deterministically safe even with pessimistic estimates. The PSA, on the other hand, is carried out on the best estimate basis to demonstrate probabilistically that the risks to the workers and public are acceptable.

## 7.5 Radiological Protection

Radiological protection in nuclear decommissioning work is an extremely important element in nuclear safety. Regulatory issues associated with radiological protection have already been described in detail in Chapter 6. This section deals with the implementation of those regulatory requirements.

### 7.5.1 Operational Protection for Exposed Workers

Operational protection of the workers against ionising radiation is based on the following principles:

- Identification of the nature and extent of the radiological risks and optimisation of radiation protection.
- Classification of workplaces into different areas, in accordance with the expected annual doses.
- Classification of workers into different categories depending on the likely exposures or risks undertaken.
- Implementation of control measures and monitoring of different areas and monitoring individuals.

#### 7.5.1.1  Nature and Extent of Hazards

Whenever any work involving radiation is undertaken, the nature and extent of the hazards arising from potential exposures of the workers need to be assessed. This is the primary reason why site or facility characterisation is carried out at the beginning of the decommissioning work. The more thorough or detailed this characterisation is, the better is the assessment of workforce exposures and risks. However, the extent of detailed site characterisation has to be balanced against the cost, particularly at the beginning of the project. Following this characterisation, the optimisation of exposures in all working conditions needs to be carried out.

#### 7.5.1.2  Classification of Workplaces

The workplaces should be classified whenever there is a possibility that workers working normally will be exposed to radiation doses in excess of 1 $mSv.y^{-1}$. In such a situation, the workplace is to be demarcated into a supervised area and a controlled area, according to the levels of risk in those areas.

The **supervised area** is to be physically demarcated and properly labelled indicating the type of area, the nature of radiation sources and their inherent risks. Working instructions appropriate to the nature and extent of radiological risks in the area are to be produced. Unclassified persons may work in this area under a system of work. The annual effective dose to a worker in the supervised area is likely to be 1 mSv or more.

A **controlled area** is one where the radiological risks are significantly higher than in the supervised area. This is the area from which contamination may spread to other areas, if not properly controlled. The area needs to be physically demarcated and properly labelled indicating the type of area, the nature of the source and the inherent risk. Barriers are to be erected and access to such areas is restricted. Monitoring and surveillance of the area for radiological risks are to be instituted. No person will be allowed to enter this area unless he/she has received appropriate instructions. The workers in this area must work according to written procedures.

## 7.5.1.3 Classification of Workers

Workers are classified into categories A and B according to the risks they are likely to face.

**Category A**: Those who are likely to receive an effective dose greater than 6 $mSv.y^{-1}$ year or an equivalent dose greater than 3/10 of the dose limits for the lens of the eye, skin and extremities. They are likely to work mostly in the controlled area.

**Category B**: Those who are not category A workers but are likely to receive effective doses in excess of 1 mSv per year. They will mostly be working in the supervised area.

## 7.5.1.4 Control Measures, Area Monitoring and Individual Monitoring

Control measures are mainly management procedures to ascertain that radiological protection principles, as required by regulatory standards, are adequately met. They are implemented by setting up local rules and procedures to ensure protection and safety for the workers and others. The work involving occupational exposures is to be adequately supervised and all reasonable steps should be taken to ensure that the rules, procedures, protective measures and safety provisions are observed. The workers are given adequate training in radiation protection.

Area monitoring of the workplaces, both supervised and controlled areas, is to be carried out. This area monitoring would include measurements of external dose rates, measurements of air activity concentration and surface contamination level. In controlled areas where airborne β-rays or γ-rays may be present, β/γ-monitors are installed at strategic points. In areas where α-activity may be present, alpha-in-air monitors are installed. In addition, workers may carry portable α- and γ-monitors. These monitors would provide a measure of the radiation dose to the workers. Based on these measurements, protective actions are to be taken. These measurements will also provide estimation of individual doses.

Individual monitoring for both category A and B workers will be carried out using approved dosimetric devices (TLDs, film badges, QFDs etc.). This monitoring for the category A worker needs to be systematic and comprehensive. The results of such individual monitoring will be retained until the worker reaches the age of 75 years or at least for 30 years from the termination of work involving exposure, whichever comes later.

### 7.5.1.5 Personal Protective Equipment and Respiratory Protective Equipment

The main emphasis in personnel protection should always be directed towards preventing, eliminating and controlling hazards, in that order, by engineering and/or administrative methods. For the protection of workers in a radiological environment, Personal Protective Equipment (PPE) and Respiratory Protective Equipment (RPE) are required, depending on the levels of hazard involved.

**PPE**: Even in well-managed radiological areas, there may be enhanced levels of surface contamination and airborne contamination. Workers in those areas would be required to wear protective clothing such as coveralls, caps or safety helmets, gloves and overshoes or safety footwear. If the activity concentrations are high or very high, a pressurised suit, which may be made of PVC, would be needed.

**RPE**: Respirators should be worn for work in areas where there may be airborne contamination. These may be passive respirators like gas masks where the wearer breathes through a canister of suitable filter material or positive pressure respirators where filtered air is supplied to the face mask by a battery operated pump. Pressurised suits carry their own air supply, which can be via an airline or in the form of pressurised air bottles in a backpack.

### 7.5.1.6 Dose Prediction Tools

Several dose prediction tools (computer code) have been developed to assist the ALARA/ALARP consideration in job planning. By modelling the work environment and various radiation sources, one can estimate the dose rates at various locations. This estimated dose rate can be used as an input data. Following the input of the details of work, i.e. route, duration etc., of the workers in the workplace, the computer code can estimate the doses workers would receive during the work. By calculating the dose for different scenarios (e.g. changing the sequence of activities, the use of shielding, and the number of workers involved) one can search for the task arrangement for the lowest dose and thereby meet the ALARA requirement.

These dose prediction tools are not only suitable for radiation protections purposes, but are also an aid in worker training and communication, and in public relations.

It should be remembered that the dose prediction is also a regulatory requirement to fulfil the ALARA requirement. This predicted collective dose is compared with the actual dose accrued by the workers in the completion of the work and this comparison gives an indication of how well the project had been managed from the ALARA point of view.

# 7.6 Application of the ALARA/ALARP Principle

The application of the ALARP principle in risk reduction strategy is the cornerstone of nuclear safety in the UK. It is intricately intertwined with the statutory regulation of the UK. This ALARP principle is synonymous with the internationally adopted ALARA principle and is based on a pragmatic and holistic approach. It is recognised in the UK that what is achievable by allocation of a disproportionate amount of resources may not always be either practicable from commercial considerations or justifiable from the overall risk reduction strategy. An allocation of a disproportionate amount of resources to achieve marginal benefits in the reduction of risk in one specific area may well entail diversion of resources from other areas which may, as a consequence, suffer from higher risks. The ALARP principle adopts a holistic approach, taking the reduction in risk in totality as the overall objective and at the same time ensuring the most cost-effective reduction in constituent risks arising from either radiological or non-radiological causes.

The ALARP principle may outwardly appear simple, but its full implementation to satisfy regulatory demands is quite involved. The difficulty arises primarily from the fact that a number of parameters implicit in the ALARP principle are judgemental, not quantified, and consequently they are prone to be misinterpreted or misunderstood in the process of implementation. This section attempts to identify those parameters, quantify them as far as possible, and thereby remove the judgemental elements from their implementation.

It must be said at the outset that the rigorous application of this methodology could be quite demanding and expensive and hence its rigorous application may be unnecessary or even unjustified where the risks are not particularly high. In such cases, industrial best practice and a degree of judgement on the part of the assessors may well be the best approach.

## 7.6.1 Tolerability of Risk

In order to maintain a balanced approach to risk, the HSE has defined two limiting conditions in its tolerability of risk document [3].

- There may be a situation where the risk is so high or the outcome is so dreadful that this risk cannot be tolerated on any grounds. The risk that is likely to be acceptable to society must be lower than or at most limited by this upper boundary. This is known as the Basic Safety Level (BSL).
- There may be a situation where the risk is small or the adverse outcome is insignificant such that any further efforts in reducing risks may not be strictly necessary. This level of risk is broadly acceptable to society and to the regulators. This is known as the Basic Safety Objective (BSO). However, the regulators in the UK require that the

licensee should make further efforts to reduce risks below
the BSO level, if possible.

Having defined these two boundary levels of risk, HSE set out to assign numerical values to these levels in its Reducing Risks, Protecting People (R2P2)
document [4]. But before doing so, the regulators needed to delineate the whole
human population into somewhat congruous groups as different groups may
be subjected to, either voluntarily or otherwise, different levels of risk. Two
distinct groups of the population in this context are: (i) the workforce who
would undertake risks at the workplace to reap benefits; and (ii) the members
of the public who would receive no such benefits from the practice producing
risks and consequently should not be subjected to the same level of risks as the
workforce.

The HSE also specified in the R2P2 document that the maximum tolerable
level of risk of fatality to workers in any industry that the society would tolerate
is about $10^{-3}$ per year (1 in 1,000 fatality per year). This level of risk is specified in the document as the BSL. The BSO for workers should be much lower:
roughly in the range of $10^{-6}$ per year is deemed reasonable. These are the risk
levels applicable to the workers from normal operations of a nuclear facility.
Within the boundaries of these two levels of risk lies the ALARP region which
is shown in Figure 7.1.

During normal operations, the workers may receive low or very low levels
of radiation which would only have stochastic effects. The ICRP in its publication 60 [5] has analysed the risk of fatality from radiation exposures based on
the UNSCEAR report [6] and has concluded that an annual whole body dose
of 20 mSv to an adult would approximately correspond to a risk of $10^{-3}$ of

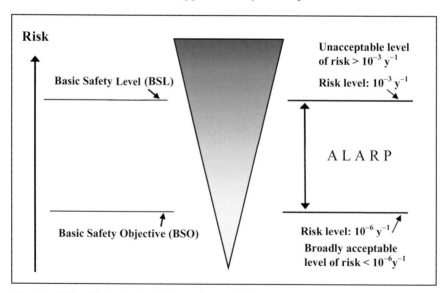

**Fig. 7.1** *Tolerable levels of risk for workers and the ALARP region in normal operations*

lifetime probability of death. Consequently, the HSE in the SAPs [7] specified the dose level of 20 mSv.y$^{-1}$ as the BSL (LL) under normal operations, where LL stands for Legal Limit. The corresponding BSO level is specified as 1 mSv. y$^{-1}$ based on the multiplicative risk projection model of the ICRP [5], which would confer a risk of 10$^{-6}$ lifetime probability of death.

In accident conditions, workers may be exposed to doses ranging from low doses where stochastic effects may be applicable to high doses where deterministic effects are relevant. In cases of high doses causing deterministic effects, the dose–risk conversion based on the nominal probability coefficient would be misleading and erroneous. Consequently, regulations controlling accident situations specify an upper limit on annual accident/incident frequency as a surrogate measure of risk, which could expose workers to various levels of radiation doses. Some accident sequences may entail very high doses, with a probability of death of 1, in which case the risk of fatality from these sequences would simply be the frequency of accidents. In other cases, the doses may be much lower and the risk of fatality can be estimated as the frequency of exposure multiplied by the anticipated dose converted to a lifetime probability of death (delayed fatality) based on 4% Sv$^{-1}$ of the nominal probability coefficient. The unit of this product is the same as the frequency (y$^{-1}$). The frequencies of all of these accident sequences affecting the workforce are summed together on an annual basis to arrive at the annual risk of fatality for comparison with the regulatory criteria [7].

The BSL and BSO values for workers under normal operations and accident conditions are given in Table 7.1.

The general member of the public in the vicinity of a nuclear facility should have a risk no more than a small fraction of that of a worker under normal operating conditions. The tolerability of risk document [3] mentions that the risk of fatality of a member of the public should not be higher than 10$^{-4}$ per year from normal operations. This is the BSL for the public. The BSO under normal operating conditions is stated to be 10$^{-6}$ per year. When these risk factors are converted to dose values using the multiplicative risk projection model, the corresponding values would translate to approximately 1 mSv.y$^{-1}$ and 0.02 mSv.y$^{-1}$, respectively, as shown in Table 7.2. These values have been specified in the SAPs [7].

**Table 7.1** *BSL and BSO levels for workers*

|  | BSL(LL) | BSO |
|---|---|---|
| Normal operation (parameter: dose) | 20 mSv.y$^{-1}$ (legal limit) | 1 mSv.y$^{-1}$ |
| Accident conditions (parameter: risk of fatality) | 10$^{-4}$ y$^{-1}$ | 10$^{-6}$ y$^{-1}$ |

**Table 7.2** *BSL and BSO levels for the public under normal conditions*

|  | BSL (LL) | BSO |
|---|---|---|
| Normal operation (parameter: dose) | 1 mSv.y$^{-1}$ (legal limit) | 0.02 mSv.y$^{-1}$ |

**Table 7.3** *Total predicted frequencies for decade bands of consequences to the public in accident conditions*

| Maximum effective dose (mSv) | Total predicted frequency (y) | |
|---|---|---|
|  | BSL | BSO |
| 0.1–1.0 | 1 | 10$^{-2}$ |
| 1.0-10 | 10$^{-1}$ | 10$^{-3}$ |
| 10–100 | 10$^{-2}$ | 10$^{-4}$ |
| 100-1000 | 10$^{-3}$ | 10$^{-5}$ |
| >1000 | 10$^{-4}$ | 10$^{-6}$ |

In accident situations, risks to the public are assessed in terms of predicted frequency of accidents/incidents rather than the dose. On the basis of the Probabilistic Safety Assessment (PSA) a wide variety of accident scenarios may be anticipated with various levels of consequences. The SAPs document uses these accident frequencies as the yardstick for indicating BSL and BSO over decade bands of consequences, which are shown in Table 7.3. Figure 7.2 shows the predicted frequencies of accidents against effective doses to the public to indicate the BSL and BSO in accident conditions, which are similar to the staircase structure.

## 7.6.2 ALARA/ALARP Methodology

The regulatory position in the UK is that it is incumbent on the licensee of a nuclear site to take steps to drive risks lower and lower for the workers as well as for the public to such an extent that the costs of any further measures would be grossly disproportionate to the risks they would avert. This overarching principle incorporates a number of implicit steps which need to be clearly defined and properly described to make the ALARP methodology easily understood and applicable. The steps which need to be considered are as follows:

(i) Identify all possible options or measures which would reduce doses to the workers as well as to the public in normal operating conditions, and risks in accident conditions and then estimate the associated committed cost for each option. This is the monetary value for that option.

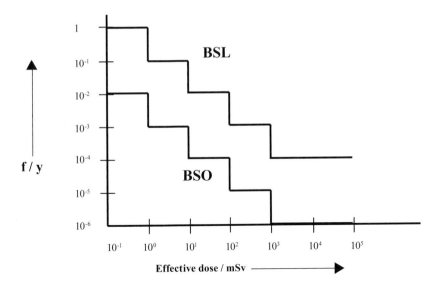

**Fig. 7.2** *Predicted frequencies against effective doses to the public in accident situations*

(ii) Convert workers' and public's averted dose to averted risk following an approved methodology.

(iii) Convert the averted risk to monetary value following an approved or acceptable technique. This is the monetary value of the benefit.

(iv) Compare the cost of the option or the measure identified in item (i) with the monetary value of the benefit (averted risk) estimated in item (iii). This comparison would give an estimate of the ratio between the cost (item (i)) and benefit, i.e. averted risk (item (iii)). The regulatory requirement is that this ratio, called the disproportion factor, must be sufficiently high to be acceptable.

(v) Carry out the sensitivity and/or uncertainty analyses of those options which are within the acceptable range of disproportion factor in order to identify the chosen option.

All of these steps need to be addressed methodically in order to demonstrate that the ALARP principle has been systematically applied and the correct conclusions drawn.

It should, however, be noted that in this context the establishment of the gross disproportionality in an option between the cost of an identified measure and the cost of the averted risk is one way of establishing the ALARP. There are other complementary ways such as the use of good practice and modern standards which could be equally acceptable to regulators as a demonstration of the ALARP principle on a qualitative basis. However, this section concentrates on the quantitative methodology of estimating proportionality between the cost of a measure and the monetary value of the risk averted.

The above-mentioned five items will now be addressed sequentially and systematically.

### 7.6.2.1 Identification of all possible measures and quantification of committed costs

Various possible measures which would improve the plant conditions and thereby reduce the dose burden on the workers under normal operations or frequencies of likely accident/incidents in fault conditions are identified at this stage. These measures are distinct programmes of work to improve plant/ process conditions from the present state to the final state. The present state may be regarded as the base case. These distinct, separate measures are the various options. The initial selection of options should ideally be done jointly by the plant managers, system designers, safety analysts and other stakeholders. Options which would reduce the doses to the workers and the public from the present level to very low levels are given priority.

Traditionally, the Cost-Benefit Analysis (CBA) was the preferred technique to identify the most cost-effective option. The CBA is based almost exclusively on the monetary costs of implementing options and comparing benefits in terms of dose reduction. Recently, more powerful techniques such as the Multi-Criteria Analysis (MCA) or Multi-attribute Utility Analysis (MUA) have become increasingly acceptable for assessing decision alternatives or options against multiple, often competing, objectives. One major difference between a criterion and an attribute is that an attribute is a quantified criterion. Each criterion/attribute of a decision alternative is measured numerically by a defined performance measure and hence it can be given a score on the strength of the performance measure. Numerical weights are then assigned to each criterion/attribute to reflect the relative valuations or preferences of the stakeholders. These scores and weights are then multiplied and aggregated to a single quantity to estimate the overall worth or utility of the option. (A full description of the MUA technique can be found in [8]). A tentative list of parameters which may be used in assessing decision alternatives or options during normal and accident conditions is given in Table 7.4. This list of parameters is only indicative; it should not be regarded as exhaustive.

A short list of options which are considered viable and meet the stipulated end-point is selected at this point from a large number of possible options. Identified options may consider some or all of the parameters shown in Table 7.4. An option would have a capital cost to implement and an operational cost. The capital cost may include the costs for design and development, project management, procurement, construction and commissioning. It may also include the plant non-availability cost. The operational cost may also include

**Table 7.4** *Some suggested decision parameters when selecting options*

| Normal Operation | Accident Condition |
|---|---|
| Individual dose | Individual risk |
| Collective dose | Collective risk |
| Capital cost | Capital cost |
| Operational cost | Operational cost |
| Inherent safety features | Inherent safety features |
| Robustness | Robustness/mitigation |
| Predicted life span of the component/system | Predicted life span of the component/system |
| Waste arising | Waste arising |
| Environmental impact | Environmental impact |

the running and maintenance costs over the lifetime of the proposed modification. The summation of these two costs is specified to be the committed cost for the option and is the parameter for comparison with the assessed benefit to demonstrate compliance with the ALARP principle.

### 7.6.2.2 Conversion of averted dose to averted risk

The implementation of an option would obviously reduce the dose burden to the workers and to the public from normal operation and risks in accident conditions. The term 'dose' means the effective dose arising from exposures to all types of radiation and from all possible pathways. The difference between dose burdens before and after the implementation of an option is the averted dose.

At the normal condition of the plant, the dose to the worker is likely to be low enough to be in the stochastic range. These dose levels can then be converted to risks on the basis of Linear No Threshold (LNT) dose–response characteristic taking the nominal probability coefficient of 4% $Sv^{-1}$ of fatal cancer to a worker (or 5% $Sv^{-1}$ of fatal cancer to a member of the public) as the unique conversion parameter. (The nominal probability coefficient had been estimated by the ICRP [5] on the basis of the epidemiological studies of Hiroshima and Nagasaki bomb survivors exposed to high doses and high dose rates. The dose–response curve was then extrapolated to low doses and low dose rates using a DDREF of 2). For example, an individual dose of 10 $mSv.y^{-1}$ to a worker is equivalent to a lifetime risk of fatality of $4 \times 10^{-4} y^{-1}$.

In anticipated accident/incident conditions, there may be a number of situations which may lead to doses which are very close to each other. These doses are normally banded together, for the sake of simplicity, in decade bands, such as 0.01–0.1 mSv, 0.1–1 mSv, 1–10 mSv and so forth. The probabilistic failure frequencies that would result in doses in a certain band are summed together to obtain the failure frequency for that band. If these summed frequencies in the decade bands are in excess of the BSO values quoted in Table 7.3, then an ALARP assessment is required.

The nominal dose values, if they are in the stochastic range, can then be converted to the probability of fatality using the nominal probability coefficient of 4% $Sv^{-1}$ for workers and 5% $Sv^{-1}$ for members of the public. The product of this probability of delayed fatality (from cancer) and the band frequency gives the actual risk of fatality $(y^{-1})$ for that band. The summation of these products for all of the dose bands in the stochastic range gives the stochastic risk. In the case of a deterministic dose culminating in early fatality, the frequency of accidents is taken as the surrogate of the risk (frequency × probability of early fatality of 1). The sum of all these risks is the risk from accident situations.

When an option is carried out, doses to the workers and the public from normal operations and risks to these population groups in accident conditions would change. The difference between the risk values before and after the implementation of the option is the averted risk.

### 7.6.2.3 Conversion of averted risk to monetary value

The conversion of risk to monetary value is quite straightforward. However, complications arise with the proper interpretation of the significance of the 'averted risk'. The significance of the 'averted risk' in the context of monetary valuation is important. The conversion can be carried out using the methodology suggested by the HSE/NII in the Technical Assessment Guide [9], which is given below.

$$\text{Monetary value} = \text{probability of fatality at a dose} \times \text{monetary value of}$$
$$\text{preventing a fatality} \times \text{number of people receiving the dose} \qquad (7.1)$$

The first term on the right-hand side of equation (7.1) the probability of fatality at a dose is, in fact, the estimated individual risk of fatality carried out in section (ii). However, in section (ii) the averted dose and averted risk were mentioned. An averted dose is the dose a worker would avert as a result of implementing an option which improved the plant/process conditions. Normally an option is chosen whose end-point dose is very low and consequently the risk at the end-point dose would also be very low. Thus the risk existent at the start of the option is the averted risk from the implementation of that option. But this may not always be the case. For example, if the end-point dose remains significant ($\sim 10 \ \mu Sv.y^{-1}$ or above), the corresponding risk cannot be ignored and hence the averted risk would effectively be the difference between the risks at the start and the end of an option. The monetary values for these two end-points need to be estimated separately to arrive at the monetary value for the averted risk.

It must be stressed here that the fatality under consideration here is the fatality of a statistical human being, not an identified individual. Consequently the value assigned here to prevent a fatality refers to preventing a statistical fatality. The monetary equivalent of a fatality has been suggested by the HSE/

NII as £2,000,000 [9]. The number of people stated above would be those involved in the lifespan of the plant under consideration.

### 7.6.2.4 Comparison of Costs and Benefits

Having identified the cost of an option and the corresponding monetary value of the averted risk as the benefit, a comparison of these two quantities can be made. It is the regulatory requirement for the purposes of the ALARA demonstration that this ratio (ratio of the cost of an option to the cost of averted risk) must be sufficiently high (grossly disproportionate) if any further expenditure is not to be justified. The quantification of this disproportion factor is subjective, but it must reflect society's willingness to undertake improvements to avoid risks to human beings. It should be noted that the disproportion factor is not a constant quantity. The higher the risk, the higher should be the disproportion factor. The nuclear industry's agreed methodology incorporating the disproportion factor as a function of risk is specified in the UKAEA's Safety Assessment Handbook Methodology [10]. The values specified are as follows:

- for an individual dose: 2 at 0.05 mSv, 5 at 0.5 mSv and 10 at 20 mSv
- for risk consideration: 1 at BSO and 10 at BSL

The disproportion factors at any other intermediate value can be evaluated by the interpolation method. If the estimated cost–benefit ratio is lower than this disproportion factor, then regulators may consider that ALARA has not been fully implemented. In other words, further resources should be allocated to drive down the risks, unless the risk is already at or below the BSO. On the other hand, if the estimated cost–benefit ratio is at or above the disproportion factor and the BSO had not been reached, it is incumbent on the licence holder to justify that any further allocation of resources would be unreasonable.

### 7.6.2.5 Sensitivity and Uncertainty Analysis

The process of estimating the committed cost and risk involves a number of input parameters which are not precisely known to the analysts or which are likely to vary in time. This aspect of uncertainty or likely variation in input parameters resulting in variability in the output parameter is an important consideration.

An assessment should be carried out by varying the values of input parameters by a certain amount and checking the impact of such variations on the assigned output parameter (which may be the cost). If the output parameter of an option changes significantly resulting in significant changes in the estimated disproportion factor, then a judgement may be made as to the precision of that estimation. Similar variations are made in other options and relative variations in the disproportion factors are made and then the final conclusions are drawn with regard to the optimum option. This method of changing an input parameter to check the variation in output parameter is known as the sensitivity analysis. The uncertainty analysis is similar; but the difference is that in this case the uncertainty of the input parameters is taken into account and varied (rather than varying the input parameters purposely by a certain amount), and then the variation in the output parameter is assessed.

# 7.7 Safety from Chemical Hazards

## 7.7.1 Chemical Hazards Consideration

In addition to radiological hazards, materials arising from the decommissioning process may be hazardous to human beings and to the environment due to their chemical properties. The term 'chemical' is used here specifically to denote all non-radioactive substances (although radioactive substances are also chemicals), products, mixtures, preparations etc. which may arise during the decommissioning activity. The hazardous characteristics of chemicals may be classified under physical hazards, health hazards and environmental hazards.

### 7.7.1.1 Physical Hazards
- explosives
- flammable gases, aerosol
- flammable liquids, solids
- self-reactive substances
- pyrophoric liquids, solids
- oxidising liquids, solids
- corrosive substances

### 7.7.1.2 Health Hazards
- toxicity
- skin corrosion
- germ cell mutagenicity
- carcinogenicity
- reproductive toxicity

### 7.7.1.3 Environmental Hazards
- acute aquatic toxicity
- chronic aquatic toxicity

The chemical substances may be variously defined by different regulatory bodies in different countries as either hazardous or non-hazardous even under the same physical conditions and that may cause genuine practical problems of trans-boundary movement of chemicals. For example, the Occupational Safety and Health Administration (OSHA) of the USA defines a flammable liquid as one whose flash point is below 37.8°C (100°F), whereas the EU defines the flash point to be 55°C (131°F). Beyond the flash points are the combustible liquids. So what is considered in the USA as combustible (>37.8°C) is considered as flammable in the EU and thereby requiring stricter regulatory controls. There are many such differences. In order to harmonise the regulatory controls, an international attempt has been made under an international mandate adopted in 1992 to produce The Globally Harmonised System of Classification and Labelling of Chemicals (GHS) [11] (see also Section 6.2.4). The GHS is a harmonised classification and labelling of all hazardous chemicals in order to improve communication between nations and even states within a nation. The GHS document (referred to as 'The Purple Book') establishes an agreed hazard classification and labelling of chemicals, which the regulatory authorities of all the countries would use to develop national programmes to specify hazardous properties of chemicals and prepare safety data sheets as appropriate. The full official text of the system is available online [11].

Brief definitions of the hazards as per the GHS [11] are given below.

**Explosive:** An explosive substance (or mixture) is a solid or liquid which is in itself capable by chemical reaction to produce gas at such a temperature and pressure and at such a speed as to cause damage to the surroundings.

**Flammable**: A flammable liquid is one whose flash point is not more than 93°C.

**Self-reactive substance**: A substance which is a thermally unstable liquid or solid and is liable to undergo a strongly exothermic thermal decomposition even without the participation of oxygen (air). This definition excludes materials classified under the GHS as explosive, organic peroxides or as oxidizing.

**Pyrophoric liquid**: A liquid which, even in small quantities, is liable to ignite within five minutes after coming into contact with air.

**Oxidising liquid**: A liquid which, while in itself not necessarily combustible, may, generally by yielding oxygen, cause or contribute to the combustion of other material.

**Skin corrosive substance**: A chemical which causes visible destruction of, or alterations in, living tissues by chemical action.

**Mutagen**: An agent giving rise to an increased occurrence of mutations in populations of cells and/or organisms. It may induce genetic mutations in human germ cells.

**Carcinogenicity**: A carcinogen is a chemical substance or a mixture of chemical substances which induce cancer or increase its incidence.

**Reproductive toxicity**: This includes adverse effects on sexual function and fertility in adult males and females, as well as developmental toxicity in offspring.

**Aspiration toxicity**: This includes severe acute effects such as chemical pneumonia, varying degrees of pulmonary injury or death following aspiration. Aspiration is the entry of a liquid or solid directly through the oral or nasal cavity, or indirectly from vomiting, into the trachea and lower respiratory system. Some hydrocarbons (petroleum distillates) and certain chlorinated hydrocarbons have been shown to pose an aspiration hazard in humans. Primary alcohols, and ketones have been shown to pose an aspiration hazard only in animal studies.

Using chemicals or other hazardous substances at work can put peoples' health at risk. Controls of such risks are effected by the COSHH (Control of Substances Hazardous to Health) Regulations 2002 (as amended) [12] in the UK. The list of substances under these regulations does not include asbestos, lead or radioactive substances as safety provisions for these substances are enforced under separate regulations. The control of chemical hazards from storage and handling is carried out in the UK by the application of the Control of Major Accident Hazards (COMAH) Regulations [13]. These regulations are applicable where threshold quantities of dangerous substances identified in the regulations are kept or used. These quantities are specified in two levels: top tier where the entirety of the regulation applies and lower tier where part of the regulation applies. Even lower tier quantities are quite substantial. For example, 5 tonnes of very toxic or 50 tonnes of toxic or 10 tonnes of explosive substances can be stored under the regulation in individual facilities.

## 7.7.2 Chemical Hazards during Decommissioning

Chemical hazards encountered during decommissioning arise from
- old chemicals or chemical wastes stored or abandoned in the laboratory, often unlabelled.
- hazardous materials used as part of the facility fabric, e.g. asbestos or as part of the process operation, e.g. sodium in fast reactors or organics used during fuel reprocessing.
- hazardous chemicals introduced as part of the decommissioning process, e.g. decontamination reagents such as acids and alkalis.

In order to avoid difficulties during the decommissioning process or subsequent management of hazardous waste, it is important that
- a careful characterisation of all chemically hazardous materials present in the facility is made.
- a comprehensive assessment and understanding of risks including reactive properties of materials is carried out.
- there is a clear understanding of the hazards to workers and the environment and that proper protective measures are planned.

- methods of work comply with the local operating procedures and national and international guidance and regulation.
- adequate emergency arrangements are put in place.

Whenever practicable or feasible, considerations should be given to the recovery, re-use or recycling of materials. However, in most cases, the opportunities for recycling materials are likely to be small, particularly where the substances are also radiologically contaminated. Details about the nature of hazards, recovery, treatment and disposal of a range of 'problematic' wastes can be found in [14].

The following sections identify chemically hazardous materials that are typically encountered during decommissioning and highlight some of the main safety aspects associated with these materials.

### 7.7.2.1 Asphyxiants

Asphyxiants are chemicals that deprive the body tissues of oxygen. Breathing air with an oxygen concentration lowered by the presence of gaseous asphyxiants can result in insufficient oxygen in the blood and tissues. Symptoms range from headaches to unconsciousness and death as the concentration of asphyxiant increases. Examples of asphyxiants are inert gases such as argon and helium, hydrogen, nitrogen, carbon dioxide, propane, methane and acetylene.

The major decommissioning activities where there are potential hazards from the use of asphyxiants are

- cutting in confined areas using gas torches, e.g. acetylene, propane
- areas where nitrogen or argon is used as a fire suppressant, e.g. glove boxes or specially enclosed areas

### 7.7.2.2 Chemical Solvents

Chemical solvents, such as strong mineral acids or complexing agents, are often introduced into the decommissioning process to decontaminate items or structures prior to dismantling or disposal. The major hazards arising from the use of these agents are related to their corrosivity. Often sequences of chemical treatments are used to improve decontamination. In these cases great care needs to be taken to ensure that there are no adverse reactions between the chemical treatments. Prior to the use of chemical decontaminants, consideration of waste disposal routes should also be undertaken since complexing agents such as EDTA can interfere with the efficiency of the effluent treatment process.

Organic solvents are often present in facility laboratories. These may be either proprietary solvents, usually labelled, but more commonly the major hazard is represented by the presence of abandoned sample pots/containers where the content is less clearly or not known.

Substantial quantities of reprocessing solvents (Tri-Butyl Phosphate (TBP)) or waste oils may build up in facilities. Accumulation of organic wastes typically occurs when disposal routes are not developed or straightforward. The major hazard arising from all organic waste is their potential flammability at elevated temperature.

### 7.7.2.3 Alkali Metals

Liquid sodium (Na) and sodium–potassium alloys (NaK) are closely associated with the development of fast breeder reactors, where they were used as coolant for transferring thermal energy from the core. Both Na and K are very reactive. Reactions with water, air and oxygen are generally violent and produce hazardous by-products such as hydrogen or caustic products. On contact with water, the generation of hydrogen and the highly exothermic nature of the reaction will result in spontaneous fire/explosion within hydrogen concentrations of 4–76%. Great care must therefore be exercised during decommissioning operations involving either bulk removal or cleaning of trace Na residues from vessels or pipe work.

A further complication arising from the ageing of NaK stored in air is that K can form dangerous peroxide crusts, super-oxides or hydroxide monohydrates which can cause very hazardous reactions at elevated temperatures.

### 7.7.2.4 Asbestos

Asbestos had been used in many nuclear installations for thermal insulation of reactor vessels and many steam and process pipes from the beginning of the nuclear industry up to the late 1960s. In the UK, significant quantities of asbestos had been used in the Magnox plants, in some cases reaching thousands of tonnes. Many forms of asbestos are carcinogenic. The EC Directive [15] requires that buildings containing asbestos cannot be demolished or renovated until it has been safely removed.

The removal and disposal of asbestos is highly regulated and generally involves manual removal under highly controlled conditions. More recently, robotic techniques have been developed to remove asbestos from the outside of pipes.

### 7.7.2.5 Lead

Lead (Pb) is widely used in nuclear facilities as shielding material in the form of lead bricks, sheets, wool or lead shots. It is also found in paint and primers routinely used during the construction of early facilities.

The main hazard of Pb during decommissioning is by inhalation of particulates arising from cutting or physical decontamination (e.g. shaving) operations. It will initially accumulate in the lungs and subsequently disperse into bones, teeth and tissues. The principal toxic effects of Pb are on the central nervous system, blood and kidneys.

### 7.7.2.6 Mercury

Mercury (Hg) may be found in small quantities in many laboratories, but was also used on a much larger scale during the early experimental fast reactor programmes. In the UK, almost two tonnes of Hg were used in the seals of the Dounreay Demonstration Fast Reactor (DFR) and Prototype Fast Reactor (PFR). Metallic Hg can enter the body by ingestion or through the skin. Because it is highly volatile, it also represents an inhalation hazard.

# 7.8 Safety from Industrial Hazards

During the operational phase of a nuclear facility, prime attention is given to radiological safety; industrial safety is considered as an addendum to it. However, in the decommissioning phase, industrial safety receives a much more prominent role as activities such as decontamination, dismantling (cutting, lifting, handling, removing) etc. are all industrial activities. Additional safety problems may arise when cutting and handling plant components containing, for example, asbestos in the thermal insulation or in cement.

## 7.8.1 Tools and Machines

Because of the unique requirements of nuclear facilities, readily available 'off-the-shelf' tools and machines may not always be available. Therefore custom-built tools and machines are to be built in many cases. Machinery, both custom-built and mass-produced, must be constructed that fits the purpose. The EC Directive 98/37/EC of 22 June 1998 [16] gives the framework for safety measures and criteria for the design and use of tools and machines. The measures aim to eliminate any risk of accident throughout the lifetime of the machinery, including during the assembly and dismantling phases. If risks cannot be eliminated, the equipment manufacturer must inform the users of the residual risks due to any shortcomings of the protection measures adopted, indicate whether any particular training is required and specify any need to provide PPE. To show that a machine complies with the directive, it incorporates the year of construction and the CE marking.

Portable handheld and/or hand-guided machinery cannot comply with all requirements applicable to large machines. The directive also provides adapted health and safety requirements for portable handheld and/or hand-guided machinery.

## 7.8.2 Maintenance

During decommissioning, regular maintenance of equipment and machinery must be carried out. This is also important during a period in which no actual decommissioning work takes place such as deferred dismantling.

## 7.8.3 Training

During decommissioning, conditions for working with tools and machines can be difficult. High dose rates and the presence of hazardous materials complicate the working procedures and shorten the stay time for the workers in those high radiation areas. Often custom-built remote control tools are used in such situations. In order to achieve the desired safety level and to do these jobs effectively, workers need to be trained to work with these tools. In complex situations, training may be carried out on a mock-up installation.

### 7.8.4 Hoisting and Lifting

Hoisting and lifting can create hazards due to the collapse of a crane, dropping of a load, hitting obstacles and meeting collisions. Hoisting and lifting equipment must comply with European and national safety regulations and needs to be checked regularly. Also each length of lifting chain, rope or webbing not forming part of an assembly must bear a mark identifying the maximum load and give reference of the relevant certificate. All relevant safety data (e.g. checks and testing of equipment) need to be recorded.

### 7.8.5 Working at Heights

When a facility is being dismantled, work often has to be done at heights. The use of scaffolding and lifebelts can prevent workers from falling. Care must be taken that tools and dismantled pieces of equipment do not fall. Hard hats must be mandatory for every worker in the area.

### 7.8.6 Electrical Equipment

When working with electrical equipment and machinery, one must comply with national and international guidelines and legislation. Additional hazards may occur when an installation is being decommissioned. Such hazards include: tangled cables, loose uncovered cables, open switch boxes etc.

### 7.8.7 Personal Protective Equipment

Examples of PPE include: respirators, safety boots, hard hats, face shields and chemical resistant clothing. In recent years PPE has become more sophisticated and specialised. Care should be taken in the selection of the most suitable type of PPE for each job. One should keep in mind that PPE has its limitations. It is essential that all persons involved in the management and use of PPE are aware of the capabilities and limitations of PPE to ensure the delivery of effective personnel protection.

## 7.9 Safety Documentation

As nuclear activities are very much regulatory driven, it is imperative that any nuclear activity, whether it is commissioning, operation, maintenance, modification or decommissioning, is properly recorded and documented. In order to demonstrate that a nuclear activity can be carried out in a safe and secure way, a 'safety case' needs to be produced by the site licensee for scrutiny and approval by the regulators. In the UK, it is a regulatory requirement under various licence conditions (LC14, LC15, LC23 and LC35) that safety cases be produced at various stages of plant life-cycle (see Section 6.3.2). These safety cases form the basis of nuclear safety and create confidence amongst regulators that the licensee can carry out the nuclear activity in a safe and secure way.

A 'safety case' is not a single document but a series of documents covering various phases of the plant's life. The 'safety report' is a top-tier report that presents safety argument and all the key safety issues of a plant under consideration. A 'safety report support file' contains all the supporting details of a safety report, namely, detailed plant or operation description, technical data, calculations, assessment etc. This support file can be a single document or a number of separate documents depending on the complexity of the plant.

There are a number of safety reports that are required to be produced at various stages of a plant's life and these are

- Preliminary Safety Report (PSR)
- Pre-Construction/Pre-Commencement Safety Report (PCSR)
- Pre-Commissioning Safety Report (PCmSR)
- Pre-Operational Safety Report (POSR)
- Operational Safety Case (OSC)
- Decommissioning Safety Case (DSC)

The main objectives and the point at which these safety reports are produced are described briefly below.

### 7.9.1 Preliminary Safety Report

The PSR is produced when there is a need for a new facility or modification to a facility to demonstrate that the proposed design can be carried out safely. It is prepared on completion of the outline design and leads to the detailed design.

### 7.9.2 Pre-Construction/Pre-Commencement Safety Report

The PCSR builds on information already presented in the PSR and provides justification that the project can be implemented safely.

### 7.9.3 Pre-Commissioning Safety Report

The PCmSR provides a justification for commissioning of a plant where safety issues are of concern and demonstrates that commissioning can be carried out with proper control of the relevant hazards. The PCmSR should be based on the PCSR but should be updated to reflect additional information.

### 7.9.4 Pre-Operational Safety Report

The POSR is designed to demonstrate that, following the safe construction and commissioning of the plant, it can be operated safely.

### 7.9.5 Operational Safety Case

The OSC dictates the whole of the plant's operational life. It is a live document and would incorporate plant incident/accident situations, modifications that may be carried out etc. A Periodic Safety Report (PSR) would be a part of this OSC.

### 7.9.6 Decommissioning Safety Case

Under the site licence condition 35 (LC35), the UK regulator (ND/NII) requires that the licensee 'makes and implements adequate arrangements for the decommissioning of any plant or process which may affect safety'. These arrangements need to be documented and approved by the ND/NII. The documents which are required to be prepared under this licence condition are

- Decommissioning programme
- Decommissioning Safety Case (DSC)
- Post-Decommissioning Report (PDR)

## 7.10 Quality Assurance

It is a regulatory requirement under LC17 that the licensee makes adequate arrangements for quality assurance in all matters which may affect safety. This quality assurance arrangement must be in place for each and every stage of nuclear activity, from design and construction to decommissioning and site remediation.

Many organisations working in the UK nuclear industry use the IAEA 'Quality Assurance for Safety in Nuclear Power Plants and Other Nuclear Installations' [17] as the basis for their quality assurance system. Another quality assurance programme entitled 'Quality Management Systems – Requirements' [18] by the ISO is also used by the nuclear industry. However, the IAEA code contains some specific nuclear safety issues which are not covered in the ISO standard.

As mentioned above, it is the responsibility of the site licensee to ensure that an acceptable level of quality assurance in relation to nuclear safety is followed. This responsibility is demonstrably carried out by the licensee by ISO QA system certification. The licensee may also require that its suppliers are also appropriately QA qualified. The most important aspects of the QA system are the audit trail, assessment and monitoring provisions.

# Revision Questions

1. What is the 'general nuclear safety objective'? How is this objective realised in practice?

2. What is the 'technical safety objective'? How is this objective implemented in practice?

3. What is meant by the term 'defence in depth'? Describe the various layers that are used in the implementation of this principle. Explain how it is technically implemented in the design of safety systems.

4. What is meant by the Design Basis Accident (DBA) analysis? What are the criteria (in terms of frequencies and radiological consequences) which would prompt the DBA analysis? How would you demonstrate that the DBA criteria have been met?

5. What is meant by the Probabilistic Safety Assessment (PSA) methodology? Briefly describe the levels of the PSA technique.

6. Compare and contrast the DBA and PSA methodologies and show how they complement each other.

7. What are the various operational steps that are taken to protect radiation workers at the workplace? Define the area classification of workplaces from a radiological point of view.

8. Describe the principle of ALARA/ALARP in the context of risk reduction. Show diagrammatically the region in the risk diagram where ALARA/ALARP is applicable, clearly labelling the BSL and BSO levels.

9. Describe the various steps in the implementation of ALARA/ALARP methodology. Explain the significance of the term 'disproportion factor'.

10. What are the main hazardous characteristics from chemical substances that are encountered in decommissioning work?

11. Describe briefly how hazards arise from the following substances:
    (i) asphyxiants
    (ii) asbestos
    (iii) alkali metals
    (iv) lead

## REFERENCES

[1] International Atomic Energy Agency, Safety of Nuclear Power Plants: Design Requirements, No. NS-R-1, IAEA, Vienna, Austria, 2002.

[2] Rahman A., PSA Methodology for Societal Risk Assessment, IBC Conference on Recent Developments in Probabilistic Safety Assessments in Nuclear Safety, IBC Conference Proceedings, 11–12 December 1996.

[3] UK Health and Safety Executive, The Tolerability of Risk from Nuclear Power Stations, The Stationary Office, 1992.

[4] UK Health and Safety Executive, Reducing Risks, Protecting People, HSE's Decision-making Process, 2001.

[5] International Commission on Radiological Protection, Recommendations of the International Commission on Radiological Protection, ICRP Publication 60, *Annals of the ICRP*, 21, No. 1–3, 1990.

[6] United Nations Scientific Committee on the Effects of Atomic Radiation, Sources, Effects and Risks of Ionising Radiation, UNSCEAR 1988 Report to the UN General Assembly, with Annexes, United Nations, New York, USA.

[7] UK Health and Safety Executive, Safety Assessment Principles for Nuclear Facilities, 2006 edn.

[8] Rahman A., Multi-attribute utility analysis – a major decision aid technique, *Nuclear Energy*, 2003, 42(2), 87–93.

[9] HM Nuclear Installations Inspectorate, Technical Assessment Guide, Demonstration of ALARP, HM NII, T/AST/005, July 2002.

[10] UK Atomic Energy Authority, Safety Assessment Methodology, UKAEA/SAH/M9, August 2001.

[11] United Nations Economic Commission for Europe, The Globally Harmonised System of Classification and Labelling of Chemicals (GHS), http://www.unece.org/trans/danger/publi/ghs/ghs_rev00/00files_e.html.

[12] UK Government, Control of Substances Hazardous to Health Regulations 2002 (COSHH) (as amended), www.hse.gov.uk/pubns/indg136.pdf.

[13] UK Government, The Control of Major Accident Hazards (Amendment) Regulations 2005, Statutory Instrument 2005, No. 1088, HMSO, 2005.

[14] International Atomic Energy Agency, Management of Problematic Waste and Materials Generated during Decommissioning of Nuclear Facilities, Technical Report, IAEA, Vienna, Austria, 2003.

[15] EC Council Directive 96/61/EC on Integrated Pollution Prevention and Control, 1996.

[16] EC Council Directive 98/37/EC of the European Parliament and of the Council of 22 June 1998, on the Approximation of the Laws of the Member States relating to Machinery, *Official Journal of the European Union* L 207/1, 23 July 1998.

[17] International Atomic Energy Agency, Quality Assurance for Safety in Nuclear Power Plants and other Nuclear Installations, Code and Safety Guides Q1-Q14, Safety Series No. 50-C/SG-Q, IAEA, Vienna, Austria, 1996.

[18] International Organization for Standardization, Quality Management Systems: Requirements, ISO Standard 9001:2000, Geneva, Switzerland, 2000.

# 8

# FINANCIAL ASPECTS OF DECOMMISSIONING

## 8.1 Introduction

Financial management is one of the most important elements in the entire decommissioning project. The availability of funds dictates, notwithstanding the regulatory requirements, when and how decommissioning can be carried out. It should be appreciated that when decommissioning of a nuclear facility is undertaken, the revenue stream to that facility had already come to an end and the licensee /owner/operator faces the liability of decommissioning costs. So an estimation of the total cost (or liability) for decommissioning as well as an annual budget requirement is essential for the financial management of the project. A prudent owner/operator may make financial provisions for decommissioning during the period when the facility generated revenue; but, more often than not, no such financial arrangements exist. Nonetheless, it is a requirement under the terms of the site licence (under the operational licence covering decommissioning in the UK or under a separate decommissioning licence in some EU member states) that decommissioning must be carried out as soon as possible and in a manner that ensures safety and security for the workforce and the public.

The financial aspects of decommissioning include two major factors: cost estimates and funding mechanisms. The purpose of the decommissioning cost estimate is to obtain an estimate of the cost of a decommissioning project from the beginning right up to its completion. The completion point, which may vary from project to project, is defined in the decommissioning plan. Sometimes achieving 'restricted' or 'unrestricted' release criteria for the site and/or buildings, or at other times a pre-defined goal such as the safe-storage or entomb conditions is the end-point.

There is another reason for the production of a robust cost estimate. In most of the countries, when the licensee/owner/operator fails to undertake decommissioning because of lack of fund or due to organisational restructuring, the government of the country is unwittingly drawn into the problem and is forced to provide public finance for the sake of the safety and security of the public. This public financing demands that the fund provider justifies the allocation of public funding only after proper scrutiny and thorough examination of its cost-effectiveness. This cost-effectiveness covers the optimisation of the operating sequence so that the total project cost is minimised. Thus, in order to procure public finance, the licensee or the owner/operator must produce a detailed cost estimate which will withstand the scrutiny of the fund provider.

This chapter sets out the techniques of cost estimation, highlighting the main factors influencing cost estimates and their associated uncertainties. It also discusses the funding mechanisms and fund management. As decommissioning is always a time-extended activity, the costs are not all incurred at one time or over a short period of time. This aspect is considered in the funding provision and leads to the 'discounting technique' which will also be addressed in this chapter.

## 8.2  Overall Decommissioning Cost Estimation

Decommissioning cost estimation is inherently a very involved process, not only because there are significant uncertainties in the cost projection far into the future but also the range of activities to be included in the estimate is extensive. To begin with, there is ambiguity as to what constitutes the start of decommissioning. It may start from the closure of an operating plant or it may begin after a clean-out phase which may be covered within the operational phase. The decommissioning may also involve construction of new facilities to facilitate treatment, conditioning and storage of wastes arising from the proposed decommissioning facility. As an example, in the UK intermediate level wastes arising from PFR and DFR at Dounreay were kept at interim storage wet silos and in geological ILW shafts. None of these facilities would meet the modern safety standards for a disposal site and hence the wastes needed to be retrieved, treated, conditioned, packaged and disposed of. All of these activities would require the building of new facilities. These costs may be incorporated within the decommissioning costs of the PFR or DFR, or they may be shown as separate new costs. An additional practical problem is that if a newly built facility is designed to cater for a number of decommissioning facilities, what will the mechanism for sharing this cost amongst the decommissioning facilities be? All of these considerations require clear definition of which items are to be included in the decommissioning cost estimate. Overall, decommissioning involves expenditure which may include

- planning, designing and building new facilities, if required, and provision for equipment
- refurbishment of existing facilities
- continuing operation and maintenance of facilities during decommissioning
- decontamination and dismantling plant items and building
- treatment and conditioning of waste
- waste storage and disposal

It may be said that without well-defined boundaries of the decommissioning activities, it is difficult to reach a credible cost estimate. However, techniques of cost estimation have developed sufficiently to give reasonable estimates. But there are areas of uncertainties and overlap between activities which may make the cost estimates unreliable.

# 8.3 Methodologies and Techniques for Decommissioning Cost Estimation

The cost estimate for the decommissioning depends on the type of facility, its physical state, radiological and other hazardous material inventory and local factors.

There may be four broad types of costs [1]
- Costs related to engineering and safety studies
- Work- or volume-related costs
- Time related costs
- Other costs

## 8.3.1 Engineering and Safety Studies

The cost for engineering studies involves a review of the state of the facility, identification of the total project, delineation of work packages, requirement for preparatory works and/or use of special decommissioning techniques. Alongside this, safety studies to meet the regulatory requirements need to be undertaken and they have significant cost elements.

## 8.3.2 Work- or Volume-related Costs

These costs are associated with the physical work of decontamination, dismantling, waste packaging, transport and eventual disposal of waste. These costs can be estimated using the unit cost methodology.

The unit costs technique is used for elementary repetitive activities such as cutting lengths of pipes, removing valves/pumps, or removing concrete of a building structure etc. The unit costs may be in terms of £/m (€/m), £/m² (€/m²), £/kg (€/kg), £ per item (€ per item) etc. To improve the accuracy of this method, a detailed list of unit costs should be set up to take account of all possible categories and sub-categories of work or work activities. A detailed list in the form of a database is necessary for this costing method.

## 8.3.3 Time-related Costs

These costs are associated with routine maintenance work, safety and security provisions, insurances, taxes, fees to local authorities and to the regulators as well as general administrative and on-site management costs.

## 8.3.4  Other Costs

The other costs, separate from volume- and time-related costs, are the costs of capital expenditure such as the purchase of specialised tools for decommissioning, setting up decontamination and dismantling workshops or test laboratories, R&D facility, radiological survey, training of personnel, contingency provision etc.

These costs are generally incurred at the time of carrying out the work. The advantage of this on-going costing is that the market conditions can be judged at the time of proposed expenditure and cost-effective solutions can be found. Inflation-related costing based on a discounting technique can be applied for funding mechanism.

The cost of a specific task may be considered to be composed of six basic elements, and these are

$$C_{task} = C_{labour} + C_{services} + C_{consumables} + C_{investment} + C_{sec.waste} + C_{contingency} \quad (8.1)$$

where $C_{labour}$ is the labour cost covering workers wages, allowances, overhead costs; $C_{services}$ is the cost for subcontracting or outsourcing the work; $C_{consumables}$ is the cost of consumable items such as protective clothing, tools, utility bills etc.; $C_{investment}$ is the cost of interest payment on borrowed capital; $C_{sec.waste}$ is the costs for managing secondary waste; and $C_{contingency}$ is the contingency cost covering all aspects.

Contingency costs are difficult to specify. However, an indication of the contingency costs can be found in the guidelines of the National Environmental Studies Project of the Atomic Industry Forum [2] and are quoted in Table 8.1.

The most common estimating techniques that are applied for various tasks and sub-tasks of a decommissioning project are

- Bottom-up Technique: This involves subdividing the whole project work into discrete and identifiable tasks and sub-tasks and then estimating costs for each of them. These elemental costs can then be added together to arrive at the total cost.
- Comparison Technique: This involves deducing the cost of a specific task or sub-task from a previous decommissioning project after allowing for the identifiable differences in the projects due to differences in radiological conditions, complexity, accessibility, local conditions and economic situation.
- Parametric Technique: This is based on the use of a model taking key driver parameters and deducing the costs from previous decommissioning experiences.
- Expert Elicitation: This is based on taking opinions from recognised experts iteratively until a consensus cost estimate is reached.

**Table 8.1** *Contingency cost as a percentage of the activity cost*

| Activity category | Contingency (%) |
|---|---|
| Engineering | 15 |
| Utility costs | 15 |
| Decontamination | 50 |
| Contaminated component removal | 25 |
| Contaminated concrete removal | 25 |
| Steam generator/pressuriser/circ. pump removal | 25 |
| Reactor removal | 75 |
| Reactor packaging | 25 |
| Reactor shipping | 25 |
| Reactor burial | 50 |
| Conventional radioactive waste packaging | 10 |
| Conventional radioactive waste shipping | 15 |
| Conventional radioactive waste burial | 25 |
| Clean component removal | 15 |
| Supplies/consumables | 25 |

# 8.4 Factors Influencing Decommissioning Cost Estimate

Decommissioning cost estimates and funding mechanisms are influenced by legal, social, economical and technical factors. This section aims not so much to give an exhaustive list of all the factors, but to give some indications of how such factors may impact on the decommissioning costs estimates.

## 8.4.1 Regulatory Factors

The decommissioning activities have to satisfy the conventional and the nuclear safety requirements defined by the national legislation. The safety requirements concern the protection of the workers, the public and the environment. Specifically, nuclear safety requirements cover the annual dose limits, free release criteria for radioactive materials, limits on discharges of gaseous and liquid effluents from plants. Changes in acceptable limits can have a significant impact on the cost estimates. So, a decrease on the free release level by a factor of 10 generates roughly the following increase in costs [3]

- nearly 5% of the costs
- nearly 17% of the waste volume
- nearly 7% of the dose uptake by workers

The decommissioning of a nuclear facility may require the availability of three types of disposal facilities

- a site for the burial of Very Low Level Waste (VLLW)

- a near-surface disposal facility for short-lived nuclides
- a geological disposal facility for long-lived waste and High Level Waste (HLW)

In many countries, the disposal routes for decommissioning wastes are not available and in many cases, even the availability of disposal facilities has not yet been planned. Therefore, arrangements have to be made on the site or in its vicinity to store the waste in a safe manner. The regulatory body can possibly require the conditioning of the waste prior to its future storage so as to reduce the risks of dissipation. Changes in the waste regulations may affect the cost of processing, containers (for example, different container qualifications) or the unit disposal cost.

The back-end solutions for the spent fuel and nuclear material can also severely disrupt the decommissioning operations and have an impact on the cost estimates. Some countries have decided to set a moratorium on the reprocessing of spent fuel and other nuclear material. Spent fuel and nuclear materials are then stored in existing or new built nuclear facilities awaiting conditioning and repackaging, if required.

Premature shutdown of a facility due to political/legal decision or for technical reasons also impacts on the decommissioning cost estimates as

- The transition from operation to decommissioning of a nuclear facility may not yet have been developed.
- Workers may not be sufficiently trained to tackle issues associated with decommissioning.
- There may be insufficient budgetary provision to launch the decommissioning project.
- Workers and public may have a negative attitude towards decommissioning.

## 8.4.2 Social Factors

Significant factors influencing the cost estimate are the local social, economic and technological bases. These conditions are important in deciding whether or not the work should be outsourced, whether to use a new or high technology and whether there should be an immediate or deferred dismantling etc.

The involvement of all stakeholders (the general public, the regulatory bodies and fund managers) is crucial. Information dissemination, stakeholders' liaison committee, public meetings etc. are organised over the duration of the decommissioning project and these costs need to be taken into account.

## 8.4.3 Economic Factors

An aggregate cost estimate is composed of labour costs, service costs, investment costs, management costs, waste costs and contingencies. Each cost component has its own inflation rate over the decommissioning period. In many countries, the rise in the annual waste cost significantly exceeds the inflation rate for labour and consumables. For example, if the cost estimates for the Biblis reactor made in 1977 are compared with those of 1991 [4], one observes that

- in 1991 the waste cost represented 18.5% of the aggregate cost instead of only 6.5% in 1977.
- in 1991 the management cost (including licensing) represented 33.5% of the aggregate cost instead of only 6.5% in 1977.
- the overall decommissioning cost increased (numerically) by a factor of 2 from 1977 to 1991.

These economic factors have to be carefully evaluated, certainly in the case of a deferred decommissioning strategy.

### 8.4.4 Technical Factors

The radioactive inventory of a facility may be considered to be the main factor influencing the decommissioning cost. Knowledge about the facility at the time of final shutdown, its operational history, any cases of incident/accident etc. are very important. A detailed analysis of the records, drawings, technical documents would give an accurate picture of the facility. Another aspect to be considered is the overall state of the infrastructure and equipment. If the infrastructure and/or equipment need to be refurbished or if their capacity needs to be extended, then that will have implications for the cost estimation.

The Waste Acceptance Criteria (WAC) and the waste management costs will influence the techniques to be used in order to optimise the decommissioning costs. As for example, the physical size of the waste acceptable for disposal may dictate the extent of the cutting operations for large components. High disposal costs may justify more decontamination efforts in order to have more free release materials.

Nuclear sites may contain several nuclear installations. Some of them can be of the same type such as the Pressurised Water Reactor (PWR) or Gas-Cooled Reactor (GCR) etc. If their decommissioning can be planned such that the same staff and same decommissioning tools and equipment can be used (with minor adjustments), then costs can be saved on staff training, safety assessment, equipment, project management, and licensing procedures [5] etc.

## 8.5  Cost Estimation Guidelines

Four principal steps are involved in the production of any detailed cost estimate, as described below.

The first step in preparing a decommissioning cost estimate is to gather all relevant information about the facility, its operational history and its present state. To perform these activities, licensing documents, safety documents, plant drawings, health physics surveys and incident/accident reports etc. will be very useful. The quality of such information should be verified by interviews with current or former employees and with visits to the facility. This first step should also cover a detailed assessment of the physical inventory of the radiological and other hazardous materials.

The second step concerns the clear identification of the boundaries for the decommissioning project such as the free release criteria, discharge authorisation for gaseous and liquid effluents, end-points of the decommissioning project, site remediation objective (green field or brown field site), availability of disposal sites for the various categories of waste.

The third step is to identify the preferred decommissioning option and to define the waste management strategy in order to optimise the project in terms of radiological and industrial safety and cost effectiveness.

The fourth and final step in the cost estimate consists of setting up the planning details in terms of work packages, tasks and sub-tasks taking into account all possible items for costing purposes. This step would also define overall project duration, the critical path and annual budgets. Methodical identification of the tasks and sub-tasks is quite a difficult undertaking at the outset of the project. A proposed cost estimation methodology produced by the EC in order to harmonise costing methods across the whole of Europe is given in [6].

## 8.6 European Cost Estimate Methodology

Various international studies on the cost estimates of the decommissioning of nuclear facilities that had been carried out recently have shown that there are significant variations in cost estimates in various countries. Although some variations in cost estimates in different countries are almost inevitable, even if the same or similar plant is considered, the variation was estimated to be excessively high. There are, of course, uncertainties in costing methods, but those could not account for the discrepancies. Studies attempting to understand the basic reasons for such differences have been somewhat thwarted by the different costing methods used in different countries with different data requirements. Problems with interpreting estimates can be encountered and invalid conclusions can be drawn when making cost comparisons if the context in which the various cost estimates were developed is not taken into account.

A study by the Nuclear Waste and Decommissioning Group of Experts of UNIPEDE has examined how different boundary conditions affect the costs of decommissioning [7]. To do that, different boundary conditions were collected from 12 countries, and costs for decommissioning a reference nuclear power were estimated adopting the same methodology. Final results of the cost estimation vary by about a factor of 6. The most important parameter is the scope of the calculation. The decommissioning timing, waste management system, administrative factors (labour rates, regulatory demands) and financial factors (discount rates) have all been recognised as important factors for the differences between decommissioning cost estimates.

The difficulties mentioned above are partly due to the lack of a standardised costing method that includes well-structured and well-defined cost items and an established estimation method. Such a structure and coherent method would be useful not only for project cost comparisons, but would also be an appropriate tool for more effective cost management.

A joint effort by three organisations (NEA/OECD, the IAEA and the EC) produced a standardised list of cost items for nuclear installation decommissioning projects [6]. The standardised list is based on the identification, definition, and verification of general and specific decommissioning tasks and related cost items. These identified and harmonised cost items for decommissioning projects have been put together in cost groups that are related to activities carried out with similar emphasis, whether or not tied to a similar time schedule. Altogether 11 cost groups have been defined in the standardised list and these are

- pre-decommissioning actions
- facility shutdown activities
- procurement of general equipment and material
- decontamination and dismantling activities
- waste processing, storage and disposal
- site security, surveillance and maintenance
- site restoration and landscaping
- project management, engineering and site support
- research and development
- nuclear material removal
- other costs

The details of these cost groups giving cost items are described below.

## 8.6.1 Pre-decommissioning Actions

This cost group includes all activities carried out in preparation of actual decommissioning

- Decommissioning planning: This involves strategic studies, conceptual planning, detailed planning, safety and environmental studies.
- Radiological surveys for planning and licensing: In some European countries a separate licence is required for decommissioning.
- Authorisation: Licence application, public consultation and

public inquiry, regulatory approval/licence approval.
- Hazardous material surveys and analyses.
- Selection of prime contractor.

## 8.6.2 Facility Shutdown Activities

This group covers all activities associated with shutdown operations of the facility and its immediate aftermath. In the UK, these activities are normally put together as the POCO. This group of activities includes
- Plant shutdown and inspection
- Removal of nuclear fuel: defuelling, transfer of fuel materials to temporary storage
- Drainage and drying of systems not in operation
- Removal of stored nuclear material
- Sampling of contaminated material for radiological characterisation following defuelling and removal of stored nuclear materials
- Removal of system fluids
- Decontamination of systems and removal of wastes
- Isolation of power equipment
- Asset recovery: resale/transfer of equipment and components to other licensed sites, if possible

## 8.6.3 Procurement of General Equipment and Material

This group covers all activities relating to purchasing of general equipment and materials at site levels. The cost items are
- General site dismantling equipment
- Equipment for personal and/or equipment decontamination
- Radiation protection and health physics equipment
- Security and maintenance equipment

## 8.6.4 Decontamination and Dismantling Activities

This group covers all activities related to actual decontamination and dismantling operations of plants, buildings, structures and components and includes
- Decontamination of areas of building and equipment to facilitate dismantling
- Fuel pool drainage and decontamination of lining
- Preparation for dormancy
- Sampling of prepared areas for radiological characterisation and site boundary reconfiguration; construction of temporary structures, enclosures to support site remedition
- Radiological characterisation for decontamination

- Preparation of temporary storage areas
- Design, procurement and testing of equipment or special tools for remote handling
- Removal of primary and auxiliary systems
- Dismantling of reactor pressure vessel and internal: this activity may be carried out immediately or delayed depending on the levels of radiation and hazards involved
- Removal of biological/thermal shield
- Removal and disposal of asbestos, if present
- Final radiological survey
- Characterisation of radioactive materials: for reuse or recycling
- Decontamination of equipment and material for reuse or recycling
- Asset recovery: sale/transfer of decontaminated materials, metal and equipment

## 8.6.5 Waste Processing, Storage and Disposal

This group comprises a large number of activities aimed at preparing systems, structures and components for dismantling and preparing wastes either for final disposal or long-term interim storage, or for release for unrestricted or restricted use or recycling. This group primarily includes

- Preparation of safety case for waste processing, storage and disposal
- Construction of supporting structures and/or services for waste processing and storage
- Waste characterisation: radioactive or non-radioactive; chemical, toxic or carcinogenic; solid, liquid or gaseous; combustible or non-combustible; special form fluids $-D_2O$, Na etc.
- Processing of waste, packaging and transport
- Waste storage (temporary or long-term) and disposal

## 8.6.6 Site Security, Surveillance and Maintenance

This involves

- Site security operation and surveillance
- Regular inspection of buildings, structures and systems in operation
- Periodic radiation and environmental survey

## 8.6.7  Site Restoration and Landscaping

This involves

- Demolition or restoration of buildings: this depends on regulatory compliance requirements and the initial end-point specification
- Land remediation
- Independent compliance verification for site use or release
- Landscaping

## 8.6.8  Project Management, Engineering and Site Support

This group covers project management, engineering services and site support and includes

- Mobilisation and preparatory work: mobilisation of personnel and construction equipment, setup/construction of temporary facilities, temporary relocation
- Project management and engineering services: deployment of project manager and supporting staff, planning and cost control, QA and QC, documentation and records control, engineering support
- Health and safety: health physics, radiation protection and monitoring, industrial safety
- Public relations
- Support services: housing, office equipment, site services, computer support
- Demobilisation

## 8.6.9  Research and Development

This group covers all costs associated with the development of decontamination and dismantling techniques.

- Research and development: decontamination techniques, specification of dismantling and cutting tools, radiation measurements
- Simulation of work model: computer simulation of work practices for the ALARP/ALARA principle, practices on innovative tools/equipment

## 8.6.10  Nuclear Material Removal

This group covers costs related to the removal of nuclear material from the site.

- Preparing temporary/interim storage facility: design, construction, maintenance and periodic inspection of

storage facility
- Transfer of fuel, nuclear material to temporary/interim storage
- Final disposal
- Dismantling/disposal of temporary/interim storage: decontamination of the facility, dismantling/disposal of the facility

### 8.6.11 Other Costs

This group covers all other costs that cannot be specifically assigned to the above groups. It may include

- Owner's costs: capital expenditure, interest on borrowed capital
- General overall costs: consulting costs, regulatory fees, inspection, certification, reviews etc.
- Taxes
- Insurances
- Contingency
- Asset recovery: resale/transfer of material, equipment, site etc.

## 8.7 Discounting Technique

A decommissioning project may take, depending on the nature and complexity of the work, a considerable period of time to complete. Consequently the total fund would not be required at the start of the project; there would be a time distribution of the fund requirement. This aspect of fund requirement leads to a methodology known as the 'discounting technique' and interest rate estimates. These techniques are used in many financial calculations, ranging from a simple task of calculating repayments on mortgages to investment appraisal, project cost estimates etc. In particular, the technique is very useful when dealing with a project which may have a number of options spread over a period of time. The principle is to transpose future cost estimates to the present value on the basis of certain assumptions and then compare the costs of various options on a like-for-like basis.

### 8.7.1 Net Present Value of Money

The value of money is not stagnant. In fact, money depreciates at the rate of inflation and that is why it is important that money is invested in such a way that gives a rate of return higher than that of the inflation. However, the inflation rate is not static and cannot be predicted for future times with accuracy. The future growth of money can be estimated by assuming a fixed interest rate.

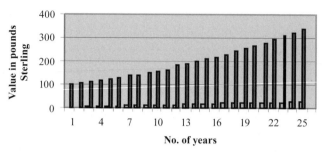

**Fig. 8.1** *Future value of £100*

Let us assume that £100.00 is invested in a bank with an interest rate of 5% (£5 interest per year per £100). At the end of the first year, the amount will become £100.00 × 1.05 = £105.00. If the amount is left in the bank at the same interest rate, at the end of the second year, the total amount will become £105.00 × 1.05 = £100.00 × 1.05 × 1.05 = £100.00 × $(1.05)^2$. This is known as the growth at a compound interest of 5%. At the end of the $n$th year, the total amount will be £100.00 × $(1.05)^n$.

We can now generalise this compound interest rate concept. If $V_0$ is the initial amount at $t = 0$ year, $x$ is the present per cent interest rate and $V_n$ is the amount at the end of the $n$th year; then

$$V_n = V_0\left(1+x\right)^n \qquad (8.2)$$

This equation can be used to estimate the amount that would be available after a certain number of years when the initial amount and interest rate are known. For example, the future value of £100 growing at a nominal rate of 5% is shown in Figure 8.1 for up to 25 years.

Now we can do the inverse calculation. If we know that an amount, $V_n$ will be required after $n$ years, we can then estimate the amount, $V_0$ that would be needed to be invested now. From equation (8.2)

$$V_0 = \frac{V_n}{\left(1+x\right)^n} \qquad (8.3)$$

$V_0$ is known as the Net Present Value (NPV) of the amount $V_n$ with a discount rate of 5%. The factor $1/(1 + x)$ is called the discount factor. Figure 8.2 shows how the present value of money ($V_0$ = £1) is eroded over time in terms of real value, when an interest rate of 5% is assumed.

To all intents and purposes, in decommissioning projects the funds are not required at the end of $n$ years but continuously, year after year, until the project ends. Let $C_0$ be the cash requirement at the beginning of the project, $C_1$ is the cash requirement at the end of the first year, $C_2$ is the cash requirement at the end of the second year and so on until $C_n$ at the end of $n$ years. On the basis of equation (8.3), we can translate these amounts to the net present value, $V_0$

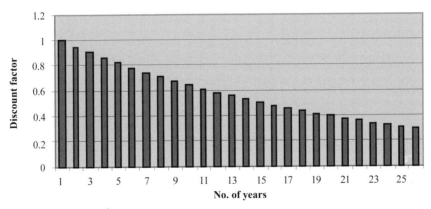

**Fig. 8.2** *Discount factor*

$$V_0 = C_0 + \frac{C_1}{(1+x)} + \frac{C_2}{(1+x)^2} + ... + \frac{C_n}{(1+x)^n}$$

$$= \sum_0^n \frac{C_n}{(1+x)^n} \tag{8.4}$$

This shows that we can estimate the net present value of the fund if a series of cash requirements is envisaged over a period of time. The advantage of the present value technique is that it makes comparison of costs of various options easier.

There are other parameters which can be utilised in fund management, some of them may be applicable to nuclear decommissioning projects while others are not. However, a list of such parameters and description is given in [8].

---

### Example 8.1

While planning to decontaminate and dismantle a nuclear facility, two options have been identified and these are: Option A which requires an expenditure of £4 M in the first year, followed by £2 M in both second and third years and then £1 M both fourth and fifth years; Option B which requires £1 M each year for 12 years. If the discount rate is assumed to be 5%, which option is financially attractive?

### Solution

Let us convert all of these expenditures, in both option A and option B, to NPV by using the discounting technique.

### Option A

| Year | Cost (£ M) | Discount factor $1/(1+x)^n$ | NPV (£ M) |
|------|------------|------------------------------|-----------|
| 1 | 4 | 0.9524 | 3.8095 |
| 2 | 2 | 0.9070 | 1.8141 |
| 3 | 2 | 0.8638 | 1.7277 |
| 4 | 1 | 0.8227 | 0.8227 |
| 5 | 1 | 0.7835 | 0.7835 |
| Total | 10 | | 8.9575 |

### Option B

| Year | Cost (£ M) | Discount factor $1/(1+i)^n$ | NPV (£ M) |
|------|------------|------------------------------|-----------|
| 1 | 1 | 0.9524 | 0.9524 |
| 2 | 1 | 0.9070 | 0.9070 |
| 3 | 1 | 0.8638 | 0.8638 |
| 4 | 1 | 0.8227 | 0.8227 |
| 5 | 1 | 0.7835 | 0.7835 |
| 6 | 1 | 0.7462 | 0.7462 |
| 7 | 1 | 0.7107 | 0.7107 |
| 8 | 1 | 0.6768 | 0.6768 |
| 9 | 1 | 0.6446 | 0.6446 |
| 10 | 1 | 0.6139 | 0.6139 |
| 11 | 1 | 0.5847 | 0.5847 |
| 12 | 1 | 0.5568 | 0.5568 |
| Total | 12 | | 8.8631 |

This shows that option B having the NPV of £8.8631M is somewhat less expensive than option A whose NPV is £8.9575M. However, other factors like continued employment for larger number of workers in option A (costing £4M in the first year), early release of the site from nuclear activity etc, may be taken into consideration when cost difference is marginal.

## 8.8 Funding Mechanisms

Under the provision of a nuclear site licence in the UK, the liability for cleaning up the site remains with the site licensee until the regulatory body declares that there ceases to be any danger from ionising radiation from anything on the site. In other words, the decommissioning liability lies solely with the licensee and consequently provisions for decommissioning need to be made by the licensee. This is the general principle not only in the UK but throughout the whole world. As mentioned previously, decommissioning not only involves decontamination and dismantling of the facility buildings and structures, but it also involves the management of waste (primary and secondary) as well as site remediation and restoration. All these activities require funds and provision needs to be made when the facility is generating income. As a general rule, the decommissioning cost is nearly the same as the original construction cost in real terms.

There are number of ways in which a decommissioning fund may be accumulated. Basically, it aims to claw back a certain amount of money, year after year, during the operational period of the facility so that sufficient funds can be generated when, at the end of that period, decommissioning is required. This principle may, however, be applicable only for commercial nuclear activities, not for military nuclear facilities which are funded exclusively by the government. Even in commercial nuclear activities public funds may be required, as adequate provisions for decommissioning were not made by the nuclear companies in the early days of nuclear activities. Provisions for such public funding may vary from country to country.

In the case of power plants, some operators may impose a decommissioning levy, which could be a small percentage of the unit generation cost of electricity (kWh) by nuclear means. Others who have nuclear and non-nuclear generating capacity may impose a smaller percentage of the cost right across the board. The imposition of a levy to defray the decommissioning costs in nuclear power plants, or the additional costs to control carbon emissions in conventional fossil fuel power plants, is becoming a practical means of raising capital. Other owner/operators may make payments annually to the fund. Whatever the modality of contribution to the fund, the level of that contribution is revised periodically to take account of

- the revised decommissioning costs estimate (usually revised every 3–5 years)
- the variability of the electricity unit price and the generating capacity of the facility (normally revised each year)

The following equation can be used to calculate the annual fund requirement:

$$FY_n = \frac{(TDF - FY_{n-1})}{Y_s - Y_n} \tag{8.5}$$

where $FY_n$ is the fund to be secured at the end of year $n$; $TDF$ is the total decommissioning fund to be made available for decommissioning (this is the estimated amount). $Y_{n-1}$ is the year when funding has already been collected; and $Y_s$ is the lifespan of the plant.

The amounts in the fund, i.e. $FY_n$ and $TDF$ mentioned in equation (8.5) are values calculated on the basis of the discounting technique. This is because the fund which is to be collected at the end of the first year of operation of the facility will have the remaining 30 or so years of operation in which to grow. So the amount at the end of the first year $(FY_1)$ and that of the penultimate year $(FY_{n-1})$ cannot be same numerically.

## 8.9 Fund Management

As funding is being raised during the operating phase, provision for managing the decommissioning fund needs to be in place. This fund management must be separate from and independent of the company management. This institutionalised segregation of the decommissioning fund, more like a pension fund, is essential to ensure that the fund will not be sucked into the company finances. In a non-segregated fund situation, the fund may very easily disappear. For example, in the UK the erstwhile CEGB (a public body) had accumulated a certain amount of nuclear decommissioning fund within the company scheme, but when the company was broken-up and privatised by the government, the whole fund disappeared, although the decommissioning liability remained with the nuclear segment of the company. Another example of misappropriation of non-segregated fund is the famous Maxwell Communication Corporation pension fund. In an attempt to shore up the sagging finances of the company, Robert Maxwell raided the pension fund and when the company collapsed, the pension fund disappeared with it. A segregated fund under a properly constituted trustee ownership could not have faced such a fate.

The whole objective of the fund manager is to achieve growth of the fund in line with, or in excess of, the nominal discount rate which has been built into the fund estimation. This nominal discount rate is normally taken to be equal to or slightly higher than the inflation rate. A fund manager should undertake a proper evaluation of risk and reward as part of the investment strategy. A cautious but steady rate of return is considered to be a prudent approach. The performance of the fund can be benchmarked against the inflation rate. At the moment, a 5% discount rate is considered an acceptable figure.

The investment strategy for the fund may involve investing in

- treasury bonds (gilts)
- international currency bonds
- national and international equities
- high yielding bank deposits
- investment in real estate

There are other investment opportunities which the fund manager may consider. But the ultimate aim is to achieve a minimum rate of return equal to the rate of inflation without taking undue risks.

## Revision Questions

1. The financial provision for decommissioning depends on two aspects: name them and briefly explain them. Briefly describe the various items used in cost estimation.

2. What factors influence decommissioning cost estimates? Briefly describe each of them.

3. What is the 'standardised list of cost items' and who produced this document? Give a brief summary of the cost items it covers.

4. What is the discounting technique? Explain its usefulness in cost estimation. Also explain the significance of the Net Present Value (NPV) of money and how it is calculated from the future value of money?

## REFERENCES

[1] EUNDETRAF II Training Course, Financial Aspects of a Decommissioning Project, EC Funded Project, 2004.
[2] Atomic Industries Forum, National Environmental Studies Project, Guidelines for Producing Commercial Nuclear Power Plant Decommissioning Cost Estimate, AIF/NESP-036, May 1986.
[3] Havard P., Reduction of radioactive waste production: where is the optimum? Proceedings of 8th International Conference on Environmental Management (ICEM'01) 30 September–4 October 2001, Bruges, Belgium.
[4] Adler J. and Petrasch P., Decommissioning Costs of Light Water Nuclear Power Plants in Germany from 1977 to date, NIS, Contract FI2D-0051, Final report, EUR-14798-EN.
[5] International Atomic Energy Authority, Planning, Managing and Organizing the Decommissioning of Nuclear Facilities: Lessons Learned, TECDOC 1394, IAEA, Vienna, Austria, May 2004.
[6] NEA/OECD, EU, IAEA, A Proposed Standardised List of Items for Costing Purposes in the Decommissioning of Nuclear Installations, Interim Technical Document, OECD, Paris, France, 1999.
[7] UNIPEDE, Cost Estimates for Decommissioning Nuclear Reactors. Why do They Differ so Much? 1998-211-0002, Brussels, Belgium, April 1998.
[8] Global Financial Management, September 1999, http://www.exinfm.com.

# 9

# PROJECT MANAGEMENT

## 9.1 Introduction

The management of a project involves planning, organising, controlling and directing resources (money, materials and people) to accomplish a clearly defined objective or objectives. The larger the project, the more involved the project management is. In the management of a nuclear decommissioning project, there are even more demands as it requires additional safety and security considerations under national, multi-national and international regulatory regimes. Implicit within the safety and security considerations is the protection of workers, public and the environment from both radiological and non-radiological hazards resulting from nuclear activities. All of these requirements impose a considerable burden on the management of a nuclear decommissioning project that extends throughout the whole of the project. Key issues specific to the decommissioning of nuclear facilities have been described in Chapter 2 of [1].

Nuclear decommissioning project management requires an extensive array of knowledge and expertise, from project planning to managerial skills, financial management, familiarity with and understanding of nuclear engineering, safety principles and legislative requirements. Obviously, such a vast array of expertise and knowledge is unlikely to be available in an individual or even in a handful of individuals. Consequently nuclear project management is very much a team effort, although someone with suitable qualifications and experience assumes the authority and responsibility of the project manager.

It should be noted that project management refers to the management of a specific project such as the decommissioning of a nuclear power plant or decommissioning of a fuel storage facility or dismantling a decontaminated building or the remediation of a contaminated site and so forth. The project is generally well defined, the scope of work is well specified in terms of timescale, staffing requirements, cost, deliverables etc. The programme management, on the other hand, is an overarching activity covering a number of projects. These projects may have some association with each other, or they may be completely independent. The NDA (see Section 5.7 and Annex 4) is the organisation set up by the government of the UK to manage the programme of decommissioning and cleaning up of civil nuclear sites in the UK. Under this programme, there are 20 major projects, each of which may be divided into a number of smaller, more congruent projects. The management of such projects is considered in this chapter.

## 9.2 General Project Management

In general, the project management is accomplished through the use of the following processes [2]:

- Initiation
- Planning
- Control and execution
- Closure

It must be appreciated that these processes are not isolated; they are iterative in nature. Planning provides the basis for the execution of a plan, and execution may lead to updates to the original plan as the project progresses. During the life-cycle of the project, changes may be necessary due to changed boundary conditions. However, retrospective changes and updates to the original plan should be kept to a minimum, otherwise there may be an endless regressive iteration of the project plan, which may hinder progress.

### 9.2.1 Initiation of a Project

A project is initiated when the authorisation for the project is given by the management of the company. However, even before this stage, a signifi-cant amount of work needs to be carried out in preparing the broad outline of a project, preparing a business case for management consideration and approval, and the allocation of resources to the project. However, in a nuclear decommissioning project, this initiation process, as understood in the usual sense of a general project, is somewhat different. Nuclear decommissioning is very much a regulatory-driven process (subject to the availability of resources by the decommissioning company) and consequently right from the start compliance with the regulatory requirements is essential. In the UK, decommissioning must be conducted as soon as possible after the cessation of operation and hence other issues such as the business case preparation etc. must precede this stage. In many European countries, decommissioning requires a separate licence and preparation for such a licence is part of this initiation process.

### 9.2.2 Project Planning

Project planning is crucial to the management of a project. The extent and complexity obviously depends on the size and scale of the proposed project. It involves project plan details in terms of time and money, selection of manage-ment and technical personnel, contract specification, contract administration, the QA programme, communication plan etc. All of these aspects are put together as a project plan. The adage for a project plan is

**Failing to Plan = Planning to Fail**

## 9.2.3 Project Control and Execution

Control and execution of a project is the central part of the management work. The main areas of project control are

- Overall control
- Schedule control
- Cost control
- Quality Control (QC)
- Risk monitoring and control

In today's management practice, project control is understood to cover more than the QC process. The project control must ensure that times, cost and resources are utilised as efficiently and closely as possible to the plan. For a control system to be meaningful a project communication system should be in place.

The coordination of the work is carried out at the Work Order (WO) implementation level. The engineering preparation of these detailed WOs may present a huge proportion of the project management workload depending of the project stage.

## 9.2.4 Completion of the Project

On completion of the project, a review of the performance of the project is carried out and a report is produced for the top management to study. This report identifies any shortcomings in the project plan and execution, lessons to be learnt and improvements to be made in future projects. At the end of this exercise, the project team may be disbanded.

# 9.3 Nuclear Decommissioning Project Management

The management practices mentioned above serve as a common basis for the management of projects and are generally applicable to nuclear decommissioning projects. However, there are significant variations in nuclear decommissioning project management as nuclear activities are very much driven by regulatory requirements and safety considerations. The very word 'decommissioning' refers to the administrative and technical actions taken to remove some or all of the regulatory controls of a nuclear facility [1] and hence the management of such a project must reflect this emphasis.

The management of a nuclear decommissioning project may be delineated into following areas:

(1) Definition of a decommissioning strategy
(2) Specification of Project Management Plan (PMP)
(3) A well-developed schedule management

(4)   Cost management
(5)   Fully developed Project Quality Management (PQM)
(6)   Specification of management structure and delegated responsibility
(7)   Information management system
(8)   Risk management
(9)   Contract and procurement management

All of these management issues are now discussed in sequence.

## 9.4 Decommissioning Strategy

When a nuclear facility comes closer to the end of its operational life, the licensee/owner/operator must put forward a decommissioning strategy. In fact, in the UK a broad outline of the decommissioning strategy needs to be incorporated in the original licence application to construct the facility. Nearer the end of its operational life, the strategy needs to be further expanded with specific details, project specification, financial arrangements, boundary conditions, objectives and end-points. Similar outline details need to be incorporated in the application for a decommissioning licence in some European countries.

## 9.5 Project Management Plan

The Project Management Plan (PMP) of a decommissioning project sets out the project description, objectives, methodology, organisation, timescale and budgetary provisions for the whole of the project based on the previously specified outline strategy. In the UK under the management of the NDA, this PMP is separated into Life-cycle Base Line (LCBL) where the overall plan over the life-cycle of the project is specified and the Near Term Work Plan (NTWP) where a three-year rolling plan is specified.

The PMP is the most important document in the overall planning, monitoring and implementation of a project. It is an approved guide to both project implementation and control. The primary aim of the document is to describe the planning assumptions and decisions, facilitate communication among the parties involved, and define the scope, cost and schedule baselines. Other items such as technical, commercial, organisational, personnel and control issues may well be included here. The essentials of a PMP are as follows:

- A summary of the project giving the essential information about the project. It should briefly state what has to be done and mention the methods and techniques to be used.
- Specification of milestones defining identifiable segments of the project with specific budgets in order to provide adequate monitoring.

- A Work Breakdown Structure (WBS) that is detailed enough to provide meaningful identification of tasks, plus all the higher-level work groupings.
- From the milestone list and the WBS, an activity network that shows the sequence of the segments of the project and their inter-relationship – which activities can be done concurrently, which ones are to be sequential etc. This is clearly more useful than just marking end-points on bar charts.
- Separation of budgets and schedules for all the segments of the project and the identification of the responsible individuals.
- A communication plan that shows how the project manager communicates with stakeholders, the contractors, the staff and other organisations that are involved.
- An indication of document requirements and review processes – which documents are needed; who reviews them; when, in what time frame and for what purpose.
- A list of key project personnel and their assignments in relation to the WBS. Key personnel are those responsible for the various phases of the project.

## 9.5.1 Work Breakdown Structure

A fundamental aspect of effective project planning is the process of defining the scope of the project and breaking it down into manageable pieces of work.

The Work Breakdown Structure (WBS) is a task-oriented detailed breakdown which defines the work or tasks to be performed. It initiates the development of the Organisational Breakdown Structure (OBS), and the Cost Breakdown Structure (CBS). Thus, the WBS is the primary planning and analysis tool used in almost all projects because it addresses two issues

- What is to be accomplished?
- What is the necessary hierarchical relationship of the work effort?

The WBS also aids the project management process by

- Providing a survey of the whole extent of work that must be performed.
- Defining responsibilities, specifying personnel, cost, duration and risk.
- Providing an easy-to-follow numbering system to allow hierarchical tracking of the progress of the project.

Thus, the WBS divides the project into manageable blocks of work for which costs, budgets, and schedules can more readily be established.

The formation of the WBS into a family tree begins with the subdivision of the project work load into smaller work blocks until the lowest level to be supported and controlled is reached. This tree-like structure breaks the project work scope down into manageable and independent units that are assigned to the various experts responsible for their completion. The WBS links the company resources with the work to be performed.

The detailed breakdown structure is project specific, even a simple WBS for a project cannot normally be applied to another similar project. There are several ways of presenting the organisation of the tasks and processes necessary to complete the project. Whether the WBS organises the project in phases or functional deliverables, the primary aim of the process is that the management team should have a clear understanding of the project scope and deliverables.

The following rules in management practice are recognised to be important:

- Always prepare a WBS.
- A team consisting of project specialists, technical experts etc. with relevant expertise and experience should be brought together for the project.
- Any element of the project against which funds are expected must be included in the WBS. It should be noted that the breakdown is a partition into functional blocks or project activities. It is not a breakdown by organisation or discipline.
- The WBS is tree-like. Therefore, an element at, say, level 3, must break down into at least two elements at level 4. All the work specified at level 3 is performed at level 4.
- Because the structural breakdown is hierarchical, no two elements in levels 2, 3 or 4 can be connected to an element at a lower level.

After the WBS has been built up, the process would be finalised by establishing the work package (WP). The WP is a specific description of work to be performed by an assigned person or persons within a specified timeframe. The WP is always prepared for each bottom-level element of the WBS. Thus a level 4 element, if not partitioned, would constitute a WP.

After establishing the WBS, the next step in executing the project plan is to develop a schedule.

## 9.6 Schedule Management

The three elements – planning a project, developing a budget and scheduling the tasks are intricately linked. Planning, budgeting and scheduling are parts of the same basic management process. For example, a budget cannot be prepared without knowing the activities to be performed and the time span within which they are to be performed.

Having defined the activities to be performed at the WP levels, the timescale for the WPs can be assigned and the budget can be estimated. Based on this information and taking into account the available resources, a project schedule can be drawn up. The outcome of this first iteration may have to be adjusted in view of changed boundary conditions, such as stakeholders' needs, regulatory demands, budgetary constraints etc. An acceptable project schedule would emerge from considering these aspects. The time management elements would be completed by controlling the schedule against the performance of the project.

Complex decommissioning projects involve a series of activities, some of which are sequential in nature while others can be performed in parallel. This collection of series and parallel activities can be modelled as a network. An activity is a task that is required to be performed and an event is a milestone marking the completion of one or more activities. The basic aim of the project schedule is to put these activities together in a way that is easy to understand and is effective for monitoring and controlling. Basically there are three types of project schedule presentation

- arrows and precedence networks
- Programme Evaluation and Review Technique (PERT) and Critical Path Method (CPM)
- simple milestones, bar or Gantt charts

The PERT is a network model where activities are represented by the lines and milestones by the nodes. An arrow on the line indicates the progression of the activity from the initial milestone to the final one. An activity is specified by a letter and the expected time duration is shown alongside it. The milestones are generally numbered, usually the ending node has a number incremented by ten to that of the beginning node. Incrementing numbers by ten allows any additional nodes to be inserted without changing the previous nodes. A complex project may have a large number of activities and milestones and consequently the PERT chart would also be quite elaborate and may require several pages. A simple PERT chart is shown in Figure 9.1.

The CPM is also a network model, but it is based on a deterministic method such that it utilises a fixed time estimate for each activity. While it is easy to understand and use, it does not take into consideration time variations that may take place in complex projects. The critical path is determined by adding the times for the activities in each sequence and determining the longest path in the project. For example, in Figure 9.1, the sequence time from node 10 to node 50 via node 20 is 6 weeks; whereas another sequence from nodes 10 → 30 → 40 → 50 is 7 weeks. So the critical path is 7 weeks. If activities outside the critical path speed up or slow down, the project time does not change. The

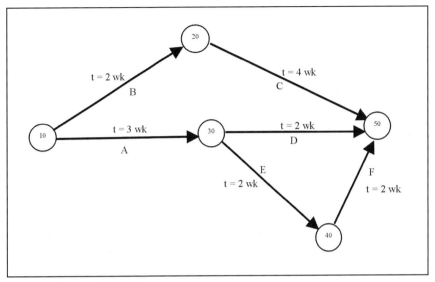

**Fig. 9.1** *A simple PERT chart*

time difference between the critical and another sequence is called the slack time for that sequence. The slack time for sequence $10 \rightarrow 20 \rightarrow 50$ is 1 week, whereas for sequence $10 \rightarrow 30 \rightarrow 50$, it is 2 weeks.

The Gantt chart, named after Henry L. Gantt (1861–1919), is a graphical representation of project activities in the form of bars placed horizontally showing the duration of the activities (tasks) and schedule dependencies against the progression in time. The Gantt chart has become a common tool for representing a project with WBS and associated dependencies. It is very useful for small projects where the project schedule can be clearly shown. But for large projects the Gantt chart may become too unwieldy and computer presentation may become too difficult.

### 9.6.1  Schedule Development

Based on the dependencies between the activities and the duration estimates, a project schedule can now be established. The project schedule is developed by determining the start and finish dates for each of the activities and events.

Determining the start and finish dates should also take into account the following factors:

- Risk management plan
- Resource availability in terms of time and conditions (working days, holidays, working time directives limiting the number of hours a worker can work etc.)
- Other external influences such as licence provisions, stakeholders' interests, social and budgetary aspects

Such external constraints might limit the management team's options. Time constraints such as 'finish no later than' or 'start no earlier than' are the most common ones.

### 9.6.2  Schedule Control

It is fair to say that projects rarely run exactly to schedule. The schedule is not, however, a monolithic document. Changes occur during the lifetime of the project. The schedule control involves the recognition that deviations from the original planning dates, programmes etc. will take place and therefore steps to rectify changes in order to return to the original programme are required. It is also necessary to take into account the extent to which schedule changes will affect the scope of the work, cost, risk, quality, and staffing requirement. Responses will have to be made if milestones will be affected. Such corrective actions need to be taken to bring the performance of the activities in line with the project plan or to ensure that the least possible delay is encountered. In complex projects, management software able to support schedule control should be used.

## 9.7  Cost Management

Alongside schedule development and control, the cost management of a decommissioning project must also be considered. This management topic is likely to address issues such as how much is it going to cost, how can the project be completed in the most cost-effective way etc. In short, cost management deals with those financial aspects which have implications for resource planning, budgeting etc.

The decommissioning cost usually includes all the costs from the point of cessation of operation of the facility, after the POCO, right up to the termination of the regulatory controls on the site. There are obviously diverse approaches to cost management with different data requirements. In order to compare the costs of decommissioning projects, it is necessary to take into account the extent to which various data are available and applicable. Chapter 8 gives the details of the standardised cost items for decommissioning that have been produced by NEA/OECD [3].

### 9.7.1  Resource Planning

Resource planning involves determining what resources (human, equipment, tools and materials) and what quantities of each should be used and when they will be needed. Such planning is primarily based on the WBS and the activity duration estimates.

## 9.7.2  Cost Estimating

Cost estimating is the process of projecting the financial requirements to achieve the objectives specified in the decommissioning plan. Although the primary activities or tasks involved can be better described and understood if they are discussed separately and sequentially, in practice they are closely related and are often carried out concurrently. The primary cost estimating tasks are based on the WBS and consist of

- selecting the WBS for preparing cost data
- collecting, evaluating, and applying the necessary cost and cost related data
- applying the proper estimating methods
- documenting the estimate in enough detail, so that it can be reviewed, evaluated, and used in the decision-making process

When the primary estimate has been completed, uncertainties, limiting assumptions, and constraints should be identified. Changes in the basic rules, schedules, quantities, system upgrade, and concepts can significantly affect the cost data.

## 9.7.3  Budgeting

A budget is simply a plan in terms of costs for allocating resources to the project activities. The project planning process has been described above, as a set of steps that began with the overall project plan and then dividing and subdividing the plan's elements into smaller and smaller pieces that could finally be sequenced, assigned, scheduled and cost estimated. Hence, the project budget is nothing other than the project plan, based on the activities or WBS, expressed in monetary terms.

Budgeting a project should involve allowing for some contingencies in order to be able to manage unexpected changes during the project. The risk of variations in project activities is inherent in project management and so budgeting should consider this aspect. Risk management is discussed in some detail in Section 9.11.

Once the budget has been established, it acts as a tool for the higher management to monitor and steer the project to time and cost. Appropriate data must be collected and reported in a timely manner. This collection and reporting system must be carefully designed in the initial project plan in order to avoid late and inaccurate reporting.

## 9.7.4  Cost Control

Cost control is achieved through monitoring, analysing, reporting and exercising controls over the commitments and expenditures with due regard to the schedule. A key element is the transparency in the method of forecasting the final cost of the project and that may lead to corrective actions ahead of time to control costs and commitments.

One of the important elements of the cost control is the implementation of an adequate accounting system. Project accounting deals with the control and historic recording of actual cash payments within the project organisation itself as well as to outside organisations. A number of contractors or sub-contractors may be involved, a number of suppliers may deliver goods or items. Proper and timely utilisation of such products is fundamental to cost control. Definitive and detailed procedures are essential for this controlling function to ensure the financial integrity and transparency of the project.

## 9.8  Quality Management

The QA begins at the project conception stage and runs through all the stages of the project. It affects cost, availability, effectiveness, safety and the environment. Therefore, the QA aspects should be given a high priority, from preparation, through implementation to completion of the project.

### 9.8.1  Elements of Quality Assured Management

The quality requirements and the QA activities considered necessary to accomplish the project objectives must be laid down in the QA handbook and QA procedures for the project. Consideration should be given to the following elements for their appropriate inclusion in the QA program:

- QA organisation
- QA plan
- Procurement control
- Document control
- Control of purchased material, equipment and services
- Identification, control, and traceability of materials, parts and components
- Control of special processes
- Inspection
- Handling and storage
- Inspection, test and operating status
- Corrective action
- QA records
- Audits

### 9.8.2  Quality Assurance and Control

QC is the process used by the project team to meet the standards required by the organisation's quality policy. The process consists of observing the performance, comparing it to the standards specified and taking the necessary actions to correct any deviation observed. It is widely accepted that keeping mistakes out of the system is far less expensive than the corrective actions which may cover expensive process interruptions, lost production or even human injury.

The basic tools used to control quality include inspections, control charts, flowchart and trend analysis.

Whereas QC is concerned with monitoring specific results and eliminating causes of unsatisfactory performance, QA is concerned with evaluation of the overall project performance to provide confidence that the project will satisfy the relevant quality standards. Therefore, to guarantee performance, the QA process must address all the interfaces in the upstream operations or processes that are both internal and external to the organisation. This includes managing internal forces and external suppliers. The project team should document in the QA manual the requirements against which each supplier will be evaluated. The team should also inform the suppliers how and when the supplied products will be utilised and monitor their functional capability and performance.

Figure 9.2 illustrates the inter-relationship of the inputs and outputs associated with QA and QC.

## 9.9 Human Resource Management

Human resource management focuses on all the processes necessary to make the most effective use of the people involved in the project. In a Nuclear Power Plant (NPP) decommissioning project, it is considered best to use the operational personnel of the NPP to the maximum extent. The organisational planning, the use of expertise and the development of skills to improve the project performance are given below.

### 9.9.1 Project Roles and Responsibilities

Project roles, responsibilities and reporting relationships should be developed at the very early stages of the decommissioning project. This organisational planning is an important part of the regulatory approval or licensing application scope. The licensing authority carefully scrutinises, and wants to be reassured that, the proposed project structure covers all the necessary skills required to carry out the decommissioning in a safe and secure way. In the UK, Licence Condition 36 demands that the licensee submits an organisational structure for regulatory approval.

The organisational planning must be documented in the Organisational Manual (OM). Throughout the project performance, the organisational structure needs to be regularly reviewed in order to ensure adequate applicability to the actual project phase.

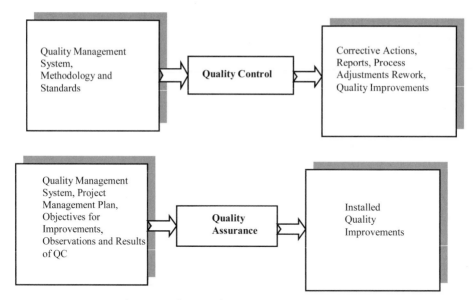

**Fig. 9.2** *Inputs and outputs of QC and QA systems*

## 9.9.2 Staffing the Project Team

Having worked out the organisational structure and the staffing requirements, the management team is responsible for identifying and deploying suitable human resources. It is advantageous to deploy the existing operational staff of the plant as much as possible. Part of the operational staff can be directly deployed to the decommissioning project with, of course, some training and skills development in areas such as radiological protection, radiation measurement and monitoring techniques, post-operation, waste management, decontamination techniques etc. Project procurement management is discussed in Section 9.12.

## 9.9.3 Training and Development of Skills

The project team will identify the skills required for the project. The initial approach should involve a review of available staff within the company who are capable of and experienced in the multi-disciplinary activities required for decommissioning work. While the number of such candidates may be limited, there are advantages associated with the utilisation of a company's own resources. Such staff members are already familiar with the company structure, processes and working practices. They should be able to form a management team quickly because they are familiar with each other and have an understanding of the various departments and their roles.

## 9.10  Information Management System

This deals with the significance of software routines to facilitate the decommissioning project management processes. Although the project management is fundamentally driven by human beings, there are opportunities when system automation may expedite the process and consequently lead to a cost-effective project operation.

### 9.10.1  Basic Considerations

Before considering the implication of project management process by automation, the objectives of the process must be defined. The key word here is 'automation', which means that the process can function without human intervention. The question then immediately arises: is it possible to have project management without human intervention and if it is, to what extent is it useful? The automation here must not be misunderstood as the complete replacement of project management by the use of software. It is totally unrealistic to expect that an automated system will function properly in a project management role without any human interface as any project management requires the ability to think critically, to respond to changed circumstances, to negotiate with individuals or solve unforeseen problems.

Nonetheless, it is useful to have the repetitive activities automated as long as they run according to certain set rules. By doing this, human error probabilities and substantial labour costs may be reduced and performance improved.

A large variety of software tools is available in the market which may cover one or more of the other project management processes. These software tools can be divided into the following basic classes:

- word processing software
- spreadsheets
- accounting software
- scheduling and tracking software
- charting software
- software development tools
- Computer-Aided Design (CAD) software
- multi-media software
- communication software
- specific waste/material tracking software

The operating systems, netware and database software, which form the basis for the classes specified above should also be mentioned. Numerous software packages on the market combine these classes, but the real challenge is the control of the overall decommissioning process. The larger and more complex a project is, the higher are the requirements on project management and the more attention has to be paid to the question of software use.

# 9.11 Risk management

## 9.11.1 Risk Management Process

It is generally accepted that risk management is an essential and integral part of project management. Traditional project management may involve specifying the project scope and carrying it out according to the project plan to time and cost. However, actual projects, particularly those associated with the decommissioning of nuclear facilities, are rarely as simple as that. One has to deal with uncertainties embedded in the project plan (so-called variability or category 1 risk) and the uncertainties that may require modification of the project plan (impact risk or category 2 risk). Thus it is imperative to recognise that a project plan must incorporate provisions to cope with identified and somewhat unidentified eventualities that may occur during the lifetime of the project. These eventualities constitute project risks. It should, however, be noted that project risk management (PRM) is not about avoiding risk, but rather recognising its existence and managing it accordingly.

PRM involves deliberately taking every risk that may arise and then proactively managing that risk to maximise the advantage for the stakeholders.

PRM procedures have been published in various texts. It is relatively straightforward and involves a series of steps normally described as follows [4]:

- Risk identification
- Risk assessment and risk quantification
- Risk response planning (mitigation and contingency planning)
- Risk analyses
- Risk monitoring and control

Risk management is a process that commences with the identification of risks and links this through to the resolution of the individual risks. This process is shown in outline in the generic level drawing of the risk management process (see Figure 9.3) [4].

The PRM process operates within the project management process throughout the life-cycle of the project. This process is complicated only by the interaction of the processes and not by the complications of the individual functions. Training of personnel in the risk management process is an important element of the risk management, as well as the design and utilisation of the databases.

# 9.12 Contract and Procurement Management

In planning for a complex decommissioning project, it needs to be recognised that all the required services, products, transport operations, disposal routes etc may not be available within the parent organisation, no matter how large that organisation may be. This requirement for product and services leads to

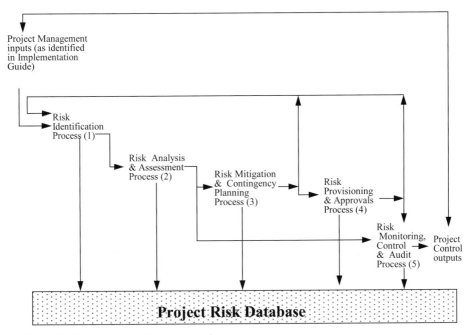

**Fig. 9.3** *Generic process for project risk management*

procurement and contract activities. Depending on the value, one is required to follow one of the EU procurement processes. Within the EU, an organisation is required to follow a European tendering procedure if the contract value is above a certain amount (~€200,000). This requirement imposes a burden on the project management to be familiar with the procurement and tendering processes within the EU. Unfamiliarity may cause project delays, legal problems and increased costs.

During the management of a decommissioning project, equipment and/ or services would be purchased and procured through competitive tendering. To ensure that the best product is bought at the best possible price, it is essential that the project team follows the standardised procedures that have been established by their companies to select and manage external suppliers. Usually, there is a procurement department within the company that leads the procurement process. Although project teams may become deeply involved in issuing and formatting the technical and preliminary contract documents, it is imperative to involve the procurement department and the legal division to negotiate and draw up the final agreement.

However, it should be noted that the quality of the services and products to be purchased should always be the top priority. Cheaper material may save some money at the outset but it may lead to an inferior finished product, delay in implementing the task due to higher failure rates, frequent refitting works etc. which would push the project cost inexorably upwards.

## 9.12.1 Procurement Management Process

Procurement is the process of acquiring services and/or products required for the project from external sources. There are several activities associated with the project procurement management and these are

- Procurement planning: determining what to procure and when
- Tender planning: procurement documents and proposal's evaluation criteria
- Tendering: obtaining the seller's response (bid, proposal)
- Source selection: pre-qualification procedures, invitation to tender and final selection
- Contract administration: managing the relationship with the contractor
- Contract close out: completion and settlement of the contract

The procurement process could be formally divided into three phases

- planning phase (pre-acquisition)
- execution phase (acquisition)
- life-cycle phase (post-acquisition)

The planning phase contains four sequential but interrelated process steps

**Step 1** Definition of need
**Step 2** Specification of product or service to fulfil the defined need
**Step 3** Tendering and tendering response
**Step 4** Establishment of the contract with the vendor selected to satisfy the need

## 9.12.2  Contract Administration

Contract administration is the process of ensuring that the contractors' performance meets the contractual requirements. In other words, contract administration includes all the management activities necessary to integrate the contract performance process in the project's overall management process.

### 9.12.2.1  Contract

A contract is a mutually binding legal agreement to establish, cancel or change a specified arrangement which is achieved by a concurring declaration of intent and acceptance between two or more parties. It differentiates between the seller (contractor), who is obligated to provide a specified product and/or service and the buyer (client), who is obligated to pay for this product and/or service. A contract includes all the aspects of a legal agreement made between the contracting parties covering a proposal, the scope of work, the terms of payment, reporting, legal recourse etc.

### 9.12.2.2 Contract Strategy

Besides taking into account the significance of the terms and conditions that appear in the project contracts, it is important for the project managers to understand and appreciate the contract strategy. In fact, understanding the contract strategy and the mode of execution of the contract are the keys to a successful outcome for the project.

Any project incorporates a degree of risk which once initiated may be countered by effective change control, producing a revised clear scope and work definition, rescheduling both the programme and the cash flows, advance payments and retentions etc. as appropriate. The type of contract used to complete a project should take into account the associated risks and then apportion them to those involved, such as the client and/or the contractor, who are best able to manage them.

The three parameters – time, cost and quality (safety, environmental impact, public perception etc.) – are the primary driving factors in a project. These three parameters may be inter-related in triangular fashion as shown in Figure 9.4. Individual project managers may emphasise one or the other aspects of the triangle, depending on the project boundaries and priorities. For example, one project may have the time constraint, i.e. completion requirement by a specific date while another one may require quality to be given top priority.

Alongside these three driving factors, risk elements to the client and contractor are also shown in Figure 9.4. The seller (contractor) will try to maximise the profit at minimum risk, whereas the buyer (client) may aim to minimise the risk to them and achieve minimum capital expenditure. These diametrically opposing aims (shown in Figure 9.4) quite often lead to an adversarial relationship between the client and contractor and that may lead to a problem in project execution. The strategies and expectations of both parties need to align so that the risk–reward criterion is jointly carried forward. If there is a high level of risk or uncertainty in the scope of the contract, as in most nuclear decommissioning work where the uncertainties could be the extent of contamination present, unforeseen radionuclides etc., the contract should have a higher contract value or a different type of contract to reflect such eventualities. On the other hand, straightforward decommissioning work with a low risk would have a lower contract value.

### 9.12.2.3 Types of Contract

The type of contract that is drawn between the client and the contractor is crucial to the success of the whole process. Obviously, contracts will vary from project to project depending on the level of complexity, degree of uncertainty, project duration etc. But there are some essential elements which must be incorporated in order to achieve a successful outcome of the project.

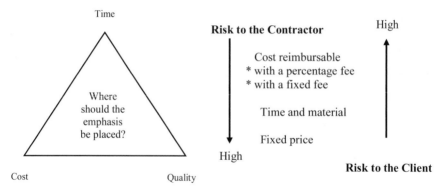

**Fig. 9.4** *Three driving factors and risk distribution*

- Project scope and deliverables must be clearly specified in the contract.
- Uncertainties, if present at the initial specification of the contract, must be quickly removed by consultation, discussions between the client and the contractor or by undertaking further studies if technical issues are involved.
- Proper risk assessment should be carried out by the contractor and accepted by the client.
- Project management by the contractor and project supervision by the client should be properly carried out.

A number of different types of contract are available for use on projects. The important point, however, is to carefully select the most appropriate type of contract for the project. In a large project, there may be separate pieces of work and each piece of work may have a different type of contract. The types of contract are given below.

**Cost Reimbursable Contract**

A Cost Reimbursable (CR) contract is also known as cost-plus-fee contract. Essentially it involves specifying: (i) an estimate of the contract cost; (ii) provisions for reimbursing contractor's expenses; and (iii) provisions for paying a fee as profit. The actual cost and contractor's expenses are most likely to be substantiated by receipts, time-sheets, travel costs etc. The third element, i.e. the fee as a profit leads to two different types of contract: (i) CR with a percentage fee; and (ii) CR with a fixed fee.

In the CR with a percentage fee, risk is very much borne by the client and the contractor's risk is minimal. The contractor may make the project cost as high as possible in order to maximise profit, as the profit is based on a percentage of the cost. This type of contract is drawn for research activities when the outcome of the work is not known at the outset or for projects where there are high levels of uncertainty or the scope of work is vague. For effective performance in this type of contract, the client and contractor should work in harmony. The CR with a fixed fee is similar to the above type of contract, but instead of a percentage fee, there is a fixed fee. So, no matter how long the contract takes to complete, the contractor receives the same amount of fee and that puts a pressure on the contractor to complete the work within a reasonable time. This type of contract is applicable to engineering design organisations, safety consultancy groups etc.

**Time and Material Contract**

In the Time and Material (T&M) and unit price contract, the client and contractor negotiate hourly rates for the specified labour of the contractor and obtain an agreement on the cost of materials. This type of contracts is frequently used to procure equipment maintenance and other support services, particularly when the time estimate to complete the work, carry out the repair, or overhaul the equipment is uncertain. The client receives a bill based on the agreed hourly rate for the labour and the cost of the materials utilised.

A unit price contract is an arrangement by which the supplier is paid on the basis of units of measurable output. A base floor and ceiling can be set and adjustments made to reflect price changes in the marketplace. These types of contracts are advantageous to both the seller and buyer because they are based on measurable costs.

**Fixed Price Contract**

The fixed price contract is the simplest and most common form of business contract. In this type of contract, the contractor undertakes the major part of the risk and cost and hence it is usually preferred by the client. However, fixed-price contracts may also provide the contractor with a greater opportunity to secure a substantial profit. In this type of contract, it is imperative that the client monitors the quality of the contractor's work.

# Revision Questions

1. What is involved in nuclear decommissioning project management? Describe the array of skills that is required for such an undertaking.

2. What are the various areas that require addressing in the management of a decommissioning project? List these areas sequentially and briefly describe them.

3. Write short notes on
   - (i)   Project Management Plan (PMP)
   - (ii)  Work Breakdown Structure (WBS) and the Work Package (WP)
   - (iii) PERT chart
   - (iv)  Gantt chart

4. What is meant by Project Risk Management (PRM) and how it is carried out in practice?

5. What is meant by 'contract and procurement management'? Briefly describe the procurement management process.

6. What is meant by a 'contract'? What are the various types of contract that are drawn between a client and a contractor? Briefly describe each of them. Show diagrammatically the risk distribution to the client and the contractor for various types of contract.

## REFERENCES

[1]  International Atomic Energy Agency, Decommissioning of Nuclear Power Plants and Research Reactors, Safety Guide, Safety Standards Series No.WS-G-2.1, IAEA, Vienna, Austria, 1999.

[2]  Massaut V., Project Management, EUNDETRAF II Training Course, EC Funded Project, 2004.

[3]  Nuclear Energy Agency/OECD, EU, IAEA. A Proposed Standardised List of Items for Costing Purposes in the Decommissioning of Nuclear Installations, Interim Technical Document, OECD, Paris, France, 1999.

[4]  Carter B., Hancock T., Morin J.-M. and Robins N., *The European Project Risk Management Methodology*, Blackwell, Oxford, UK, 1994.

# 10

# PLANNING FOR DECOMMISSIONING

## 10.1 Introduction

The need for proper planning to decommission a plant, large or small, is paramount. As decommissioning is the ultimate end of the life of a plant, when regulatory controls from the plant are going to be withdrawn, thereby releasing the licensee from the licence responsibilities, it is in the interest of the licensee to conduct this operation in a responsible and cost-effective way. A proper planning for decommissioning at an early stage is essential. The adage is

**Failing to Plan = Planning to Fail**

It should also be noted that the production of a report detailing the decommissioning plan and the submission of the plan for regulatory approval are required under the licence requirements in the UK. These licence conditions also specify that the approval of the regulatory body is required for the implementation of a decommissioning project and no alteration or amendment to the approved project can be made. If alteration to the project is required, regulatory approval for such alteration or amendment must be obtained.

The regulatory control of decommissioning is done in the UK by a single overall licence spanning from the operational phase to the decommissioning. In many European countries, a separate licence is required for the decommissioning operation (see Appendix 5 for the details of institutional framework and regulatory controls in some of the EU countries for of decommissioning and radioactive waste management). The organisation holding the licence for decommissioning becomes the operating organisation. Whatever the regulatory infrastructure, the operating organisation is responsible for producing a programme detailing the course of action and the end-point to be achieved in decommissioning for regulatory review and approval.

## 10.2 Decommissioning Options

There are a number of possible options for decommissioning a nuclear facility. It may range from immediate dismantling and removal of all radioactive materials and wastes from the site leading to site release to an option of in situ disposal by safe enclosure of the highly active areas and maintaining a care and maintenance programme. This in situ disposal of a facility is also known as deferred dismantling. (It should, however, be noted that a radioactive waste repository does not undergo decommissioning, it is simply closed.) For immediate dismantling and site release (whether partial or whole site release), a decommissioning plan is needed. Even if the chosen option is deferred dismantling, studies of appropriate methods and approaches need to be carried out in preparation for the eventual dismantling (see also Section 5.8).

## 10.3 Detailed Planning Description

The main components of the plan for decommissioning are shown in Figure 10.1 and described in detail below [1].

### 10.3.1 Initial, Ongoing and Final Planning

An initial plan for the decommissioning of the facility should be prepared at the time of designing the facility and submitted by the owner/operator to the regulator in support of the licence application. However, it should be noted that many older facilities were designed, constructed and operated with little or no regard to decommissioning. For facilities where an initial plan was not prepared, it should be made at the earliest opportunity. This initial plan does not need to be elaborate. It should only state the overall decommissioning strategy, the resources necessary for such a plan and the waste management provision.

A proper consideration for decommissioning at the planning stage of new plants would help to reduce decommissioning costs considerably. In addition, proper design features help maintenance and inspection during the operational lifetime of the facility. Specific factors should include

- In reactors, careful selection of materials would reduce activation which would help waste minimisation in decommissioning.
- Careful design of the reactors may reduce the circulation of activated corrosion products.
- In glove boxes, fume cupboards etc., use of non-absorbent materials such as steel, plastic-coated material instead of wood would reduce absorption of radioactive liquor.
- Plant design, layout and access routes would facilitate removal of large components, decontaminate components etc.

Following the preparation of the initial plan, it needs to be reviewed periodically and updated, as and when necessary, while the facility in operation. The operating history of the facility, significant abnormal events, modifications and improvements that had been carried out in response to the regulatory requirements etc. should be incorporated in the ongoing plan.

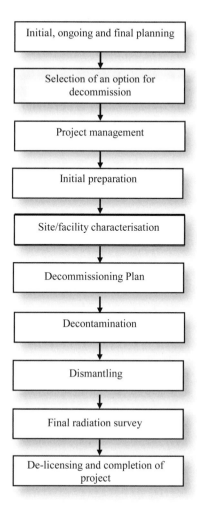

**Fig. 10.1** *Flow chart of a decommissioning project*

The final decommissioning plan should be prepared before the final shutdown. The operating records, records of all abnormal incidences and accidents should be preserved and handed over to the operating organisation. If the facility was shut down because of an abnormal situation, the final plan should be completed as soon as possible and submitted to the regulator for approval. A detailed description of these stages of the plan can be found in [2] and [3].

## 10.3.2 Selection of an Option for Decommissioning

As mentioned above, the full range of options at the end of the operational life of a nuclear facility is quite wide. It may vary from immediate dismantling leading to site release to deferred dismantling. The decision to decommission a nuclear facility varies from case to case. The major factors which drive the decision to decommission are

- National policy and regulatory requirements
- Future use of the site or the facility
- Occupational, public and environmental safety
- Skill resources
- Cost considerations including availability of funding
- Technology requirements
- Structural deterioration
- Inter-dependence with other on-site activities
- Availability of waste storage and disposal facilities

On the basis of all of these considerations, a decision to decommission the facility and an outline strategy for its implementation are taken.

The selection of a preferred decommissioning option may take the following aspects into consideration:

- Safety considerations, including an assessment of hazards and risks involved
- Regulatory requirements concerning radiological, environmental and industrial regulations and government policy
- Financial considerations
- waste treatment and storage facilities and waste disposal routes
- Balance between detrimental physical deterioration of structures and beneficial radioactive decay
- Availability of staff with specialist skills and knowledge
- Interaction with other on-site facilities
- Confidence in the technology

The advantages and disadvantages of each of these decision parameters should be listed and the associated risks assessed. High on the list of decision parameters is safety and risk assessment. The risks here should include not only operational risks associated with decommissioning activities but also any accidents that may arise during these activities. In other words, a full safety assessment should be carried out to asses the full extent of risk, and that assessment should include occupational exposure from normal decommissioning activities as well as risks from faults/failures and accidents. If the facility happens to be a major nuclear installation which may require an extensive decommissioning operation, the risks to the public may also need to be assessed. Other decision parameters should also be considered and a list of possible options should be drawn up.

On the basis of this assessment, it may be possible to eliminate some of the options on safety, environmental, economic or technical grounds. The options which pass these initial screening tests are then subjected to detailed analysis. Until recently, analyses of the options were restricted to a financial appraisal. A more robust option study can now be carried out using techniques such as the Multi-Criteria Analysis (MCA) or Multi-attribute Utility Analysis (MUA). These techniques have been fully described in Section 7.6.2.

The outcome of this option study is a report which recommends a preferred option with a detailed justification. In this justification, the advantages/disadvantages and possible risks of the preferred option along with other closely contested options are set out. The main options are then subjected to sensitivity and uncertainty analyses to establish their robustness.

## 10.3.3 Project Management

The management of a decommissioning project is a complex, multifaceted and multi-disciplinary activity. Once the decision to decommission a facility has been taken, a project management team needs to be formed to oversee the operation. In some cases, the project management team may be formed even before the decision to decommission is taken, and that team is given the task of identifying the most cost-effective decommissioning option, taking into account regulatory, technical, financial, legal and other parameters. Depending on the complexity of the project, the team should include people with expertise in the following areas:

- project management with schedule control and cost control
- risk monitoring and control
- nuclear and conventional safety
- radiation protection
- QA and QC
- plant system and operational experience
- safety and security

The project management, its functions and responsibilities are discussed in detail in Chapter 9.

## 10.3.4 Initial Preparation

The initial preparation of the site or facility to be decommissioned involves organising the administrative set-up, forming a decommissioning team, setting up the QA team and organising training requirements. This stage also involves the removal of operational radioactive materials and waste, ensuring continuation of site infrastructure, essential safety provisions etc. This phase of work may encompass the activity which is generally known in the UK as POCO. A description of the POCO phase (along with the IAEA transition phase) can be found in Section 5.6. The POCO phase is, however, generally considered to be outside the decommissioning operation.

## 10.3.5 Site/Facility Characterisation

The site or facility characterisation is a vital element of the decommissioning operation. The radiological characterisation of the site provides a reliable database of information on the quantity and type of radionuclides, their distribution and their physical and chemical conditions. Characterisation involves a survey of the existing data, in situ measurements, sampling and analyses.

This characterisation is essentially sequential in nature. The initial objective of characterisation, at the planning stage, is to collect sufficient information to assess the radiological status of the facility and the nature and extent of the problem areas. This information may be used to prioritise and determine the sequence of the decommissioning activities. As the planning process progresses, a more detailed characterisation covering the physical, chemical and radiological conditions of the site/facility would be required. Using this database, the decommissioning planner produces various decommissioning options covering

- Decontamination techniques and dismantling procedures to be followed – manual, semi-remote and fully remote
- Radiological protection of workers, public and the environment
- Waste estimation
- Cost estimation

The characterisation of the site or facility is described in Chapter 11.

## 10.3.6 Decommissioning Plan

The Decommissioning Plan (DP) provides a strategic overview of the decommissioning project from the beginning right up to its final end-point. The end-point may be the unrestricted use of the site (green field site) or some other agreed condition which may be a brown field site. The DP includes

- Description of the facility
- Description of regulatory requirements
- Decommissioning strategy
- Project management
- QA programme
- Decommissioning activities
- Environmental Impact Assessments
- Safety assessment
- Final radiation survey proposal
- Final decommissioning report

The description of the facility should include physical description of the site and facility. It should also include the operational history and any abnormal incidents/accidents that may have an impact on the decommissioning. A description of the systems and equipment would be needed.

The decommissioning strategy should include the following items:

- Decommissioning objectives
- Decommissioning options
- Types and volumes of waste that are likely to arise
- Dose and cost estimates
- Financial arrangements
- Selection of the preferred option and its justification

The safety assessment should cover the following items:

- Operational limits and conditions
- Dose predictions for tasks
- Demonstration of ALARA/ALARP for tasks
- Radiation monitoring and protection systems
- Control of physical security and materials
- Management of safety
- Risk analysis

## 10.3.7 Decontamination

The term decontamination covers those activities associated with the removal or reduction of radioactive contaminants, either fixed or loose, from the bulk of the body or from the surface of structures, components, tools, equipment etc. The process of decontamination can be carried out before, during or after dismantling. It is carried out inside as well as outside a structure or a system so that the exposures to radiation of the workers carrying out subsequent operations are reduced. The other purpose of decontamination is to reduce the volume of waste. A higher category waste can be decontaminated to a lower category waste or a lower category waste can be decontaminated so that it can be released under the clearance criteria.

There are various decontamination techniques: chemical, mechanical and other techniques. The suitability and effectiveness of a technique depends on a number of factors such as the type and size of the material, the distribution and accessibility of contaminants etc. Full details of the decontamination processes and techniques are given in Chapter 13.

## 10.3.8 Dismantling

A number of dismantling techniques are available for nuclear decommissioning work. Depending on the level of radiation where the operation is to take place, either remote dismantling or manual dismantling can be chosen. Various tools can be used such as mechanical cutting tools, thermal cutting tools, electrical cutting tools and new techniques.

The primary objective of the dismantling operation is to reduce the volume of contaminated systems, structures and components in a safe and environmentally acceptable way. Minimisation of secondary waste is also a major requirement. All of these diverse and incompatible requirements can only be accommodated by careful selection of a technique which makes a balanced compromise. For example, the risks to workers can be significantly reduced by using remote dismantling techniques. But there will then be an increase in secondary waste, from the cutting operations and contaminated remote handling machines. Also underwater cutting of large thick steel plates would reduce worker exposure, but that would increase the amount of radioactive slurry. Taking all of these issues into consideration, dismantling techniques are utilised. Full details of the dismantling techniques are given in Chapter 14.

## 10.3.9  Final Radiation Survey

The decommissioning plan must provide for a final radiation survey. The purpose of this survey is to ensure that the radiation protection objectives, specified at the beginning of decommissioning operation, have been fulfilled. This survey should be conducted by a competent, independent organisation which reports to the regulatory body. On the basis of this survey the regulatory body decides the outcome of the decommissioning operation. The survey report may highlight certain concerns or hazards which need to be addressed by the licensee before the regulators can make a final decision. For example, the final radiation survey by the NRPB at the JASON reactor building at the erstwhile Royal Naval College at Greenwich identified tritium migration into the adjoining building areas which the MoD had to remove before unrestricted use of the site was allowed.

## 10.3.10  Delicensing and Completion of Project

Delicensing is the final outcome of a decommissioning operation. Delicensing can either be with some restrictions or without any restrictions. If the site is released from regulatory controls with some restrictions, then the site is called a brown field site. Such a site can be used for further nuclear activities or for industrial purposes, a warehouse, car park etc. But it cannot be used for housing, schooling or agricultural purposes. On the other hand, if a site is released by the regulator without any restriction, then it can be used for any purpose. In the UK, a site released without restrictions from nuclear activity cannot be called a green field site, as nuclear activity is deemed to have removed green field status from a site for ever.

On completion of the decommissioning work, a final decommissioning report is prepared. It provides confirmation that decommissioning is complete. It should contain the following information:

- Description of the facility
- Decommissioning objectives
- Radiological criteria used for the removal of the facility from regulatory controls

- Inventory of radioactive materials produced during decommissioning and the present location in storage or disposal of the waste
- Summary of abnormal events or problems during decommissioning
- Summary of occupational and public doses from the decommissioning operation
- Results of the final radiation survey
- Lessons learnt

Appropriate records should be retained once the decommissioning has been completed.

## Revision Questions

1. Describe how the preparation for decommissioning is made from the design stage to actual decommissioning of a plant.
2. What are the various stages of decommissioning planning? Show them in a flow diagram and briefly describe each of these stages.
3. What are the prime drivers in the selection of a decommissioning option? Briefly describe them.
4. What are the possible end-points of a decommissioning project? Briefly describe them.
5. What is the purpose of a final radiation survey? Who carries out this survey? What criteria must be fulfilled to have the site released from regulatory control?

## REFERENCES

[1] International Atomic Energy Agency, Decommissioning of Nuclear Power Plants and Research Reactors, Safety Guide, Safety Standards Series No.WS-G-2.1, IAEA, Vienna, Austria, 1999.
[2] International Atomic Energy Agency, Decommissioning Techniques for Research Reactors, IAEA, Vienna, Austria, 1994.
[3] International Atomic Energy Agency, Decommissioning of Small Medical, Industrial and Research Facilities, Technical Reports Series No. 414, IAEA, Vienna, Austria, 2003.

# 11

# SITE/FACILITY CHARACTERISATION

## 11.1 Introduction

The characterisation of a nuclear-licensed site or a nuclear facility involves the estimation of the type, amount, extent and distribution of radioactive substances as well as non-radioactive contaminants in the site or facility. It is an important and essential step in the process of decommissioning as it can directly affect the start and/or delay of work between the stages of decommissioning. An estimate of the inventory of radionuclides is a prerequisite in the planning to ensure that decommissioning is carried out in a safe, economic and timely manner [1]. Non-radiological hazards may arise during decommissioning from carcinogenic materials such as asbestos, PCB etc. or from chemo-toxic materials such as Pb, Hg and other heavy metals. This chapter, however, focuses specifically on the radiological characterisation of site buildings, structures and equipment. It also includes a broad description of the methods to detect radioactivity and other methods to characterise nuclear facilities.

Following the completion of the decontamination process, a survey of the area that has been cleaned up is carried out in order to demonstrate that the initial objectives have been achieved by adhering to the stated criteria. The stated criteria depend on the end-point specified at the beginning. For example, if the end-point is the delicensing of the site, the criteria for the characterisation process would be to demonstrate that there is 'no danger' from ionising radiations on the site. (The significance of the criterion of 'no danger' from the regulatory point of view is described fully in Section 6.4.) On the other hand, if the end-point is the reuse of the site for other nuclear activities, the criteria are likely to be less rigorous. In any case, a final site survey is required. The techniques and methodology for the final site survey and the pre-decommissioning characterisation processes are very similar. This chapter addresses these survey techniques and identifies specific areas of application such as the pre-decommissioning survey, post-decommissioning survey, and the soil characterisation associated with decommissioning or remediation of contaminated land.

## 11.2 Characterisation Objectives

The main objective of radiological characterisation is to provide a reliable database of information on the type and quantity of radionuclides, their distribution and their physical and chemical status in order to obtain an overview of the levels of hazard involved in the decommissioning process. This process involves the evaluation of historical data in order to identify the present conditions of the facility, conduct radiological and non-radiological surveys, design and implement various measurement and estimation techniques. The information is used to plan and manage the decommissioning operation as well as to estimate the waste arising. This initial estimate of waste arising is needed not only to meet the regulatory requirements but also to carry out a cost estimation for the decommissioning work.

## 11.3 Process of Characterisation

The process of site/facility characterisation needs to be approached systematically and logically. If a facility which handles small quantities of low radiotoxic materials comes to the end of its life in a normal way, then the characterisation is simple. On the other hand, if a nuclear facility happens to be a nuclear power plant which comes to an abrupt halt because of an incident or accident, the characterisation process may be arduous and hazardous with a large element of uncertainty. All these aspects need to be taken into account at the initial planning stage.

At the very beginning of the planning stage of decommissioning, the approach is to gather as much information as possible about the site/facility, its historical background and its present stage. To carry out these objectives, the following steps are taken:

- Site identification: The extent and boundary of the site to be characterised is identified. At this stage, the site can be demarcated by a radiological area classification. If the site forms part of a larger site, interactions and interfaces between the various sections of the site with regard to services such as water, drainage, ventilation etc. should be taken into consideration.

- Historical assessment: A review of the past and ongoing operations and all incident/accident reports should be undertaken. The licensing file(s) would be an extremely valuable source of information. Based on past/ongoing operations, areas or locations which are likely to be impacted and the extent of the impact should be determined. This exercise can also give some indication of the potential migration of contamination beyond the site into the surrounding areas through, for example, the ground-water.

- Scoping survey: The extent and magnitude of the radio-activity is estimated and the previous area classification is confirmed or revised. The major contaminants on the site are also identified. This is the initial phase of the survey.
- Survey design: A properly designed survey is initiated to obtain detailed information regarding the extent and magnitude of radiological and non-radiological contaminants. This chapter deals with the main elements of this detailed survey. The whole process of characterisation in the context of the decommissioning is shown in the grey boxed area in Figure 11.1 [1].

Following the cessation of operation and prior to the start of decommissioning a facility, the radiological and non-radiological conditions are characterised.

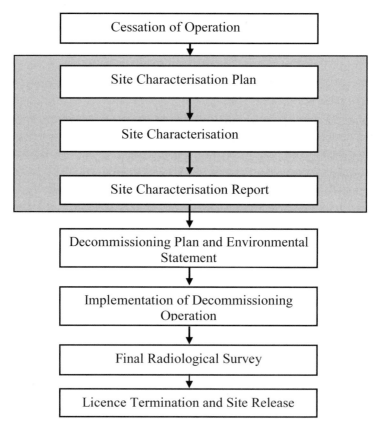

**Fig. 11.1** *Site characterisation in the context of the general decommissioning process*

Initially, a plan for characterisation detailing the systematic and chronological progress of the characterisation is produced. The process of characterisation is iterative. At the beginning of the planning stage of decommissioning, the purpose is to collect sufficient information to assess the radiological or non-radiological status of the facility and the nature and extent of any problem areas. As the planning progresses, characterisation objectives move towards developing a more detailed description of the radiological conditions of the facility. The physical and chemical states of the contaminants are evaluated at this stage. If the facility is a nuclear-chemical plant, the radioactivity is likely to be widespread and contamination may be non-fixed or loose; whereas if the facility is a nuclear reactor, most of the activity (besides irradiated fuel) is likely to be fixed as activated products. It should be noted that such a distinction between non-fixed and fixed contaminations is somewhat arbitrary. Fixed contaminants may become loose or loose contaminants may become embedded. If the contaminants remain in situ under normal working conditions, they can be considered to be fixed and they will not give rise to hazards from inhalation or ingestion.

Computational methods may be used to calculate the induced activity and its distribution, which may be followed by sampling the relevant areas. The details from these activities are then used to estimate the volume of waste, operational risks and cost. The outcome of the characterisation process and its conclusions and recommendations are included in the site characterisation report.

Once the site characterisation process has been completed, the decontamination of components, systems and buildings takes place. Methods and techniques applied during the implementation of the characterisation process are also generally applicable to surveys conducted following the decontamination activity. These techniques are briefly described in Section 11.4. Finally, site soil and groundwater surveys are carried out as part of the final radiological survey in order to demonstrate compliance with the regulatory requirements for delicensing. Guidance on the best practice for managing contaminated land is given in [2].

## 11.4 Methods and Techniques of Characterisation

The process of site/facility characterisation is essentially the collection of radiation data by performing surveys in a planned and systematic way. As stated above, the essential objective of the characterisation process is to obtain an understanding of the prevailing radiological (as well as non-radiological) conditions. Although historical information is a valuable asset, it should not be accepted without some degree of scrutiny. It may be flawed or not give a comprehensive picture of the radiological condition. The characterisation can be done by one of the following three methods or a combination of them [3]:

- A direct measurement: This is carried out by placing a detector at or near the surface or in the medium being investigated and taking direct readings of the radiation levels.
- Scanning: This is the dose rate evaluation technique which is carried out by moving a portable radiation detector at a constant speed at a specified distance from the surface. This technique is applied when a large surface area is to be covered (such as the building wall, contaminated soil etc.) at a relatively short timescale.
- Sampling: This is the process of collecting a portion of material as a representative of the entire medium. The collected material is then analysed in the laboratory under controlled conditions to determine the activity concentration.

The life-cycle of data collection essentially involves three steps: planning, implementation and assessment [4]. During the planning stage, Data Quality Objectives (DQOs) are used to define quantitative and qualitative criteria which determine when, where and how many samples or measurements are to be carried out and what should be the desired level of confidence in the results. This information along with sampling methods, analytical procedures and QA and QC procedures are put together in the site characterisation plan which itself forms part of the project plan. The collected data are then subjected to the Data Quality Assessment (DQA) process to determine if the planning objectives were achieved. The statistical techniques for the DQA have been given in Section 4.4.

Besides sampling and measuring data, there is a computational technique which allows the estimation of the radiation levels in areas where measurements cannot be undertaken either due to high levels of radiation, which may exist in the reactor pressure vessel, or due to access constraints. This computational method may also be applied to estimate dose levels which can then be used in the planning process such as the dose reduction strategy for the workforce (application of the ALARP principle) or for optimisation of the work schedule.

Each of these processes is described in Section 11.6. Table 11.1 summarises the data needs, uses and collection methods for this purpose [1].

## 11.5 Instrumentation

For the measurements of radioactivity or $\gamma$-radiation dose level, the resolution, sensitivity and accuracy of the detector must be consistent with quality of the data that is required. There are basically three types of detectors: (i) gas-filled detectors (ionisation chambers, proportional counters and Geiger–Müller (G-M) counters), (ii) scintillation detectors, and (iii) solid-state detectors [5].

**Table 11.1** *Data requirements and collection methods*

| Data needs | Specific uses of data | Data collection methods |
|---|---|---|
| Radiation (α-β-γ) dose or exposure rates | Necessary to identify radiation hazards and access limitations, to specify decommissioning procedures and methods and to estimate waste volumes | Direct radiation measurements, screening level, air monitoring |
| Amount of loose and fixed contamination on surfaces | Necessary to evaluate effectiveness of pre-decontamination, to plan protection against airborne releases and to identify personnel protection measures | Analyses of smear samples and correlated radiation measurements |
| Location of radiation sources and contamination ("hot spots") | Necessary to evaluate design sequence of decommissioning actions to specify decommissioning procedures and methods | Scanning, historic knowledge of plant |
| Contaminant penetration into walls and floors | Necessary to design sequence of decommissioning actions to specify decommissioning procedures and methods | Scanning analyses of core samples |
| Contamination levels in soils under and near the facility | Necessary to specify decommissioning procedures and methods, to assess foundation removal and excavation hazards | Analyses of soil samples, historical soil sampling data |

## 11.5.1 Gas-filled Detectors

Gas-filled detectors produce ion pairs when the incoming radiation interacts with the gas atoms in the sensitive volume and these ion pairs are then collected by charged electrodes. The central electrode, the anode, is connected to the positive polarity of the applied voltage and collects negative ions or electrons. The outer cylindrical surface, called the cathode, is connected to the negative applied voltage and collects positive ions. The amplitude of the ion pulse as a function of applied voltage can be divided into three regions: ion saturation region; proportional region; and avalanche region. The detectors operating in these regions are, respectively, the ionisation chamber, the proportional counter and the G-M counter. The filling gas could be either: (i) air; or (ii) Ar or He with a small amount of a halogen gas such as Cl or Br acting as a quenching material. Quenching is used to eliminate spurious pulses when operated at a high applied voltage such as in the G-M operating range.

## 11.5.2 Scintillation Detectors

Scintillation detectors operate on the principle that when radiation interacts with a solid or liquid luminescent material, electrons are produced which jump from stable energy states into excited energy states. These excited electrons jump back to the normal energy levels emitting photons which are taken to be proportional to the energy imparted by the incoming radiation. The emitted photon energy is then converted to an electrical signal by a photo-multiplier tube. The most common scintillating materials are NaI(Tl), ZnS(Ag), Cd(Te) and Cs(Tl) which are used in radiation survey instruments. When the scintillating material, NaI, is doped with a small amount of Tl it is denoted by NaI(Tl). The NaI(Tl) detectors are used in the detection of X-rays or $\gamma$-photons and ZnS(Ag) detectors are used in $\alpha$-surveys.

A NaI(Tl) detector is often used to carry out Low Resolution Gamma Spectrometry (LRGS). The NaI detectors have a much poorer energy resolution than the High Purity Germanium (HPGe) detectors (see Section 11.8.3), but they are suitable for use when the $\gamma$-ray spectra are relatively simple. They can be manufactured in larger volume than the HPGe and are cheaper and require less maintenance than an HPGe detector. The LRGS systems also tend to use multiple or scanned detectors to measure a rotating waste. Very high efficiency, low background LRGS assay systems can be produced either by carefully shielding a few detectors close to the measured item or by building NaI detectors into a low background shielded enclosure.

## 11.5.3 Solid-State Detectors

In solid-state detectors, incoming radiation interacting with a suitable semiconducting material creates electron–hole pairs. An electron jumping from a valence band into a conduction band and a vacancy (called a hole) left in the valence band form an electron–hole pair which can be considered to be the solid-state equivalent of an ion pair in a gas. The design and operating conditions of a solid-state detector determine the types of radiation ($\alpha$-, $\beta$- and $\gamma$-)

**Fig. 11.2** *High resolution γ-spectrometer with Ge detectors*

that can be detected and measured. The semi-conducting materials that are currently used are Ge and Si which can be either n-type or p-type in various configurations. Spectroscopic techniques using these detectors provide a marked increase in sensitivity in detection in many situations.

Spectroscopy provides a means of discriminating amongst various radionuclides on the basis of their characteristic energies. When a specific radionuclide contributes only a small fraction of the total particle fluence rate or energy fluence rate, gross measurements are inadequate and radionuclide-specific detection is necessary. In situ γ-spectrometry is particularly effective in field measurements. A large HPGe detector permits the measurement of low abundance γ-emitters such as U-238 or low energy γ-emitters such as Am-241 and Pu-239.

If HPGe detectors are used instead of the plastic scintillation detector, the signal created in the detector is more closely proportional to the γ-energy emitted by the radionuclide that is present. An example a of high resolution spectrometer with HPGe detectors is shown in Figure 11.2. The radioactive material in a box is placed inside the shielded cabinet of the spectrometer. Because of the small energy gap between the trapped sites and the conduction band of Ge (0.7 eV), it is conventional to operate such detectors at liquid nitrogen temperature (77K).

There is also a class of detector where the incoming radiation energies are integrated over a specified period of time. This class includes thermo-luminescent dosimeters (TLDs). Because these detectors can be exposed for a long periods, they can provide better sensitivity for measuring low activity levels which may be encountered in materials close to clearance levels or for surveillance purposes.

**Table 11.2** *Characteristics of some handheld contamination detectors*

| Detector type | Radionuclide measured | Emission detected | MDA (Bq.cm$^{-2}$) |
|---|---|---|---|
| Bell-type organic GM counter | C-14 | β | 0.24 |
| Cylindrical halogen GM counter | Sr-90/Y-90 | β | 0.33 |
| Air-filled counter | U-235 | α | >0.003 |
| Gas flow counter | Am-241 | α | >0.0007 |
|  | Tl-204 | β | >0.012 |
| NaI(Tl) crystal | Fe-55, Pu-238, Pu-239 | X | 0.9 |
| NaI(Tl) crystal | Co-60 | γ | 2.6–8 |

In assaying materials by spectral analysis, it is sometimes neither necessary nor desirable to measure all the radionuclides, so selection criteria must be established. In general, radionuclides with half-lives of less than one year can be disregarded since they have little bearing on the potential detriment to humans during most decommissioning operations. The selection of the remaining radionuclides will depend on the type and nature of the contamination.

Full details of all these detectors including their underlying theories can be found in [5]. The characteristics of some of the handheld contamination detectors and their Minimum Detectable Activity (MDA) are given in Table 11.2. The MDA is that level of activity which a specific item of equipment and technique can be expected to detect for 95% of the time.

## 11.6 Direct Measurement

To conduct direct measurements of surface contamination from α-, β- and γ-emitting radionuclides, instruments and techniques providing the required detection sensitivity are selected. This selection is dependent of the type of potential contamination, required sensitivity and the radiological survey objectives [3]. Direct measurements are taken by placing the instrument at an appropriate distance from the surface, taking discrete measurements for a pre-determined time interval such as 10 s, 60 s, 1 h or even several days or weeks. The integration times depend on the type of detector and the required detection limit. In general, the lower the required detection limit, the longer is the integration time. The detection limit of an instrument is specified by its MDA which is the activity level which can be detected with 95% confidence.

The instrument that can be used in direct measurements of photon emitting radionuclide concentrations is the portable Ge detector, or in situ γ-spectrometer. The in situ γ-spectrometer can discriminate between various radionuclides on the basis of the characteristic photon energies to provide nuclide-specific measurement. A calibrated Ge detector with multi-channel analyser measures the fluence rate of primary photons at specific energies that are characteristic of the radionuclide and the fluence rate may then be related to the average surface activity. A collimator may be placed in high radiation areas where activity from adjacent areas may interfere with the direct measurements. The collimator, usually lead, tungsten or steel, shields the detector from radiation fields outside the specified area of the surface.

This measurement technique may also be applied to estimate soil activity. However, the profile of the radioactivity can be assumed to be uniform in order to convert the fluence rate to a concentration of activity. This assumption of a uniform distribution is not unrealistic as soil is regularly ploughed or overturned causing homogenisation of the activity distribution.

Direct measurements of β-emitting radionuclides are performed by placing detectors at or near the surface to be measured. Because of the limited range of β-particles in air, the detector head needs to be close to the contaminated surface. The contaminated surface also needs to be relatively smooth and impermeable where activity is present as surface contamination. Direct measurements of porous materials such as wood, soil etc. cannot, in general, meet the objectives of the survey. However β-scintillators may measure, with sufficient accuracy and reliability, the concentration of β-emitting radionuclides in soil under certain conditions.

Limitations similar to those for β-emitting radionuclides apply to direct measurements of α-emitting radionuclides as α-particles have very short ranges (about 1 cm in air). Recently special instruments such as long-range α-detectors have been developed to measure concentrations of α-emitting radionuclides in soil under certain conditions.

Direct measurement may be carried out at random locations in the survey area or at predefined locations to supplement scanning surveys for the identification of small areas of enhanced activity. All direct measurement results and locations are recorded.

## 11.7 Scanning

Scanning is the process by which the presence of radionuclides on specified surfaces such as ground, walls, floors, ceilings etc. is identified by using portable radiation detection equipment. It is relatively quick and inexpensive to perform. For these reasons, scanning is usually performed before direct measurement or sampling.

This type of measurement can give an overall picture of the radiation field. It can be useful in area classification, identification of hot spots etc., but it does not give detailed knowledge of the distribution of radioactive materials, isotopic composition etc. The accuracy of this method depends on factors such as surface geometry, isotopic mixture, background radiation level and, of course, measurement procedure. Nonetheless this type of dose profile can be quite useful.

The instrument is held close to the surface and then moved systematically along the surface at a speed that is low enough to detect a reasonable number of counts. For scanning areas for γ-radiation, NaI(Tl) detectors are normally used as they are very sensitive to γ-radiation, easily portable and relatively inexpensive. The detector is normally held close to the surface (~6 cm) and moved at approximately 0.5 m.s⁻¹. The limiting speed is a function of the detector sensitivity, the type and intensity of radiation and the instrument resolving time. Speeds of 3–5 cm.s⁻¹ are normally used. However, a large area probe may allow a faster scan rate.

If the equipment used for scanning is capable of providing data of the same quality as required in direct measurement complying with the detection limit, spectroscopic distribution etc., then scanning may be used in place of direct measurements.

An interesting development of the γ-mapping has taken place over the last few years. It consists of an imaging system which displays the locations and relative intensities of radioactive sources superimposed in real time over a picture of the area on a video monitor [1]. A system of cameras or a transportable monitor has been developed. They are mainly used in the preliminary phase for the characterisation and the grouping of material per category.

This system can consist of

- an HPGe monitor cooled by nitrogen. The detector is collimated to measure the radiation emitted by the material in a specific geometrical angle. The activity level is estimated by defining the composition of the material (identification of the density of the material and the location of the activity) and then using a mathematical code. This monitor will give information on the activity level and the radioactive material present.
- plastic scintillation detector associated with an imaging system. The collimated detector scans the area to be measured. The activity level is measured at a number of points and a map of the activity level can then be created. This mapping can be associated to the picture of the area scanned in order to obtain a better visualisation of the activity distribution in the area of interest. This kind of instrument is often used in the pre-study to identify the 'hot spots'.

Alpha scintillation survey meters and thin window gas flow proportional counters are normally used for α-surveys. As α-particles have very limited range (1 cm), detectors must be kept close to the surface. Consequently, α-scanning is normally suitable for smooth, impermeable surfaces such as concrete, metal etc., not for porous material such as wood, soil etc.

For scanning β-emitting surfaces, thin window gas flow proportional counters are normally used, although solid scintillation counters are also available. Typically, β-detectors are held about 2 cm from the surface and moved at a speed which can attain the desired level of investigation.

## 11.8 Sampling

Accurate characterisation requires representative samples, which may be surface samples of contaminated materials or bulk samples of activated materials, to be taken and analysed properly in a laboratory. The spectrum of radiation energy from the sample is measured and from this measurement the constituent radionuclides and their activities may be determined. Such analysis requires equipment such as Ge detectors with multi-channel analysers, α-spectroscopic equipment or liquid scintillation detectors. The choice of equipment depends on the required data quality. Table 11.3 gives the main detection methods for important radionuclides and their MDA [1]. The half-lives of the respective isotopes are also given in the first column under the heading 'Isotope'.

There are basically two techniques of sampling: unbiased sampling and biased sampling. In cases where activities are expected to be uniformly distributed, unbiased sampling should be done. An indication of whether or not the

**Table 11.3** *Detection methods and MDAs*

| Isotope | Emission | Detection method | MDA (Bq.g$^{-1}$) |
|---|---|---|---|
| H-3(1.2E+01) | β$^-$ | Liquid scintillation | 10 |
| C-14(5.7E+03) | β$^-$ | Liquid scintillation | 1 |
| Fe-55(2.7E+00) | Electron capture, X | X-ray spectroscopy or Liquid scintillation | 10 |
| Co-60(5.3E+00) | β$^-$, γ | γ spec. | 0.5 |
| Ni-59(7.5E+04) | EC, X | X-ray spec. | 10 |
| Ni-63(1.0E+02) | β$^-$ | Liquid scintillation | 1 |
| Sr-90(2.9E+01) | β$^-$ | Liquid scintillation | 1 |
| Tc-99(2.1E+05) | β$^-$ | ICPMS(a) | 0.6 |
| Ru-106(1.0E+00) | β$^-$, γ | γ spec. | 0.5 |
| I-129(1.6E+07) | β$^-$ | ICPMS(a) | 0.007 |
| Cs-137(3.0E+01) | β$^-$, γ | γ spec. | 0.5 |
| U-235(7.0E+08) | α, γ | ICPMS(a) | 0.0001 |
| U-238(4.5E+09) | α | ICPMS(a) | 0.00001 |
| Pu-239(2.4E+04) | α | α spec. | 0.02 |
| Pu-241(1.4E+01) | β$^-$ | Liquid scintillation | 1 |
| Am-241(4.3E+02) | α, γ | α spec. | 0.02 |

* ICPMS stands for Inductively Coupled Plasma Mass Spectrometry

activity distribution is uniform can be obtained from operational and historical data. The facility to be characterised should be divided into discrete sampling areas and survey units. Survey units are discrete geographical areas of specified size and shape from which decisions can be made as to whether or not these areas have been affected by the facility's operations. For sampling following decontamination, the decision is whether or not the unit has attained the site specific clean-up standard. Survey units are generally formed by grouping contiguous site areas with a similar operational history and the same classification for contamination or activation potential. Survey units are established to facilitate survey process and statistical analysis of survey data. If the activity distribution is expected to be non-uniform (e.g. hot spots are anticipated), then a biased sampling should be carried out. Biased sampling should indicate the size and location of the sampled area. In both cases, the level of confidence in the sampling and measurements (normally 95%) should be specified.

Typical survey areas may include

- Floors where potential spills of radioactive liquor or other contaminants may be deposited
- Walls where dust, sprays etc. may settle
- Ceilings where contaminated air, vapour etc. may settle
- Other horizontal surfaces such as work surfaces, railings, external surfaces of pipes etc. where dust may settle preferentially

The sampling and the analysis of the samples are dependent on the quality of data required. If rigorous characterisation is required, then the following steps need to be carried out:

- Specification of types, numbers, sizes, locations of samples
- Methods of taking samples and performing analyses
- Specification of equipment to meet objectives
- QA requirements
- Provision for the disposal of waste generated during sampling

It should be emphasised here that one of the underlying objectives of sampling is to estimate the quantity of waste that would arise from the operation. Although the total activity concentration is an important parameter, it is not sufficient in itself. The composition of the contaminants and half-lives of individual radionuclides are required as wastes with long half-lives should be segregated from those with short half-lives. This segregation helps to reduce costs as wastes with long half-lives are much more expensive to dispose of (in deep underground facilities) than those with short half-lives which are disposed of in shallow underground facilities.

A practical example of the sampling method is given below. In the Japan Power Demonstration Reactor (JPDR) (decommissioned between 1984 and

1994), the initial rough characterisation of floors, walls and ceilings was carried out by radioactivity measurements. Samples from each of 2 m × 2 m areas where contamination was suspected were taken. Each sample was 1 cm deep and 4 cm in diameter and a NaI(Tl) detector with a single-channel analyser was used to measure γ-intensity. About 1800 samples were taken from the total area of 20,000 m² of floors, walls and ceilings. This gave a contamination profile of the building. To obtain detailed information on the penetration of the contamination, analysis was carried out by one of two methods: (i) in areas where surface contamination was detected, a thin layer of surface (2 mm deep) was repeatedly removed until contamination was undetectable. These 2 mm deep samples were analysed for radionuclide composition and activity. (ii) In areas where deep contamination was suspected, 10 cm deep cores were taken from the concrete. Samples of 1 mm thickness were taken down to a depth of 10 mm, and then samples of 10 mm thickness down to 100 mm. A γ-spectrum of each sample was taken with a Ge detector. Using these sampling methods, it was found that in over 85% of contaminated areas, the activity was confined to within a depth of 2 mm. Thus 2 mm of surface skimming would decontaminate most of the building surfaces.

In order to reduce the costs of the characterisation, one can use statistical techniques whereby a limited number of samples are taken and inferences are made about the whole of the representative area. However, in doing so one must be careful about interpreting the results and extrapolating the outcome. The method for estimating data quality by statistical means is given below.

## 11.9 Statistical Evaluation

As mentioned before, the quality of the data to be collected is specified at the initial stage of the characterisation process by the DQOs. These objectives will also determine the number of data points to be measured, equipment to be used with their MDA capability and tolerance limits etc. The various methods of data collection based on various types of sampling, e.g. probability sampling, search sampling etc. have been described in Section 4.3. Following the completion of initial sampling and estimation of data values, calculation of basic statistics and generation of graphs take place. Graphical representation may include display of individual data points, statistical quantities, temporal data, spatial data and so on. This analytical information may be used to learn more about the structure of the data and to identify patterns and relationships or potential anomalies.

A detailed data collection exercise is then carried out. Following the completion of this exercise, an assessment of data quality is undertaken by analysing the data, based on the review of the DQOs, the sampling design and the preliminary data review. At this stage the key underlying assumptions that must hold for the statistical procedure to be valid are also identified.

## 11.10  Computer Calculations

Computer calculations are most useful for estimating activation products in and around the reactor compartment of a nuclear power reactor. Neutron fluxes and energy spectra produced in a reactor are incorporated in a computer code which then models the neutron interactions with surrounding materials once the neutron interaction cross-sections have been provided. The transport of neutrons through matter can be calculated by a one-dimensional model, called ANISN [6], which solves the transport equations. For somewhat complex geometries, a two-dimensional neutron transport code, such as TWODANT [7], may be used. For complex geometries, the deterministic solution of the transport equations becomes very difficult. In such situations, the Monte Carlo method of neutron transport may be used. McBEND [8] is a very widely used and validated code for such calculations. McBEND can also model coupled neutron/γ-interactions and produce output giving total energy absorptions at various distances along the length of the track. The ORIGEN 2 [9] code can be used to calculate the neutron-induced radioactivity in materials.

## 11.11  BR 3 Characterisation

As an example of the radiological characterisation of a nuclear reactor, the characterisation of the Belgian Pressurised Water Reactor, BR3, is given below [10].

Two aspects are important during planning for the decommissioning of a nuclear reactor: (i) the dose rate aspect which, in the case of a PWR, is dominated by the γ-radiation from Co-60; (ii) the contamination aspect, for which not only the γ-emitting nuclides such as Cs-137 and Co-60 are important, but also the presence of α-contamination can present a particular problem.

For waste management, it is also important to determine the so-called critical nuclides, i.e. the nuclides which are difficult to measure (the pure β-nuclides such as Ni-63 and Ni-59, Sr-90, Nb-94, C-14, H-3 etc. and the α-nuclides such as the Am-241, Pu and U isotopes) and which are an issue during long-term disposal due to their long lives and their specific radiotoxicity. The determination of these critical nuclides in waste packages is a difficult task. Several approaches are followed to satisfy the disposal requirements. Estimations on the basis of neutron activation calculations, materials composition and irradiation history allow us to determine precisely the activation levels for the major components of the irradiated materials, such as the Ni and Fe isotopes for metals and Ca isotopes for the concrete. For elements present at trace levels such as Nb, C, H-3, Eu, the accuracy of the estimation depends strongly on the exact original content of these trace elements which is generally not known precisely.

Radiochemical measurements are the best way to determine the exact radiochemical composition of the activated materials. For BR3, during the dismantling of the highly active internals, samples were taken systematically during the cutting operations. Some swarf material was collected during the cutting operation and subjected to detailed radiochemical characterisation. The radiochemical determination implies a complex analytical work with a series of separations to eliminate the strong γ-nuclides which are present in activity levels several orders of magnitudes higher than the investigated isotopes.

A still more difficult task is to estimate the critical nuclides coming from fuel leakages which are fission products such as Sr-90 and Cs-137 or all the α-isotopes. This is quite impossible to model so only radiochemical determinations can solve the problem. This requires the estimation of mean surface contamination levels of α-, β-, γ-emitters and the determination of specific contamination isotopes such as the Cs-137 γ-emitter, the correlated Sr-90 β-emitter, the determination of the α-spectroscopic composition including the Am-241, the long-lived Pu, U and Cm isotopes as well as the β-emitter of Pu-241.

For BR3, the compositions of two mean surface contamination levels were determined:

- the first one, the high contamination level, is representative

Table 11.4 *Radionuclide vectors of different waste streams in BR3*

| Correlation | Thermal shield | 'Vulcain' internals | 'Westing-house' internals | Contamination vector |
|---|---|---|---|---|
| Ni-63/Co-60 | 1.4E+00 | 3.6E+00 | 7.1E+00 | 1.1E+00 |
| Ni-59/Ni-63 | 2.8E–03 | 1.9E–03 | 3.6E–02 | 2.0E–03 |
| Fe-55/Co-60 | 1.2E+00 | 3.8E+00 | 4.8E–02 | 1.8E+00 |
| Nb-94/Co-60 | 4.4E–05 | 1.3E–04 | 2.3E–04 | 4.0E–03 |
| C-14/Co-60 | 1.1E–04 | 4.3E–04 | 7.2E–04 | 4.2E–03 |
| H-3/Co-60 | 1.3E–04 | 5.2E–05 | 3.2E–04 | 3.2E–04 |
| Cl-36/Co-60 | 4.6E–06 | 1.3E–05 | 2.7E–05 | 3.7E–06 |
| Sb-125/Co-60 | | | | 1.9E–03 |
| Tc-99/Co-60 | | | | 5.9E–06 |
| Sr-90/Cs-137 | | | | 3.7E+04 |
| Am-241/$\alpha_{tot}$ | | | | 4.6E–01 |
| Pu-238/$\alpha_{tot}$ | | | | 3.5E–01 |
| Pu-239+240/$\alpha_{tot}$ | | | | 1.5E–01 |
| Pu-240/Pu 239 | | | | 3.8E+04 |
| Pu-242/Pu 239 | | | | 3.4E–03 |
| Cm-244/$\alpha_{tot}$ | | | | 4.0E–02 |
| Pu-241/Am 241 | | | | 4.3E+01 |
| $U_{tot}/\alpha_{tot}$ | | | | 4.0E–03 |

of the primary pieces which were never decontaminated

- the second one is representative of the items which were decontaminated during the full system decontamination of the primary loop, which was the first step in the dismantling strategy

Table 11.4 gives an overview of the radiochemical isotope vectors which were derived for different waste streams (reference date: 1 July 1998, i.e. 11 years after shutdown).

## Revision Questions

1. What is meant by 'site characterisation'? State the main objectives of radiological characterisation of a site.

2. Radiological contamination is primarily of two types. State them and explain how they arise and how they can be detected.

3. How is site characterisation carried out in practice? Briefly describe these processes in the order in which they are carried out.

4. What are the methods and techniques that are used in the characterisation of a site? Briefly describe each of these methods.

5. Describe the significance of the terms 'Data Quality Objective (DQO)' and 'Data Quality Assessment (DQA)'.

6. What are the main types of detectors used for site characterisation? Briefly describe their functions.

7. Write short notes on the following with examples:
   (i)    scintillation detector
   (ii)   $\gamma$-spectroscopy
   (iii)  minimum detectable activity

8. Why is sampling is carried out? What are the main techniques used in sampling? Describe them briefly.

# REFERENCES

[1] International Atomic Energy Agency, Radiological characterisation of shutdown nuclear reactors for decommissioning purposes, Technical Reports Series No. 389, IAEA, Vienna, Austria, 1998.

[2] CIRIA, Best Practice Guidance for Site Characterisation: Managing Contaminated Land on Nuclear Licensed and Defence Sites, CIRIA, London, UK, October 2000.

[3] United States Environmental Protection Agency, Multi-Agency Radiation Survey and Site Investigation Manual (MARSSIM), June 1, 2001 updates in Revision 1, August 2000, http://www.epa.gov/radiation/marssim.

[4] United States Environmental Protection Agency, Guidance for Data Quality Assessment, Practical Methods for Data Analysis, EPA QA/G-9, QA00 Version, Final July 2000.

[5] Knoll G.F., *Radiation Detection and Measurement*, 2nd edn., Wiley, New York, USA, 1989.

[6] Oak Ridge National Laboratory, ANISN / PC Manual, Idaho National Engineering Laboratory, EGG-2500, ORNL, RSIC Computer Code Collection, CCC-0514, ORNL, Oak Ridge, TN, USA, April 1987.

[7] Alcouffe, R.E., Brinkley F.W., Marr D. and O'Dell R.D., User's Manual for TWODANT: A Code Package for Two-Dimensional, Diffusion-Accelerated, Nuclear-Particle Transport, Report LA-10049-M, Los Alamos National Laboratory, Los Alamos, NM, USA, 1984.

[8] Chucas, S.J., Curl I, Shuttleworth T. and Morrell G., Preparing the Monte Carlo Code McBEND for the 21st Century, (Proceedings of 8th International Conference on Radiation Shielding, Arlington, IL, USA, 1994), American Nuclear Society, Langrange Park, IL, USA, 1994.

[9] Oak Ridge National Laboratory, ORIGEN 2.1, Isotope Generation and Depletion Code Matrix Exponential Method, RSIC Computer Code Collection CCC-371, ORNL, Oak Ridge, TN, USA (revised August 1991).

[10] European Commission, 6th Framework Programme Contract No. FI60-CT-2003-509070, EUNDETRAF II, 2002–2006.

# 12

# ENVIRONMENTAL IMPACT ASSESSMENT AND BEST PRACTICABLE ENVIRONMENTAL OPTION

## 12.1 Introduction

The Environmental Impact Assessment (EIA) is a key element of the national environmental regulation strategy derived mainly from the EU's environmental policy. Since the introduction of the first EIA directive by the EU in 1985 [1], the law and the implementation practices of the EIA have been further extended by the Council Directive 97/11/EC [2]. All member states of the EU are required to abide by this directive.

On the other hand, the requirement for the Best Practicable Environmental Option (BPEO) comes from the work of the Royal Commission on Environmental Pollution (RCEP) in the UK in 1976. It is basically an assessment methodology to optimise environmental damage and degradation following polluting releases to all three environmental media (land, water and air). The competent authorities to oversee the implementation of this regulatory requirement are the Environment Agency (EA) in England and Wales, and the Scottish Environmental Protection Agency (SEPA) in Scotland. Both agencies work under the direction of the DEFRA. Alongside this BPEO requirement of the environmental agencies, there is a specific radiological protection requirement embodied in the ALARP principle (see Section 7.6) which is enforced and implemented by the HSE/ND in the UK.

This chapter addresses the issues of EIA and BPEO sequentially in the context of their applications in the EU in general and in the UK in particular. The discussions on EIA are based primarily on EU guidelines on this topic. Needless to say, UK regulations are generally in keeping with the EU directives.

The EIA directive [2] requires that the member states carry out EIAs on certain public and private projects where it is believed that the projects are likely to have a significant impact on the environment. For some projects, such as the construction of motorways, airfields and nuclear power stations (commissioning and decommissioning), EIA assessments are obligatory. While for others, such as urban development projects, tourism and leisure activities, member states are required to operate a screening method to determine which projects would require assessments. They need to apply certain thresholds or criteria set by the member state and carry out a case-by-case examination to identify projects which would have significant impacts on the environment. During the EIA procedure, the public may provide input and express environmental concerns with regard to the project. The results of such consultations are to be taken into account in the EIA process.

## 12.2  Purpose of the EIA Process

The prime purpose of the EIA is to identify any significant environmental impacts of a major development project and, where possible, to take mitigating actions to reduce or remedy those impacts in advance of any decision to authorise the construction of the project. As a tool to aid the decision-making process, EIA is widely seen as a proactive environmental safeguard measure which, together with public participation and consultation, can help to meet wider EU environmental concerns, EU policy and principles.

Public participation in environmental decision-making processes is based on the belief that it is right for the public to be involved in decisions which affect them. It also ensures that if the public are aware of the environmental issues and are involved in the decision-making process, they are more likely to accept the outcome and the implementation of the decision will be easier. Another benefit of public participation is that the final decision is likely to be of better quality, as relevant information and the interests of stakeholders have all been taken into account.

### 12.2.1  Regulatory Requirements

The current EIA requirements of the EU are set out in the Council Directive 97/11/EC of 3 March 1997 [2] amending Directive 85/337/EEC [1] on the assessment of the effects of certain public and private projects on the environment. These two directives need to be consulted together to determine the obligations on the part of the member states. The basic elements of these requirements have been incorporated in the UK for the benefit of the nuclear industry in the 'Nuclear Reactors (Environmental Impact Assessment for Decommissioning) Regulations 1999' (EIADR) which came into force on 19 November 1999 [3].

The emphasis of the amended EU directive is that EIA should be undertaken for private and public projects which are likely to have significant effects on the environment. The directive also emphasises that the consent should be dependent on an adequate EIA having been undertaken. For a number of projects, an EIA is mandatory, e.g. for nuclear power stations and other nuclear reactors, decommissioning of nuclear reactors (except research reactors whose maximum power does not exceed 1 kW continuous thermal load). The EIA is also required for installations designed for the disposal of irradiated nuclear fuel, for the final disposal of radioactive waste, or for the storage of irradiated nuclear fuels or radioactive waste at a site different from the production site.

## 12.3  EIA: A Phased Process

The EIA process is time-consuming as it involves a large number of people and organisations who may be collectively called stakeholders. Typically it takes between one and three years to complete the whole process. In order to conduct this process efficiently and effectively and to ensure that all necessary input is obtained, a phased approach is required.

The main steps can be described as

- Screening: to determine if an EIA is required.
- Scoping: to establish the information to be submitted.
- Environmental impact evaluation: to assess the impacts.
- Preparation of the ES (environmental statement).

### 12.3.1  Screening

A screening procedure may be necessary to determine if an EIA is required for projects other than those specified as mandatory projects in Annex I of [2]. Annex II of [2] lists projects that on the basis of scrutiny, either on a case-by-case examination or on the basis of thresholds and/or criteria set by the member state, it is determined that they are likely to have significant effects on the environment. For projects such as extensions or alterations to nuclear decommissioning work, there may be significant environmental impacts. All of these projects require screening for the requirement of an EIA. The EU provides guidance to facilitate the screening process [4].

For case-by-case screening or criteria based-scrutiny, a set of selection criteria is provided in [2] in Annex III. Competent authorities must take these selection criteria into account when making screening decisions on a case-by-case basis and when setting thresholds and criteria for projects requiring an EIA. The screening process is shown in Figure 12.1.

#### 12.3.1.1  Checklist

In considering whether a project requires an EIA, some information about the project is needed. The following checklist should help to identify which information is required:

Checklist on information for screening

(1)  Contact details of the company
- name of the company
- main postal address, telephone, fax and e-mail details for the company
- name of the main contact person and direct postal address, telephone, fax and e-mail details

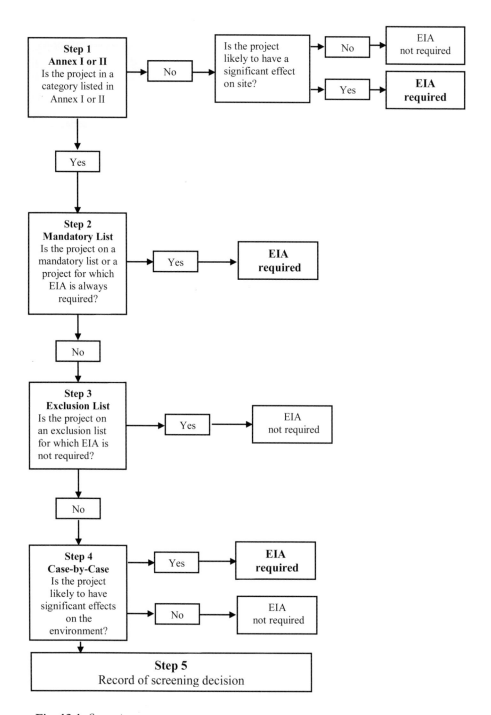

**Fig. 12.1** *Screening process*

(2) Characteristics of the project
- brief description of the proposed project
- reasons for proposing the project
- a plan showing the boundary of the project including any land required, even temporarily, during construction
- the physical form of the project (layout, buildings, other structures, construction materials etc.)
- description of the main processes including size, capacity, throughput, input and output
- any new access arrangements or changes to existing road layout
- a work programme for construction, operation and commissioning phases, and restoration and after-use where appropriate
- construction methods
- resources used in construction and operation (materials, water, energy, etc.)
- interaction with other existing/planned projects
- information about possible alternatives
- information about mitigating measures
- other activities which may be required as a consequence of the project (such as new roads, additional housing and sewage disposal provision, new water supply, generation or transmission of power etc.)

(3) Location of the project
- maps and photographs showing the location of the project relative to surrounding physical, natural and man-made features
- existing land users on and adjacent to the site and any future planned land users
- zoning or land-use policies
- protected areas or features
- sensitive areas (e.g. areas of outstanding natural beauty, areas of special scientific interest etc.)
- details of any alternative locations which have been considered

(4) Characteristics of the potential impact
- impacts on people, human health, flora and fauna, soils, land use, material assets, water quality and hydrology, air quality, climate, noise and vibration, the landscape and visual environment, historic

and cultural heritage resources, and the interactions between them

- nature of the impacts (direct, indirect, secondary, cumulative, short-, medium- and long-term, permanent and temporary, positive and negative)
- extent of the impact (geographical area, size of the affected population/habitat/species)
- magnitude and complexity of the impact
- probability of the impact
- duration, frequency and reversibility of the impact
- mitigation incorporated into the project design to reduce, avoid or offset significant adverse impacts
- trans-frontier nature of the impact

In the EU guidance on EIA [4], a screening checklist provides a list of questions to help identify where there is the potential for interactions between a project and its environment. This checklist is designed to help decide whether those interactions are likely to be significant. Those responsible for making screening decisions may find difficulties in defining what is 'significant'. A useful simple check is to ask whether the effect is one that ought to be considered and to have an influence on the development consent decision. At the early stage of screening, there is likely to be little information on which to base this decision but the following list of questions may be helpful.

### 12.3.1.2 Questions to be Considered

The following questions need to be considered:

- Will there be a large change in environmental conditions?
- Will new features be out-of-scale with the existing environment?
- Will the effect be unusual in the area or particularly complex?
- Will the effect extend over a large area?
- Will there be any potential for trans-frontier impact?
- Will many people be affected?
- Will many receptors of other types (fauna and flora, businesses, facilities) be affected?
- Will valuable or scarce features or resources be affected?
- Is there a risk that environmental standards will be breached?
- Is there a risk that protected sites, areas, and features will be affected?
- Is there a high probability of the effect occurring?
- Will the effect continue for a long time?

- Will the effect be permanent rather than temporary?
- Will the impact be continuous rather than intermittent?
- If it is intermittent will it be frequent rather than rare?
- Will the impact be irreversible?
- Will it be difficult to avoid, or reduce or repair or compensate for the effect?

There is no specific rule that can be used to decide whether the results of using the screening checklist should lead to a positive or negative screening decision, that is, whether or not an EIA is required. The questions are designed so that a positive answer will generally point towards the need for an EIA and a negative answer to an EIA not being required. A 'don't know' answer should also point towards a positive screening decision (i.e. an EIA is required) because the EIA process will help to clarify the uncertainty.

### 12.3.1.3 Thresholds

In some member states, threshold values or criteria are used to avoid a full screening procedure.

This process results in at least three levels of screening thresholds/criteria

- Exclusion thresholds/criteria: projects below a given size or at particular location or other characteristics which do not require EIA by law.
- Indicative thresholds/criteria: projects above thresholds/ criteria levels which are more likely to require EIA, though these thresholds are not always clear quantitative indicators.
- Inclusion thresholds/criteria: projects of a certain size or of particular location or other characteristics which require mandatory EIA.

If a project fulfils the indicative thresholds/criteria, a full screening process may be necessary.

## 12.3.2 Scoping

Scoping is the process of determining the content and extent of the subject matters which should be covered in the environmental information to be submitted to a competent authority for projects which require an EIA.

Scoping was introduced in Directive 97/11/EC. It was not made mandatory by the 1997 directive, but it was stated that all member states that do not have scoping in their EIA procedure are required to introduce, as a minimum, a voluntary scoping process. The minimum requirement is that competent authorities must provide a 'scoping opinion' if requested by a proposer. The 'scoping opinion' should identify the content and extent of the information to be elaborated and supplied by the developer to the competent authority.

Even if the authorities do not require scoping, it is good practice to make scoping part of the EIA process since it brings a number of benefits.

- It helps to ensure that the environmental information submitted to the decision-making process provides a comprehensive picture of the significant effects of the project, including issues of particular concern to affected groups and individuals.
- It helps to focus resources on the important issues in decision-making and avoids wasting effort on issues of little relevance.
- It helps to ensure that the environmental information provides a balanced view and is not burdened with irrelevant information.
- It stimulates early consultation between the developer and the competent authority, and with environmental authorities, other interested parties and the public, about the project and its environmental impacts.
- It helps effective planning, management and resourcing of the environmental studies.
- It should identify alternatives to the proposed project and mitigating measures which ought to be considered by the developer.
- It can identify other legislation or regulatory controls which may be relevant to the project and provide opportunities for the necessary assessment work for different control systems to be undertaken in parallel, thereby avoiding duplication of effort and costs for all concerned.
- It reduces the risk of delays caused by requests for further information after submission of the development consent application and the environmental information.
- It reduces the risk of disagreement about EIA methods (baseline surveys, predictive methods and evaluation criteria) after submission of the environmental information.

### 12.3.2.1 Identification of Stakeholders

The stakeholders in a project need to be identified at an early stage. Ideally all individuals and groups who feel affected by the decision should have an opportunity to participate in the decision-making process. Common potential stakeholders are the following:

- Local authorities
- Local residents and landowners
- Local community groups
- Non-governmental organisations and interest groups

- Trade unions
- Commercial associations
- Independent experts
- Media
- Educational institutions
- National and local government agencies with responsibilities for management of natural resources and welfare of people likely to be affected by the project (e.g. local highway authority)
- Affected public and authorities in potentially affected neighbouring countries

At the start of the scoping process, not all potential stakeholders may be recognised. A combination of several of the following means can be used in an attempt to reach as many potential stakeholders as possible:

- Initial announcements about the scoping process in local or national newspapers
- Posting notices announcing the scoping process at the site, in the neighbouring area and at the offices of local authorities
- Preparing a leaflet or brochure about the project giving brief details of what is proposed with a plan or a map, describing the EIA process and the purpose of scoping and inviting comments
- Distributing letters or questionnaires to potentially interested organisations and nearby residents requesting information and comment on the proposals
- Telephone discussions or meetings with key organisations, groups or individuals
- Articles in newspapers, on radio or on television
- Public meetings (it may be helpful to invite an independent person to chair public meetings)
- Public exhibitions. (An exhibition may be preferable to a public meeting as people who are nervous about standing up and speaking at a public meeting may feel more comfortable speaking to someone on a one-to-one basis at an exhibition. Meetings can also be dominated by a few vocal attendees, so that the full range of issues are not discussed or the most important issues might be overlooked.)
- Electronic media, e.g. website or provision of CD-ROMs

Once the potential stakeholders are known, information must be made readily available. All information needed to form an opinion and to contribute to the scoping process must be provided in a form easily understood by non-experts.

### 12.3.2.2 Outline of Feasible Alternatives

A number of possible alternatives to a project such as the decommissioning of a nuclear facility may be considered at this stage. They may include top-level options such as safe storage, entombment or full decommissioning at an early stage (see Section 5.5). The top-level options may be broken down into work packages and task levels to identify the best possible options. A selection of the proposed feasible alternatives should be undertaken as part of the public participation process at this stage. The aim of this step is to optimise resources on the evaluation of environmental impacts of realistic options only. Section 12.6 dealing with the BPEO will give further details of the feasible options.

In considering feasible alternatives, the following aspects need to be taken into account:

- national legislative/regulatory requirements
- economic factors (e.g. availability of funding, implications for local employment)
- physical status of the nuclear facility (structural integrity of buildings and equipment, radioactive inventory and levels of contamination)
- options for the management of radioactive and chemical wastes (e.g. the availability of interim storage and/or final disposal facility)
- availability of personnel with adequate knowledge of the facility
- potential future uses of the site

## 12.4 Implementation of Environmental Impact Assessment

The EIA starts with a description of the environmental baseline. The next step is to describe the environmental impacts of the selected project such as the decommissioning of a nuclear facility, and identification of various options and assessment of their impacts and then comparing those impacts with each other and with the environmental baseline to come to a final decision.

### 12.4.1 Determination of Environmental Baseline

The description of the environmental baseline must describe the environment as it existed prior to initiation of the decommissioning project in sufficient detail for it to serve as a basis for assessing the project's potential impacts. This would include not only factors relating to the natural environment (e.g. air and water quality, soil, flora and fauna, and landscape) but also socio-economic factors (e.g. land use, culture, infrastructure, population and economy). In order to reduce costs at this stage, existing data should be used as much as possible.

The typical contents of a baseline study are summarised as follows:
- air quality and meteorology
  - regional climate
  - meteorology of the site
  - atmospheric diffusion
  - air quality
- water: hydrology and hydrogeology
  - surface hydrology
    - hydrological system
    - quality of surface waters (radioactive and non-radioactive constituents)
  - hydrogeology
    - hydrogeological system
    - quality of groundwater (radioactive and non-radioactive constituents)
  - sewerage
- land and soil: geological (geotechnical studies; soil characteristics and quality)
  - regional geography
  - local geology
  - detailed geology of the plant site
  - geotechnical studies of the site and conclusions
  - seismology
  - soil on site
  - soil in surrounding areas
- flora and fauna, ecosystems
  - terrestrial species and ecosystems
  - aquatic species and ecosystems
  - marine species and ecosystems
- landscape
  - visibility analysis
  - quality of the landscape
  - visual fragility of the landscape
  - human presence
- noise and vibration levels
- land use
- cultural heritage
- infrastructure: territorial system
  - transport systems, including level of use
  - demographic density and typology of settlements
  - communications and basic infrastructures

- human aspects
- population
  - demographic evolution
  - demographic projection
  - distribution by age, religion and gender of population
  - floating population
  - health aspects
    - public health
    - dosimetric models and critical groups
- economic system
  - income and employment
  - agricultural production
  - fishing production
  - industrial activity

## 12.4.2 Impact Identification

In order to identify impacts which are likely to arise from a decommissioning project, both environmental factors such as flora, fauna, air quality, water quality etc. as well as activities associated with a decommissioning project (with the potential to cause impacts) need to be considered simultaneously. Possible activities may include: modifications to existing buildings, site, demolition of buildings and construction of new ones. Other activities are transport, handling and storage of materials. The environmental parameters to be considered are roughly the same as those used in the study of the environmental baseline.

A commonly used methodology for the identification of impacts of a project on the environment consists of drawing up an identification matrix. This methodology inter-links project activities that might cause impacts to the factors associated with physical, chemical, biological and socio-economic issues of the environment. When there is an interaction, the intersection box of the matrix is marked with a symbol, such as X, and the interaction is subsequently analysed. These marked boxes only identify the likelihood of impacts; subsequently it is necessary to estimate impacts, followed by additional efforts to identify indirect and cumulative impacts.

An example of an impact identification matrix is given in Table 12.1 [5].

## 12.4.3 Impact Assessment

In order to evaluate the effects of a decommissioning project on the environment, impact indicators have been developed. These indicators are used as markers on which efforts are to be devoted to quantify the magnitude of the impacts. Some examples of indicators are given in Table 12.2.

**Table 12.1** *Impact identification matrix [5]*

| E12 Population and Economy | E11 Human Factors | E10 Infrastructure | E9 Cultural Factors | E8 Land Use | E7 Noise and Vibration | E6 Landscape | E5 Fauna | E4 Flora | E3 Water | E2 Land and Soil | E1 Air | Environmental Factors / Project Actions |
|---|---|---|---|---|---|---|---|---|---|---|---|---|
| × | | | × | × | × | | | | | × | | A1 Modification of industrial site |
| | | | | | × | × | | | | × | | A2 Modification of industrial buildings |
| | | | | × | | | | | | × | | A3 Modification of property limits |
| | × | | | | × | × | | | | × | × | A4 Demolition of Buildings |
| | × | | | × | × | × | | | | × | × | A5 Construction of new buildings |
| | × | | | | × | × | | | | × | × | A6 Landfills and earth movements |
| | | | | | | | | | × | × | | A7 Silting and Drainage |
| × | | | | | | | | | | × | | A8 Recycling of wastes |
| | × | × | | | × | | | | | | × | A9 Transport of materials |
| | × | | | | | | | | | | | A10 Handling of Hazardous radioactive or toxic materials |
| | × | | | | | | × | × | × | × | × | A11 Emission of liquid and gaseous effluents |
| | × | | | × | × | × | × | × | × | × | | A12 Use of rubble tips or solid inert waste tips |
| | × | | | × | | × | × | × | × | × | | A13 Storage of solid radioactive wastes |
| | × | | | | × | | | | | | × | A14 Fires |
| | × | | | × | | | × | × | × | × | × | A15 Releases or leakage of contaminating liquids or gases |
| | × | | | | | | | | | | | A16 Operating failures |
| | × | | | | | | | | | | | A17 Personnel accidents |
| × | × | | | × | × | | × | × | × | × | × | A18 Structural failures due to external events |
| | × | | | | | | | | | | | A19 Monitoring and control operations |

Socio-Economic Medium (E12–E8), Physical Medium (E7–E1). Project actions having the potential to cause an impact.

**Table 12.2** *Parameters used in impact assessment*

| Environmental factors | Potential impact indicators |
| --- | --- |
| Air | Concentrations of radioactivity in the air. |
| | Concentration of dust particles in the air. |
| | Atmospheric dispersion. |
| | Level of acoustic intensity or pressure. |
| Land and soil | Concentration of contaminants in the soil. |
| | Contaminated surface. |
| | Volume of soil affected and surface restored. |
| Water | Concentrations of contaminants in surface water. |
| | Volume of contaminants released. |
| | Water consumption. |
| | Priority routes for infiltration. |
| | Changes in temperature gradients. |
| Flora | Number and identity of species (especially protected species that might be damaged). |
| | Areas in danger of contamination. |
| | Areas with increased potential for fires or flooding. |
| Fauna | Number and identity of species, especially protected species that might be affected. |
| | The existence of protected species and migration routes. |
| Landscape | Visual quality. |
| | Changes to and/or destruction of existing structures. |
| | Volume of earth movements. |
| | Type and location of embankments and cleared areas. |
| Noise and vibration | Noise levels adjacent to inhabited properties. |
| | Location of inhabited properties. |
| | Location of structures susceptible to vibration damage. |

**Table 12.2** *(continued)*

| Environmental factors | Potential impact indicators |
| --- | --- |
| Land use | Surface area of site to be reclassified for free use following decommissioning. |
| Infrastructure | Energy supply and use. Changes in the transport network and associated traffic levels. |
| Human elements | Doses due to levels of radiation in the area and from radioactive effluents. Areas with high levels of noise. The inputs of atmospheric contaminants. |
| Population | Changes in population distribution and demographic composition |
| Economy | Level of employment. |

In the discussions of the significance of various environmental factors, one must take into account the views of various stakeholder groups which may lead to variations in weighting factors. As different stakeholders may have different perception of risks, attention should be given to the assessment and presentation of risks. A comprehensive study of the decision-making process covering Multi-Criteria Analysis (MCA) and Multi-attribute Utility Analysis (MUA) can be found in [6].

## 12.4.4 Mitigation Measures

For each decommissioning alternative investigated, mitigation measures should be included which would reduce impacts or risks. The effectiveness of mitigating measures can vary from scenario to scenario. Examples of mitigating measures are minimisation of atmospheric emissions by using high-efficiency filters, minimisation of releases to inland waters by recycling of waste water, control of landfills, adequate personnel training etc.

Although these measures, and others that might be specified for a particular project can be taken, it is necessary to have sufficient details to demonstrate not only that attempts have been made to carry them out but also that further improvements may not always be justified. Once the preventive and mitigation measures have been defined, the full impact generated by the project can be re-assessed.

## 12.5 ES Preparation

When all the above mentioned information has been made available, the Environmental Statement (ES) can be written. An ES is a report which provides a full description of the proposed project and its full impact, particularly on the environment. The primary aim of the ES is to provide sufficient information to two groups: decision makers and people potentially affected by a project. The most important aspect of the ES is that it should be clear and transparent to these audiences.

The desirable characteristics of an ES are

- It has a clear structure with a logical sequence describing existing baseline conditions, predicted impacts (nature, extent and magnitude), scope for mitigation, agreed mitigation measures, significance of unavoidable/residual impacts for each environmental topic and monitoring plan.
- There is a table of contents at the beginning of the document.
- A clear description of the consent procedure and how the EIA fits within it is given.
- There is a clear explanation of complex issues.
- A good description of the methods used for the studies of each environmental topic.
- Each environmental topic should be covered in a way which is proportionate to its importance.
- It provides evidence of good consultations.
- It includes a clear discussion of alternatives.
- It makes a commitment to mitigation and to monitoring.
- It has a non-technical summary for a wider audience.

This description of the proposed plant must include, but not limited to the following:

- physical, geographic and topographic data (e.g. location of the site, access routes, geographical area)
- plant description. For a nuclear power plant, for example, it should include the number and type of the reactor and electricity generating systems, containment building, cooling system, water usage, waste treatment systems etc.
- foreseeable status of the installation at the beginning of decommissioning

In some member states of the EU, a draft ES is provided first for independent peer review. This is followed by the submission of the reviewed ES to the competent authority.

## 12.5.1 ES Review

Review is the process of establishing whether an ES is adequate for the competent authority to make its decision [7]. The decision usually involves consideration of other issues, in addition to the environmental information, but the aim of review is to check that the environmental information is adequate.

In some member states of the EU, review of the adequacy of ES before they are used for decision making is mandatory in the EIA procedure. In those cases the review may be undertaken by the competent authority itself or by an independent organisation on behalf of the competent authority. Where the ES is considered to be inadequate, the developer will be asked to provide additional information and the decision-making process will not begin until all this information has been provided. There will usually be a defined procedure for the transfer of information. On the other hand, if the developer considers that excessive demands for information have been made, there is an appeal procedure against such demands.

In other member states, there is no formal requirement for review in the EIA procedure but competent authorities usually undertake some sort of review before starting the decision-making process to ensure that the requirements of the legislation have been met. They have the power to ask for further information from the developers before the decision-making process starts, if they consider the ES to be inadequate.

Review may also be undertaken informally by the developer prior to submitting the ES to the competent authority or by consultants after it is submitted, to check that the information is adequate.

The word adequacy means completeness and suitability of the information. In particular, it is aimed at helping reviewers decide whether the information meets the two main objectives of

- providing decision-makers with all the environmental information necessary for them to make their decision.
- communicating effectively with consultants and the general public so that they can comment in a useful manner on the project and its environmental impacts.

The checklist cannot verify whether the information meets legal requirements. This can only be done by considering the national legislation of the specific member state. It is also unable to verify the technical or scientific quality of the information or the adequacy of the environmental studies that have gone into its preparation.

## 12.5.2  ES Decision

After reviewing the ES, the competent authority will decide if the environmental implications of the planned decommissioning project are acceptable. The competent authority communicates the decision to the developer as soon as possible.

# 12.6  Best Practicable Environmental Option

## 12.6.1  The Concept of Best Practicable Environmental Option

The term BPEO, originally introduced by the Royal Commission on Environmental Pollution (RCEP) in 1976, is a systematic assessment methodology for optimising pollution damage and control to the environment. The RCEP proposed that polluting releases to the environmental medium should be directed in such a way that least overall environmental damage would be done. Therefore, an integrated approach whereby the damage from polluting releases to all three environmental media (land, water and air) is taken into consideration. This concept was incorporated into Part I of the Environmental Protection Act 1990 (EPA90) [8] where the Integrated Pollution Control (IPC) authorisation was based on Best Available Techniques Not Entailing Excessive Cost (BATNEEC). Subsequently, the RCEP defined BPEO in 1988 as 'the outcome of a systematic consultative and decision-making procedure which emphasises the protection and conservation of the environment across land, air and water'. The BPEO procedure establishes, for a given set of objectives, the option that provides the most benefits or least damage to the environment as a whole, at acceptable cost, in the long term as well as in the short term.'

The BPEO concept then found its way into the management of disposal and dispersal of radioactive wastes and radioactive discharges. In October 2000, the then Department of the Environment, Transport and the Regions (DETR) (the environmental role of DETR has been taken over by the DEFRA) issued a guidance note to the EA on the BPEO. It stated: 'Radioactive discharges may arise in different physical forms, but not necessarily be discharged in the form in which they arise. The EA, before granting discharge authorisation, needs to be clear that alternatives, where they exist, are properly evaluated and the choice is made that will have a low environmental impact, i.e. BPEO is chosen'.

Thus the BPEO is an assessment methodology whereby the most effective environmental option out of a host of environmental options is chosen such that the least environmental damage is done. The quantification of environmental damage is complex and difficult. Section 12.3 has given some indication of the processes involved. In the context of radiological protection, BPEO is about achieving doses (both to the critical group as well as collective dose to the population) that are ALARP (see Section 7.6). The optimisation objective of the BPEO can be achieved by utilising the BPM for the dispersal and disposal operation of radioactive waste. Thus BPEO can be considered as a broad-brush industrial option whereas the BPM is a specific operational option within the overarching BPEO. Within a BPEO, various BPM can be applied.

## 12.6.2 Assessment Technique

As BPEO assessments involve considerations of various attributes or criteria, a methodology known as the MCA or MUA may be applied. But before going into the details of MCA or MUA techniques, it is worth decomposing the term BPEO and deciphering the meaning of the words that make up the BPEO.

(i)   First of all, the word 'Best' signifies most effective, or most desirable or optimised option.

(ii)   The word 'Practicable' identifies an option out of a host of options when considerations are given to the following aspects:

- regulatory requirements
- government policy requirements
- international obligations
- workers' health and safety
- public health and safety
- technical feasibility
- economic feasibility
- environmental impacts
- public acceptability

(iii) The term 'Environmental Option' signifies that considerations are given to the following aspects which have environmental impacts:

- minimisation of waste generation
- treatment of waste
- design and operation of waste treatment system
- conditioning of waste
- design and operation of waste conditioning process
- storage/disposal of waste

Each of these considerations constitutes an attribute or criterion. To achieve a BPEO, these attributes need to be considered, evaluated and their overall impact assessed. The assessment technique, as suggested by the RCEP in 1988, is that seven distinct stages be devoted to this task and these are

(1)   Define the objective.

(2)   Identify various options.

(3)  Evaluate these options.
(4)  Summarise and rank these options.
(5)  Select the BPEO.
(6)  Review the BPEO.
(7)  Implement and monitor the BPEO.

The objective or objectives of the BPEO exercise needs to be defined at the outset (stage 1). In stage 2, various possible options to complete the project at hand are then stated. These options are independent, separate roadmaps for the completion of the project. Stage 3 is the process of evaluating these options. If radioactive waste disposal is the project under consideration, then the evaluation of the option should concentrate on estimating the radiological consequences to the public in the short term as well as in the long term. Many other attributes such as minimisation of industrial hazards, minimisation of monetary costs, maximisation of benefits etc. may be taken into account. Following the evaluation exercise, each option is given a score against each of the attributes considered. This is a somewhat subjective process and so 'expert elicitation' may be sought. A method for weighting the attributes against each other and then aggregating them for each option is carried out in stage 4. Following this stage, these options are ranked to give an indication of their desirability. Once the identified options have been ranked, it is up to the decision makers, such as the public inquiry inspector or the top management of the developer etc., to choose the best option. The review of the selected option, if necessary, and finally implementation of the option are the last two stages of the process.

# Revision Questions

1. Describe the origins of the Environmental Impact Assessment (EIA) and the Best Practicable Environmental Option (BPEO) methodologies. In which projects is the EIA obligatory and in which is it advisory?

2. How is the EU directive on EIA incorporated in the UK in the decommissioning of nuclear reactors? Describe briefly the regulatory requirements of this piece of legislation.

3. The EIA is a phased process. What are the steps of this process? Briefly describe the main elements of these steps.

4. In some EU member states threshold criteria are applied to avoid a full screening procedure. What are these threshold criteria? Briefly describe them.

5. When implementing the EIA, an environmental baseline needs to be drawn. What is this environmental baseline? Briefly describe it.

6. What procedure is normally followed to identify the environmental impact of a proposed project? Show the structure of the impact identification matrix and describe how it helps to identify impacts.

7. Following impact identification, what steps are taken in the preparation of an Environmental Statement (ES)? Describe the basic elements of this ES.

8. Describe the concept of Best Practicable Environmental Option (BPEO) as practised in the UK. How is Best Practicable Means (BPM) applied in the BPEO methodology?

## REFERENCES

[1] Council Directive 85/337/EEC on the Assessment of the Effects of Certain Public and Private Projects on the Environment, *Official Journal of the European Union*, No. L 175, 5 July1985, p.40.

[2] Council Directive 97/11/EC of 3 March 1997 amending Directive 85/337/EEC on the Assessment of the Effects of Certain Public and Private Projects on the Environment, *Official Journal of the European Union*, No. L 73, 14 March 1997, p.5.

[3] UK Government, Nuclear Reactors (Environmental Impact Assessment for Decommissioning) Regulations 1999, Statutory Instruments 1999 No. 2892, 1999.

[4]  European Communities, Guidance on EIA – Screening, Office for Official Publications of the European Communities, Luxembourg, 2001.

[5]  Environmental Impact Assessment for the Decommissioning of Nuclear Installations, EUR 20051 EN, Final Report Volume 2, Revised edn. February 2002.

[6]  Rahman A, Multi-attribute utility analysis – a major decision aid technique, *Nuclear Energy*, 2003, 42(2), pp 87–93.

[7]  European Communities, Guidance on EIA – EIS Review, Office for Official Publications of the European Communities, Luxembourg, 2001.

[8]  UK Government, Environmental Protection Act 1990, http://www.opsi.gov.uk/acts/acts1990/Ukpga_19900043_en_1.htm.

# 13

# DECONTAMINATION TECHNIQUES

## 13.1 Introduction

Radioactive contamination may be present in materials in two forms: (i) loose contamination on the surfaces due to spillage of solid or liquid radioactive substances and/or from the deposition of airborne radioactive material; and (ii) fixed contamination arising from the activation of metallic components during the operational period. Loose contamination is easier to remove than the fixed contamination which is embedded in the material. In order to reduce activity levels on the surfaces of metal, plastic, concrete or any other material as well as in building structures, decontamination is carried out. Various decontamination techniques such as chemical decontamination and mechanical decontamination as well as novel techniques such as microwave scabbling, laser flashgun etc. are now available. The choice of a particular decontamination technique is dependent on the type of surface to be decontaminated, the nature of the contamination, amount of secondary waste likely to arise, the likely dose burden on the workforce etc. A selection of major decontamination techniques that are commercially available at the moment and their main applications are given in Table 13.1 [1]. Comprehensive descriptions of decontamination techniques are available elsewhere [1, 2].

## 13.2 Chemical Decontamination

Chemical decontamination is generally most effective on metallic and non-porous surfaces. The choice of chemical agents is dependent on the chemical nature of the contamination to be removed, the extent of contamination, type of contaminated surface as well as the volume of likely waste arising. It should be borne in mind that the secondary waste needs to be treated and, possibly, conditioned before storage or disposal and so the type of waste arising is a major consideration. Various chemical techniques and their applications are given in Table 13.2.

### 13.2.1 Strong Mineral Acids

The purpose of using strong mineral acids is to attack and dissolve metal oxide

**Table 13.1**  *Typical decontamination techniques and their main applications*

| Technique | Application | | |
|---|---|---|---|
| | Large volume and closed systems | Segmented parts | Building surfaces/ metal structures |
| **Chemical decontamination** | | | |
| Chemical solutions[a] | x | x | x |
| Multiphase treatment | x | x | x |
| Foam decontamination | x | – | x |
| Chemical gels | x | x | x |
| **Mechanical decontamination** | | | |
| Flushing with water | x | x | x |
| Vacuuming/wiping/scrubbing | – | x | x |
| Strippable coatings | x | x | x |
| Steam cleaning | – | x | x |
| Abrasive cleaning | – | x | x |
| $CO_2$ blasting | – | x | x |
| High pressure liquid nitrogen blasting | – | x | x |
| Wet ice blasting | – | x | x |
| High pressure and ultrahigh pressure water jetting | x | x | x |
| Grinding/shaving | – | x | x |
| Scarifying/scabbling/planning | – | – | x |
| Milling | – | x | – |
| Drilling and spalling | – | – | x |
| Expansive grouting | – | – | x |
| **Other decontamination techniques** | | | |
| Electropolishing | x | x | – |
| Ultrasonic cleaning | – | x | – |
| Melting | – | x | – |
| **Emerging technologies** | | | |
| Light ablation | – | x | x |
| Microwave scabbling | – | – | x |
| Thermal degradation | – | – | x |
| Microbial degradation | x | – | x |
| Electromigration | – | – | x |
| Supercritical fluid extraction | – | – | x |

x  Technique is utilised     – Technique is not utilised

(a) Chemical solutions include strong mineral acids, acid salts, organic acids, bases and alkaline salts, organic solvents and complexing agents. These chemical solutions are given in Table 13.2.

films of contamination and lower the pH of solutions to increase solubility or ion exchange of metal ions.

**Table 13.2** *Chemical decontamination techniques and their applications*

| Chemical agents | Application material/surface |
|---|---|
| Strong mineral acids | |
| Nitric acid | Stainless Steel (SS), Inconel |
| Sulphuric acid | Carbon Steel (CS), SS |
| Fluoroboric acid | Metal and metallic oxides |
| Acid salts | Metal surfaces |
| Organic acids | Metal and plastic surfaces |
| Bases and alkaline salts | CS |
| Organic solvents | Metal, plastic, concrete |
| Complexing agents | Metals |
| Multiphase treatment process (REDOX, LOMI, MODIX) | CS, SS, Inconel, Zircaloy |
| Foam decontamination | Porous and non-porous surfaces |
| Chemical gels | Porous and non-porous surfaces |

### 13.2.1.1 Nitric Acid

Nitric acid is widely used to dissolve metallic oxide films on stainless steel and inconel surfaces. As it is highly corrosive, it can cause some damage to the surface. However, during the process of decontamination in decommissioning, surface damage is of little or no concern. But this technique should not be applied to decontaminate surfaces of operational systems.

### 13.2.1.2 Sulphuric Acid

Sulphuric acid is an oxidising agent and is used to a limited extent to remove surface contaminants that do not contain calcium compounds. In France, sulphonitric acid has been tested successfully. Sulphuric acid with cerium solution has been tested successfully at the Japan Power Demonstration Reactor (JPDR).

### 13.2.1.3 Fluoroboric Acid

Fluoroboric acid reacts with nearly every metal surface and metallic oxide. However, it is not as aggressive as nitric acid and hence thin layers of contaminated surfaces can be removed by this process (DECOHA process) without much risk of surface damage.

## 13.2.2 Acid Salts

The salts of various weak and strong acids may be used on their own or, more effectively, in combination with the acids to decontaminate metal surfaces. Examples of such salts are sodium phosphate, sodium sulphate or bisulphate, ammonium oxalate, sodium fluoride and ammonium bifluoride.

## 13.2.3 Organic Acids

Organic acids are used quite extensively to decontaminate metal surfaces as well as plastic or polymeric surfaces, particularly during the operational phase. Examples of organic acids are formic acid, oxalic acid, oxalic peroxide and citric acid.

Formic acid has been used in Slovakia in its A1 NPP decommissioning project to decontaminate metal oxides by mixing it with complexing agent and corrosion inhibitor and simultaneously agitating by ultrasound. Surface decontamination from $10^3$–$10^4$ Bq.cm$^{-2}$ to below release level has been achieved [3]. Oxalic acid is effective in removing rust from iron and is an excellent complexer for niobium and fission products. Citric acid is very effective in decommissioning stainless steel following alkaline permanganate treatment.

## 13.2.4 Bases and Alkaline Salts

These compounds are used on their own or in solution with compounds to remove grease and oil films, to remove paint coatings, to remove rust from mild steel and to neutralise acids. Examples of such compounds are potassium hydroxide (KOH), sodium hydroxide (NaOH), sodium carbonate ($Na_2CO_3$), ammonium carbonate ($(NH_4)_2CO_3$) etc.

## 13.2.5 Organic Solvents

These chemicals are used to remove organic materials such as grease, oil, paint and wax from metals, plastics, building surfaces etc. These solvents include kerosene, tetrachloroethane, trichloroethylene, xylene and alcohols.

## 13.2.6 Complexing Agents

Complexing agents such as diethylene triamine penta-acetic acid (DTPA), ethylene diamine tetraacetic acid (EDTA), hydroxy ethylene diamine tri-acetic acid (HEDTA) etc. are used to form stable complexes with metal ions and prevent their redeposition on the surfaces from the solution.

## 13.2.7 Multi-phase Treatment Process

Multi-phase treatment process uses a number of chemicals and chemical processes to achieve decontamination. The main processes are described below.

The REDucing OXidising (REDOX) process is a multi-stage decontamination process. It works on the principle of initially oxidising or increasing the oxidation states of metal ions by injecting alkaline or acidic permanganate followed by a reducing stage in order to dissolve superficial metal oxide layers of contaminated metals.

The Low Oxidation state Metal Ions (LOMI) process is similar in principle to the REDOX process but less aggressive. It was developed for decontaminating operational Steam Generating Heavy Water Reactor (SGHWR) internals at Winfrith, UK. It may be applied to structural materials such as carbon and stainless steel, Inconel and Zircaloy.

Multi-stage OxiDative by Ion Exchange (MODIX) is a four-stage process which is applied quite extensively to reduce contamination at the primary circuit of a PWR. The first stage involves injecting potassium permanganate ($KMnO_4$) solution to oxidise metal ions (Cr and others in corrosion films) which would then dissolve over a wide range of pH values. The second stage involves neutralising excess $KMnO_4$ solution by injecting disodium EDTA ($Na_2EDTA$) acid. The next stage involves injecting a mixture of nitric acid and ammonium citrate to dissolve metal ions. Finally, the dissolved cations are removed from the solution by ion exchanges in the resin in ion exchange column.

### 13.2.8 Foam Decontamination

Foam, such as that produced by detergents, is used on its own or as a carrier for chemical decontaminant. It can be applied very effectively on large porous or non-porous surfaces of complex shapes, both internally and externally. The foam generation equipment is simple and cheap and it can be used manually or remotely. Foaming equipment was developed in the UK and was used in the maintenance bay at the DIDO facility at Harwell. The advantage of this process is that the secondary waste volume is low.

### 13.2.9 Chemical Gels

Chemical gels containing chemical decontamination agents are sprayed or brushed onto a component or surface and then scrubbed, wiped, rinsed or peeled off. The method requires long contact times. This technique using aggressive chemical agents can achieve a high level of decontamination. The advantage is its small secondary waste production but the disadvantage is that it requires a long contact period for the gel to work.

## 13.3 Mechanical Decontamination

Mechanical decontamination can be used for surface or sub-surface decontamination. Table 13.3 gives a list of various mechanical decontamination techniques.

### 13.3.1 Flushing with Water

Water acts by dissolving certain chemical species and so by flushing contaminated surfaces with water, decontamination can be achieved. This technique is used when the surface area is too large for wiping or scrubbing. However, a drawback of this process is that a large amount of secondary liquid waste is produced.

### 13.3.2 Vacuuming/Wiping/Scrubbing

The standard technique of vacuuming (suction cleaning) may be used to remove loose contaminants from surfaces. The technique may be applied manually or remotely. If high level of loose contaminant is present, then remote suction is preferable in order to reduce the dose burden to the workforce. For large air volumes, a ventilation system with air filters is used to remove dust, aerosol or particles in the air. Small surfaces, with or without chemicals, are wiped or scrubbed. This technique may be considered as a pre-treatment for other techniques.

### 13.3.3 Strippable Coatings

This technique consists of an application of polymer and decontaminant mixture to a surface and then stripping off the stabilised polymer layer after it has set. The removed polymer layer incorporates surface contaminants. This

**Table 13.3** *Mechanical decontamination techniques and their applications*

| Mechanical techniques | Applications |
| --- | --- |
| Flushing with water | Large area: too large for wiping or scrubbing |
| Vacuuming/wiping/scrubbing | Large surfaces and air volumes for vacuuming, small surfaces for wiping/scrubbing |
| Strippable coatings | Large, non-porous, easily accessible surface |
| Steam cleaning | Large surfaces, complex shapes |
| Abrasive cleaning | Metal and concrete surfaces. With $CO_2$ pellets, ceramics, plastics, composites, CS, SS etc. |
| High pressure water jetting | Metal, concrete |
| Grinding | Floors and walls |
| Scarifying/scabbling | Concrete surfaces |

**Fig. 13.1** *Removal of strippable coating (Courtesy IAEA TRS 395 and Argonne National Laboratory, USA)*

technique may be applied to reasonably large, non-porous, easily accessible surfaces. Figure 13.1 shows the application of strippable coatings to decontaminate a surface.

## 13.3.4 Steam Cleaning

Steam cleaning utilises the solvent action of hot water along with the blasting effect of water particles. This process can remove contaminants, even in the presence of oil and grease, from large or complex surfaces. The equipment is simple and inexpensive. The secondary waste volume may be minimised by ventilating and then condensing steam.

## 13.3.5 Abrasive Cleaning

This process utilises materials such as plastic, glass beads, steel beads, grits, aluminium oxide pellets etc. to blast onto a surface using water or compressed air as the propellant. The process can remove coatings or fixed contaminants from metal or concrete surfaces. The volume of secondary waste could be quite large as a significant amount of sub-surface material could also be scraped off, particularly from concrete surfaces.

A variation of the above process is the use of solid $CO_2$ pellets in place of grits to blast onto the surface. A reasonable thickness of surface removal can

**Fig. 13.2** *High pressure water jet being used to decontaminate structural steel at Gentilly, Canada. Courtesy IAEA TRS 395.*

be achieved, but brittle materials may shatter. The secondary waste volume is low as gaseous $CO_2$ can be removed by a ventilation system.

### 13.3.6 High Pressure Water Jetting

A high to very high pressure water jet from $10^5$ Pa (1 bar) to over $10^8$ Pa (1000 bar) may be used to remove contamination from surfaces. This technique is very useful for inaccessible places, high surface contaminated areas such as interiors of cells and caves and underwater surface contaminated areas etc. However, the secondary waste volume could be significant and hence a recirculation and water treatment system may be necessary. Figure 13.2 shows the use of a high pressure water jet to decontaminate structural steel at Gentilly-1, Canada.

### 13.3.7 Grinding

Grinding utilises diamond grinding wheels or multiple tungsten carbide surface discs to chip off surface contamination, particularly from concrete surfaces. It can be used remotely to grind thin layers from contaminated horizontal or vertical surfaces. A ventilation system is usually installed to remove dust and airborne particles. Figure 13.3 shows grinding of a concrete wall surface using a diamond tipped rotary cutting head.

### 13.3.8 Scarifying/Scabbling

Scarifying/scabbling works on the principle of using scabblers, consisting of several pneumatically operated piston heads, which strike the surface simul-

**Fig. 13.3** *Concrete decontamination using an automatic wall shaver equipped with a diamond tipped rotary cutting head. Courtesy IAEA TRS 395.*

**Fig. 13.4** *Decontamination of a floor using a scabbler at JPDR decommissioning project (Courtesy IAEA TRS 395)*

taneously. A different type of scabbler consisting of a set of uniform needles, several millimetres long, is also pneumatically driven. This process can be used to scarify concrete floors and walls. Dust and airborne particles are normally ventilated. Figure 13.4 shows the use of a scabbler in the JPDR decommissioning project.

## 13.4   Other Decontamination Techniques

### 13.4.1  Electro-polishing

Electro-polishing works on the principle of anodic dissolution of a controlled amount of a contaminated metal surface in a closed circuit. The surface must be conducting, which means that there must not be any paint or surface coating. The components can be immersed in a bath of fluid or treated in situ using closed circuit systems which can be deployed manually or by manipulators. The choice of electrolyte is important. Typical electrolytes are based on nitric acid, phosphoric acid and organic acid.

### 13.4.2  Ultrasonic Cleaning

Ultrasonic cleaning is utilised to decontaminate small metallic objects submerged in a cleaning solution. The energy of the high frequency ultrasound is converted into mechanical vibrations of the cleaning fluid particles which scrub loose deposits from the surface. Good decontamination of the surface may be achieved by using aggressive chemicals with ultrasound.

## 13.5   Emerging Technologies

The following decontamination techniques are now emerging as state-of-the-art technologies. Some have already undergone field tests whereas others are still being tested. However, they all show promise for future use.

### 13.5.1  Light Ablation

This technique relies on the absorption of concentrated light energy by a surface material and its conversion to heat to remove surface coatings, adhesives, corrosion products and deposited airborne contaminants. A laser beam has been used for etching and ablation. Three types of laser have been tested by the Idaho Chemical Processing Plant (ICPP) and these are continuous wave $CO_2$, Q-switched Nd and Excimer using krypton fluorine gas. Some laser and xenon flashgun sources are now commercially available.

Laser decontamination has advantages over other techniques as: (i) it is a dry process, so production of secondary waste is minimal; (ii) by increasing the energy of the laser beam, thicker surface layers (up to about 6 mm) can be removed from concrete surfaces; (iii) it can be selectively and remotely applied using optical fibres.

## 13.5.2 Microwave Scabbling/Scarifying

This technique depends for its operation on the absorption of microwave energy by the surface to heat up the moisture present in the concrete matrix and the moisture which has been turned into steam scarifies the surface. Depending on the absorbed energy and the moisture content, reasonable thickness of concrete surface may be removed by this technique. However, the disadvantage is that the process does not work well in a dry concrete matrix. This was found when this technique was applied to dismantle the biological shield of the DIDO plant at Harwell. The advantage, however, is that the secondary waste arising is very low.

## 13.5.3 Microbial Degradation

In this process, a microbial solution is applied to the contaminated surface. The microbes come into contact with the contaminants, consume them and penetrate the surface. After a while, a solvent wash is applied to remove the reaction products and the microbes with contaminants are removed. This technique is most effective in decontaminating abandoned process equipment, storage tanks, piping etc.

# Revision Questions

1. What is meant by decontamination? Describe the main techniques for decontamination.

2. On what surfaces is chemical decontamination most effective? Give some examples of chemical agents and describe their application.

3. Describe the principles and application procedures of
    (i)   MODIX operation
    (ii)  chemical gels
    (iii) strippable coatings
    (iv)  ultrasonic cleaning
    (v)   microwave scabbling

## REFERENCES

[1] International Atomic Energy Agency, State of the Art Technology for Decontamination and Dismantling of Nuclear Facilities, Technical Reports Series No. 395, IAEA, Vienna, Austria, 1999.

[2] European Commission, Handbook on Decommissioning of Nuclear Installations, Rep. EUR 16211, Office for Official Publications of the European Communities, Luxembourg, 1995.

[3] American Society of Mechanical Engineers, Recent trends in the area of the decontamination of nuclear power plants in the Slovak Republic and in the Czech Republic, Proceedings of International Conference on Nuclear Waste Management and Environmental Remediation, Prague, Czech Republic, 5–11 September 1993, pp. 301–306.

# 14

# DISMANTLING TECHNIQUES

## 14.1   Introduction

The dismantling operation is normally carried out after the decontamination of systems, components, shielding materials, building structures etc. in a nuclear facility. However, dismantling or cutting of some large items may also be carried out prior to decontamination in order to reduce sizes for decontamination. Dismantling thus covers segmentation of metal items (such as reactor pressure vessel, steam generators/boilers, piping etc. in a nuclear power plant or other large items in a nuclear facility) and demolition of building structures and fabrics of the facility. The whole operation is considered as the final phase of the D&D operation.

This chapter deals with various dismantling techniques which can be itemised as follows [1]:

- mechanical cutting techniques
- thermal cutting techniques
- abrasive water jet cutting
- electrical cutting techniques
- emerging technologies
- mechanical demolition techniques

The choice of technique depends on the size and complexity of the operation as well as on the end-point objectives to be achieved. In addition, there are some constraints to be met. The constraints of the dismantling process are that the techniques should produce a minimum of secondary wastes, offer a low dose burden to the workforce, be efficient in operation and be cost effective.

## 14.2   Mechanical Cutting Tools

Mechanical cutting tools are used to cut, grind, fracture or split an object by direct mechanical means. There are a number of cutting techniques available, a selection of which is shown in Table 14.1. With the exception of the explosive cutting, these techniques produce secondary waste streams which can easily be collected by extraction systems. In general, they produce much lower airborne contamination than thermal cutting, but the cutting speeds are generally lower.

**Table 14.1** *Mechanical cutting techniques and their applications*

| Cutting technique | Application | Comments |
| --- | --- | --- |
| Shears | Cutting metals | Can be used manually or remotely, in air or underwater |
| Mechanical saws | Cutting metals | Can be used manually or remotely, in air or underwater |
| Orbital cutters | Metals | Can be used manually or remotely, in air or underwater |
| Abrasive cutting wheels, blades, wires | Cutting metals, concrete | Can be used manually or remotely, in air or underwater |
| Explosives | Dismantling concrete and metal structures | Detonated in air or underwater |

## 14.2.1 Shears

The operation of shears is based on the principle of applying pressure on an object by one arm while the other arm remains fixed, rather like the action of a pair of scissors. They can be operated manually, pneumatically, hydraulically or electrically and can be used in air or underwater. In terms of construction, there are basically three types of shears:

- A two-bladed device which works like a pair of scissors. These are normally used for cutting small diameter pipes, thin metal plates etc. They are small and lightweight and hence can easily be handled by workers. They can be operated manually or remotely. Use of these types of shears had been made at the Windscale AGR (WAGR) decommissioning project where the tubes were filled with cement grout to reduce tube deformation and thereby reduce overall cutting force. Another improvement that has been effected in the technique is the development of crimp shear where a tube is cut and the ends are crimped at the same time. Thus the ends are sealed and contaminants cannot leak out of the pipe.
- A blade and anvil device where the blade forces through the work piece resting against the fixed anvil. They are bigger and heavier than the scissor-type shears and hence can cut metal pipes of larger cross-section and thickness. They can be operated manually or remotely.

- Demolition shears which can be of the anvil or scissors type but are much larger and heavier. They are used for sectioning structural steel or for crushing concrete.

## 14.2.2 Mechanical Saws

Various types of mechanical saws are being used in decommissioning work. They range from small hand-held hacksaws to very large and heavy bandsaws for cutting large structures like steam generators or turbines. There are basically three types of mechanical saws: reciprocating saws, circular saws and bandsaws.

Reciprocating saws work on the principle that as the saw blade moves forwards and backwards against an object, it splits the object. The saw blade may be supported by a frame at one or both ends. Figure 14.1 shows a reciprocating saw cutting a heavy metallic cylinder. The simplest reciprocating saw is the hacksaw. The complex and heavy duty saws designed to cut hard, thick materials such as stainless steel are normally fixed at a place and the materials are brought to it. They can be operated in air or underwater and are pneumatically driven. However, the cutting speeds for hard metals could be relatively slow compared to other techniques such as plasma arc cutting. Nonetheless reciprocating saws are very widely used.

A circular saw consists of a disc saw blade which is rotated by a motor to carry out the cutting operation. Depending on the thickness of the plate to be sheared, a circular saw can have a small disc with teeth that are a few millimetres deep up to a wheel as large as 1–2 m in diameter. They can be operated in air or underwater. The large circular saws may have diamond-tipped blades to cut reinforced concrete or even metal structures. A variation of a circular saw is the orbital cutter which can be manually operated or self-propelled as it cuts and moves inside or outside the circumference of a vessel. It is particularly useful in cutting large active surfaces, as the cutting is done automatically, but it needs to be positioned manually at the beginning.

A bandsaw consists of a loop saw blade supported by a frame which allows the loop to circulate and a motor which drives the loop. They can be small portable ones for cutting thin tubes and plates or very large ones for cutting large structures such as steam generators. They can be used in air or underwater for highly activated components and can cut horizontal as well as vertical structures.

## 14.2.3 Orbital Cutters

Orbital cutters operate on the basis that the cutter goes around the outside or inside circumference of the cutting object. The cutting object can be a pipe or a vessel with large dimensions. These units can be manually actuated or

**Fig. 14.1** *Reciprocating saw cutting a heavy metallic cylinder*

self-propelled as they go around the circumference of the cutting object such as steam generators, pressure vessels etc. A small pipe can be cut by moving a hardened wheel which compresses and shears the outside diameter of the pipe.

## 14.2.4 Abrasive Cutting Wheels, Blades and Wires

These devices use electrically, hydraulically or pneumatically operated wheels, blades or wires containing abrasive materials. Typical abrasive materials include aluminium oxide, silicon carbide or diamond. They can be used either dry or with a liquid coolant which is normally recirculated to reduce secondary wastes. The technique is widely used and is quite reliable.

## 14.2.5 Explosives

Explosives are used quite extensively to demolish or dismantle metallic or concrete structures.

Conventional explosives having low detonation speed and shock waves are used to dismantle large solid structures such as exhaust stacks, cooling towers, biological shields etc. Seed explosives are placed at pre-defined critical points and are then detonated remotely. This type of controlled blast has been used to dismantle the exhaust stack at JPDR and at the LIDO reactor in the UK.

Shaped explosives, on the other hand, having high detonation speed and shock waves are used to fracture large diameter pipes at precise points in a controlled manner.

Another use of explosive, termed as linear shaped charges, is to propel a V-shaped blade by the force of detonation into an object to cut it. This technique has also been used at the JPDR and at the LIDO in the UK.

# 14.3 Thermal Cutting Tools

Thermal cutting is used quite extensively in decommissioning work, both in air and underwater. This technique has certain advantages and disadvantages over other techniques such as mechanical cutting. The advantages are that the thermal cutting speed is generally faster and the equipment is much lighter and easily portable. The disadvantages are that it produces aerosol, dust and dross, which need to be removed so that the workplace remains clean and the workforce is not exposed to high doses of radiation. For underwater cutting, it is necessary to have a water filtration process to maintain water quality. There are two main types of thermal cutting: (i) flame cutting; and (ii) plasma arc cutting.

## 14.3.1 Flame Cutting

Flame cutting uses a flowing mixture of a fuel gas (acetylene, hydrogen and propane) or fuel vapour (gasoline) and oxygen which are mixed and ignited to produce a high-temperature flame. When cutting a mild steel surface, the flame is allowed to heat up before the cutting oxygen is injected into the centre of the flame oxidising it. As iron oxide melts at a lower temperature than the parent metal, the flame causes a cut. The technique can be used either manually or remotely in air or underwater.

## 14.3.2 Plasma Arc Cutting

Plasma arc cutting works on the principle that a direct current is set up between a tungsten electrode and the surface of a conducting metal. The high direct current ionises the intervening gas producing a plasma of ionised gas which is then blown towards the object which is to be cut. The heat of the plasma gas causes localised melting and the gas also blows the molten metal away, thereby creating a cut in the object. However, the process produces dust and aerosol in the air and hence there is a need for air filtration. The principle of plasma arc cutting is shown in Figure 14.2.

Plasma cutting is a fast process and the cutting head is light and easily manoeuvered. The unit can be operated either manually or remotely. The process can be used in air or in water. In air, the conducting gas as well as the electrode-cooling gas is air. As the cooling rate is lower than that of water, the power density in air is maintained at a lower level and hence the cutting speed is also lower than in water. Plasma arc in air technology and underwater plasma arc technology are both well developed and widely used.

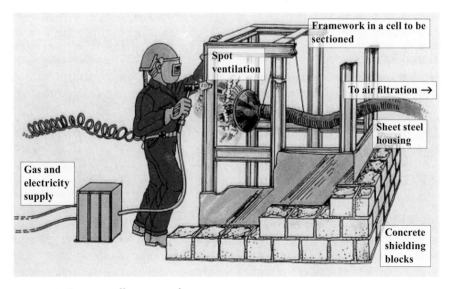

**Fig. 14.2** *Diagram illustrating plasma arc cutting operation*

## 14.4 Abrasive Water Jet Cutting

Abrasive water jet cutting involves the use of an abrasive material such as sand propelled by high pressure water. The technique is particularly effective in cutting reinforced concrete, although it can be applied to cut any material. However, this method produces a large amount of secondary waste which needs to be considered.

The efficiency of the tool can be increased by adding abrasives to the plain water jet. Two kinds of abrasive water jets are well known: The Abrasive Water Entrainment Jet or Abrasive Water Injection Jet (AWIJ) and the Abrasive Water Suspension Jet (AWSJ), which can be generated by different pumping principles. Figure 14.3 shows the variations in abrasive water jet techniques.

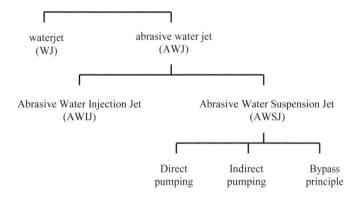

**Fig. 14.3** *Various abrasive waterjet techniques*

### 14.4.1 Abrasive Water Injection Jet

The AWIJ was developed in the late 1970s. Its main component is a mixing head, with an assembly of a water nozzle and a focusing or mixing tube. The water nozzle has a diameter of 0.2–0.5 mm and generates a plain water jet. This jet runs through the mixing chamber and generates a vacuum pressure. Abrasive particles are sucked into the chamber pneumatically through an opening. The abrasive and the water are mixed, accelerated and focused in the mixing tube, as shown in Figure 14.4.

### 14.4.2 Abrasive Water Suspension Jet

Another way to generate an abrasive water jet is the AWSJ which was developed in 1984 by the BHR-Group at Cranfield, UK. A high-concentration suspension is stored in a vessel under system pressure in the pressure circuit.

The main difference to the AWIJ is the absence of air in the jet. Part of the pressurised water is used to feed the highly concentrated suspension into the main water stream. The suspension can be transported via long high-pressure hoses to the cutting location (Figure 14.5).

As there is no air in it, this jet is much more efficient than the AWIJ. AWSJs are well known in the dismantling industry, though only a few applications for manufacturing purposes are currently known. State-of-the-art AWSJs have pressures up to 200 MPa. AWSJs at pressures of 400 MPa have been developed and are running under laboratory conditions.

### 14.4.3 Characteristics of AWIJ and AWSJ

The AWIJ consists of three phases (≈95% vol. air, ≈4% vol. water, ≈1% vol. abrasive), the AWSJ, however, only consists of two phases (80–90% vol. water,

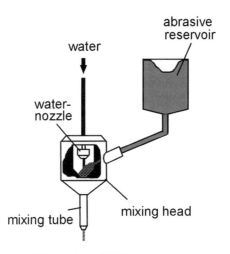

**Fig. 14.4** *The abrasive water injection jet – AWIJ*

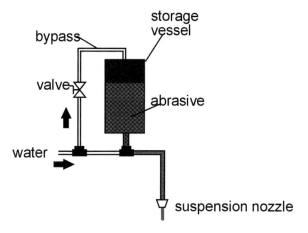

**Fig. 14.5** *The abrasive water suspension jet – AWSJ*

10–20% vol. abrasive). This leads to a better acceleration of the abrasive particles in an AWSJ. Therefore its cutting efficiency is at least twice as high as an AWIJ of the same hydraulic power and abrasive flow rate (see Figure 14.6). However, the AWIJ is commonly used with pressures as high as 400 MPa, which leads to similar performances to those of the AWSJ in terms of the depth of cut.

As AWSJs only consist of water and abrasive material, the particles are better guided than in an AWIJ. This leads to a higher jet stability and finally to an improved quality and efficiency of the cutting.

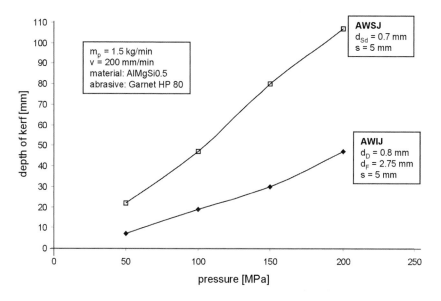

**Fig. 14.6** *Comparison of cutting efficiency of AWIJ and AWSJ*

**Table 14.2** *Advantages of the waterjet technology*

| Principles | WJ | AWIJ | AWSJ |
|---|---|---|---|
| Multi-functional tool | Cutting, drilling, turning, decoating, cleaning | | |
| Non-thermal process | No toxic reaction products | | |
| Omni-directional | Sharpness of the jet from every side | | |
| Almost all materials can be cut | 'Soft' materials | Metallic and ceramic materials | |
| | Homogeneous- and inhomogeneous material, composite materials | | |
| Small width of cut | > 0.1 mm | > 0.4 mm | > 0.3 mm |
| Achievable depth of cut | e.g. PVC 20 mm | e.g. steel 120 mm | e.g. steel 300 mm |
| Small and flexible tool | | | |
| Application in different environment | In air, under water, in explosive environment | | |
| Low reaction forces | 15–250 N | | |
| Low stand-off distance sensitivity | No focusing necessary | | |
| Natural resources | Water | Water and abrasive | |

To generate an AWIJ, a water jet leaves the water nozzle, passes the mixing chamber and then enters the focusing tube. This requires the focusing tube diameter to be at least twice (normally three to four times) as large as the water nozzle diameter. Therefore AWSJs of the same nozzle diameter, and consequently the same hydraulic power, produce lower cutting widths and higher cutting depths.

The advantages of the water jet technology in relation to thermal and conventional procedures are summarised in Table 14.2.

### 14.4.4 Abrasive Water jets for the Dismantling of Nuclear Power Plants

An example of dismantling by AWIJs is the biological shield of the JPDR in Japan. In addition to water jets, diamond saw and drill, an explosive technique was used to dismantle the reinforced concrete shield.

The performance of the abrasive water jet cutting technique combined with other new developments led to using this technique as an alternative decommissioning technique in the experimental nuclear power plant (Versuchs Atomkraft Work, VAK) at Kahl, Germany. Two research projects, the generation of higher working pressure as well as the application at VAK, Kahl, were sponsored by the German Federal Ministry of Education, Science Research and Technology. Within the project at Kahl the cutting of the lower core shroud (activated material to be cut underwater) and the development of a strategy to cut the reactor pressure vessel with AWSJs were planned.

The following two cutting strategies for the dismantling at Kahl were adopted: first, for cutting through to achieve total material separation, a setting angle for cutting of 15° was taken. Secondly, if other parts of the object were not to be cut and should only be minimally affected, the cutting angle could be increased up to 45° to minimise the effect of the gap.

For the kerfing a definite percentage of the material thickness (for example 95%) was used to prevent the surroundings being contaminated by impurities. The used abrasive and the machined (radioactive) material was collected inside the pressure vessel. Only when cutting the remaining wall thickness was a small percentage of abrasive and machined material ejected to the surroundings.

The first application of an AWSJ was at the nuclear power plant VAK, Kahl. Initially a 140 MPa cutting unit was used to cut the lower core shroud and the thermal shield. The reactor pressure vessel itself was cut by a 200 MPa unit. Further details are given in Table 14.3.

An advantage of abrasive water jet cutting compared to thermal cutting is the small amount of aerosol, the disadvantages the secondary waste. Both

**Table 14.3** *Cutting parameters of VAK Kahl*

|  | **Lower core shroud** | **Thermal shield** | **Reactor pressure vessel** |
|---|---|---|---|
| Material | X 6 Cr Al 13 | X 6 Cr Al 13 | Austenitic plated, ferritic steel, 19 Mn 5 |
| Material thickness | 51 mm (132 mm) | 32 mm | 104.5 mm (6.5 + 98) |
| Working pressure | 140 MPa | 140 MPa | 200 MPa |
| Water flow rate | 8–20 l.min$^{-1}$ | 8–20 l.min$^{-1}$ | 9.5–20 l.min$^{-1}$ |
| Abrasive flow rate | 1.3 kg.min$^{-1}$ | 1 kg.min$^{-1}$ | 1 kg.min$^{-1}$ |
| Cutting speed | 40 mm.min$^{-1}$ (13 mm.min$^{-1}$) | 65 mm.min$^{-1}$ | 25 mm.min$^{-1}$ |
| Total length of cut | 20 m | 70 m | 63.9 m |
| Total consumption of abrasive | 1000 kg |  | 2553 kg |

were quantified and analysed for kerfing and cutting through for application in air as well as underwater. Only a very small amount of waste is spread into the air as aerosols, most of the waste comprises sediment particles.

To manage the waste, a catcher and a special filtering device need to be installed. At the VAK, Kahl most of the abrasive material (97%) was directly filled into a special container. As small particles were held back by a special filter system, the water could be reused.

An optimised cutting process and the right cutting strategy are necessary to minimise the flow rates of abrasive.

## 14.5   Electrical Cutting Techniques

Electrical cutting techniques utilise metal evaporation as against thermal cutting techniques which melt metals. These processes do not generate any metal flow in the melt pool, but they do produce large amounts of aerosol in air or hydrosol underwater as compared to thermal cutting techniques.

### 14.5.1 Electro Discharge Machine

The Electro Discharge Machine (EDM) works on the principle of thermo-mechanical erosion of metals by the sparks that are generated. The technique can be used when a conducting material such as metal is to be cut and is ideally suited for underwater application. The technique was used to dismantle reactor internals at the BR3 project for cutting thermal shields (76.2 mm) and some delicate operations such as removing bolts which were difficult to access.

### 14.5.2 Arc Saw Cutting

Arc saw cutting uses a circular, toothless saw blade to cut conducting metals without making physical contacts. A high current electric arc between the blade and material being cut is maintained and as the blade is rotated, a cut is made. The process works most effectively in high conductivity materials such as stainless steel, copper, aluminium, Inconel etc. At JPDR, the technique had been applied under water and remotely operated for cutting the Reactor Pressure Vessel (RPV).

## 14.6   Emerging Technologies

### 14.6.1 Liquified Gas Cutting

Liquified gas cutting is similar in principle to water jet cutting, except that liquefied gas is used as the propellant instead of water. Liquid nitrogen has been used as the carrier medium in the USA. The advantages of this technique are that it produces hardly any secondary waste, there is no risk of fire or explosion and it can be remotely operated.

## 14.6.2 Laser Cutting

Laser cutting works on the principle that the concentrated laser beam falling on an object heats it up locally beyond its melting point, thus cutting it. The technique can be applied to cut any non-inflammable material. Both CO and $CO_2$ lasers have been tested for decommissioning work in Europe and an oxygen–iodine laser has been tested in Japan. In Japan, research work has been successfully conducted to transmit a laser beam to a remote location using fibre optics.

There are three basic cutting processes: laser flame cutting, laser fusion cutting and laser sublimation cutting. The application of the different processes depends on the material and the given demands.

With laser flame cutting, the additional exothermal reaction energy of the burning process is exploited. Thus, high cutting speeds are achieved as the release of energy during the burning process can be as high as the energy input of the laser beam. The process gas oxygen is applied, which starts the exothermal reaction. The material burns to a fluid slag, which drains off easily. The gas pressure is, therefore, set lower than that of the laser fusion cutting. The exothermal energy has to be limited in order to guarantee a controlled cutting process. An oxide layer is formed at the edge of the cut by the burning reaction. This layer only has a weak connection to the workpiece.

Laser fusion cutting is characterised by the use of an inert process gas, i.e. N or Ar. An additional exothermal reaction is prevented by the atmosphere of the shielding gas. Thus the process only works with the input of laser power and, therefore, the cutting speed is substantially lower than the laser flame cutting. In return, the cut edge is not oxidised after the processing. Further treatment of the workpiece is thus not necessary. However, the fusion resulting during the treatment is viscous. For that reason, the gas pressure has to be very high in order to expel the melt from the kerf and to realise a clean burr-free cut.

High-power densities are required for laser sublimation cutting in order to transform the material directly from the solid to the vapour phase. An inert process gas is also applied to prevent the material from burning. The gas pressure used depends on the material which is to be processed. Because of the high vaporisation temperature of metallic materials and the limitations of available power densities, only organic or plastic materials are treated using laser sublimation cutting. Metals are mainly processed by the other two cutting methods described above.

### 14.6.2.1 Use of Laser in Dismantling Techniques

Laser cutting is characterised by narrow cutting kerfs, sharp cutting contours, narrow heat affected zones, little or no distortion of the workpiece, stress-free treatment and high reproducibility. However, a high investment is necessary and the low efficiency of lasers is coupled with high energy consumption.

A considerably lower amount of particle emission is released in this process than in the plasma process. The comparison of the laser process with the plasma process relating to the Nominal Hygienic air requirement Limit (NHL) shows that less fresh air is needed to meet the tolerable limit values with the laser processing than with plasma processing. The NHL-value characterises the amount of fresh air which is theoretically necessary with a given emission mass flow in order to remain within the tolerable limit value for each hazardous substance in the air.

### Components of a Laser Dismantling System

Like conventional laser material processing, for the dismantling of nuclear power plants, a system consists of the laser source, beam delivery, beam shaping devices (lenses or mirrors), gas nozzle and process gas, as well as the handling system with controls. Because different laser types have different wavelengths, different beam delivery systems are necessary. The $CO_2$ laser beam can only be delivered to the workpiece by mirror systems, while there is the possibility of using fibre optics for Nd:YAG laser radiation. For that reason, for the dismantling of nuclear power plants, an Nd:YAG laser is much more flexible for the beam delivery. Beam shaping is influenced by optical components on the laser beam, such as the use of a focussing lens. It gives a characteristic form to the beam, in order to use, for example, the small focus of a beam for material processing.

### Use in nuclear facilities

Laser technology can be used in many areas of dismantling nuclear power plants. Experimental investigations on laser processing underwater have been carried out. First, the process gas is switched on, in order to protect the optical components and to prevent water from entering the processing head. The processing head is then dipped into the water where the cutting process is carried out. The water does not hinder the laser beam, as the gas beam displaces the water and guarantees the free propagation of the laser beam.

The laser also offers process-specific advantages for asbestos cutting. The release of cancerous fibrous aerosols can be significantly reduced by using the laser, as the asbestos is vaporised and condenses to harmless spherical particles in the air. At the same time, the cut edge of the asbestos material is glazed during the treatment. Thus, a durable sealing of the cut edge is guaranteed, and the release of remaining fibres from the asbestos material is prevented.

Emission-minimised cutting using special process parameters offers the possibility of attaching the molten material on the underside of the workpiece in the form of a burr. This reduces the release of emission products and contaminations. Tubes can also be cut by laser. Only the material and its thickness is decisive for the process. The production of a burr is also possible in this case.

When dismantling tanks or storage basins which consist of concrete walls lined with steel plates, cutting of the steel material is difficult. The metal sheets lie directly on the concrete, and it is very difficult to cut them mechanically.

Thermal cutting methods producing very deep fusion penetration in the adjacent concrete can lead to contamination of the concrete. In that case, a further treatment of the concrete would be necessary. Laser cutting offers specific advantages here. The energy input is very precise and can be controlled in depth so that the process can be adjusted exactly to the thickness of the workpiece. Fusion penetration in concrete can therefore be minimised. Moreover, there is the problem of the treatment of coated sheets. The sheets can be processed nearly without burning the coating. An additional advantage of this method is the separation of gas and laser beam. A special nozzle technique can be used to expel the molten material to the top surface of the sheet. A specific removal of the released process emissions by suction is also possible.

The mobility and flexibility of the laser is an important reason for its application in nuclear facilities. A condition for the realisation of these applications is the availability of handheld laser processing heads. For this reason, a device was developed at the Laser Zentrum (Centre in English) Hannover (LZH) which allows manual guidance. The system specifies the feed rate in order to guarantee a stable cutting process. With this device, which was specially constructed for the demands of dismantling nuclear facilities, programming and teach-in procedures are not necessary, as for example with the use of robots, which leads to significantly lower costs and saving of time. For areas with low level contamination, this system offers a useful alternative to other thermal cutting techniques, providing all the advantages of laser technology.

## 14.7 Mechanical Demolition Techniques

There are a number of techniques which can be applied to demolish building structures, solid surfaces etc. They are described below.

### 14.7.1 Wrecking Ball or Wrecking Slab

A wrecking ball is a conventional demolition technique used for demolishing lightly reinforced or non-reinforced concrete structures less than 1 m thick. As the process produces large amounts of dust and dross, it is not generally applied to radioactive buildings.

### 14.7.2 Expansive Grout

Expansive grout is used to fracture non-reinforced concrete by drilling holes and filling them with wet grout mixture. As the grout cures, it expands, thus creating an internal stress and eventual fracture of the concrete substrate.

### 17.7.3 Rock Splitting

Rock splitting has traditionally been used in the quarrying industry. It is a method of fracturing rock or concrete by hydraulically driving a wedge-shaped plug into a pre-drilled hole.

## 14.8 Developments in the EU

The EC under its R&D programme is in the process of creating a database on tools and costs to harmonise costs and share experiences amongst the EC Member States. Recently, the remit of this database, called EC DB NET2, has been extended to incorporate experiences and working practices of all 27 member states. Another recent development is that the EC, the OECD/NEA and the IAEA have jointly produced an interim document giving a standardised list of items for costing purposes in the decommissioning of nuclear installations [2]. This document categorises all decommissioning operations into 11 cost groups and lists cost items under each group. Further details of this standardised list of items can be found in Section 8.6. The plan is that all organisations within the EU will use this defined format for costing purposes and thereby generate a harmonious cost database. The EC DB NET2 is consistent with this document.

# Revision Questions

1. What are the currently available dismantling techniques? What dictates that a specific dismantling technique is chosen to perform a task?

2. List the main mechanical cutting tools and briefly describe them.

3. What are the main types of thermal cutting tools? Briefly describe them.

4. What are the advantages and disadvantages of thermal cutting techniques over the mechanical cutting techniques?

5. Describe the principle of abrasive water jet cutting. List the various types of abrasive water jet cutting technique and show them in a schematic diagram.

6. Describe the working principle of the Abrasive Water Suspension Jet (AWSJ) with a schematic diagram. Also compare the cutting efficiencies of AWIJ and AWSJ techniques.

7. What are the different types of laser cutting techniques? Briefly describe these techniques.

8. What are the various types of mechanical demolition techniques? Briefly describe them.

## REFERENCES

[1] International Atomic Energy Agency, State of the Art Technology for Decontamination and Dismantling of Nuclear Facilities, Technical Reports Series No. 395, IAEA, Vienna, Austria, 1999.
[2] OECD Nuclear Energy Agency, A Proposed Standardised List of Items for Costing Purposes in the Decommissioning of Nuclear Installations, NEA/OECD, 2001.

# 15

# CASE HISTORIES
# AND LESSONS LEARNT

## 15.1 Introduction

This concluding chapter on decommissioning presents an overview of decommissioning of some of the nuclear facilities in Europe covering procedures adopted, problems encountered, solutions found and lessons learnt. Although the fundamental requirements for decommissioning of nuclear facilities and its objectives remain the same in almost all countries, the implementation strategy differs from country to country. The details of decommissioning strategy encompassing waste management, accomplishment of the stated end-points etc. are dependent on the plant concerned and can vary even within the same country. Additionally, it should be noted that the decommissioning strategy that used to be the norm some years ago has now been modified due to changed priorities and regulatory requirements.

This chapter presents four decommissioning projects as case studies, pointing out in each case the scale of decommissioning activity, the problems that were encountered and the lessons that were learnt. Two of these projects: (i) AT-1 in France; and (ii) KRB-A in Germany were two of the four pilot projects undertaken by the EU within the Research and Technological Development (RTD) programmes. (The remaining two EU pilot projects were WAGR in the UK and BR3 in Belgium [1]). The other two decommissioning projects presented here are: (iii) decommissioning of Greifswald site in Germany, the largest decommissioning site in the world; and (iv) decommissioning of the world's first commercial nuclear plant in the UK (Calder Hall in West Cumbria).

## 15.2 EU Decommissioning Pilot Projects

### 15.2.1 Overview

The EU RTD funding provisions follow a five-year cycle under the framework programmes which started at the beginning of 1982. The primary objectives of these five-year programmes are to encourage scientific and technical activities in the EU so that the EU as a whole can develop a strong scientific and technological base and become competitive in the world arena. A key part of the EU RTD activities involves demonstrating the feasibility of decommissioning activities through pilot projects and contributing to improve advanced techniques and the reliability of decommissioning cost estimates. In particular, generation of specific data on costs, tools, dose uptake by workers and the amount of secondary waste generated are all considered to be important objectives for these pilot projects. Two of the four pilot projects that were selected under the EU RTD programmes will be described here giving their present status and the lessons that could be learnt from these activities.

## 15.2.2 AT1 Reprocessing Plant in France

The AT-1 plant, situated on the COGEMA site at La Hague near Cherbourg, France is a reprocessing plant for fast breeder reactor fuels. The plant was operated for 10 years (1969–1979) and then finally shutdown in 1979. Following the cessation of operation, there was a 12-month campaign to clean out the site and then an 18-month campaign to decontaminate the circuits.

From 1982, the CEA/UDIN (Unité de Démantèlement des Installations Nucléaires) was responsible for decommissioning the plant to the IAEA stage 3 level. Specifically the objectives were

- To dismantle and remove all contaminated circuits and equipment from the site.
- To decontaminate various shielded cells to a level that would allow unrestricted access to buildings.

### 15.2.2.1 Operational Period

During the operational period of the plant, the following reprocessing activities were carried out in concrete cells [2]:

| | |
|---|---|
| Cell 901 | spent fuel storage |
| Cell 902 | fuel cropping |
| Cell 903 | fuel dissolution |
| Cell 904 | first extraction cycle |
| Cell 905 | second/third extraction cycle |
| Cell 952 | fourth U/Pu separation cycle |
| Cells 950/951/906 | U and Pu concentration and Pu precipitation |
| Cell 907 | liquid effluent storage |
| Cells 920/908/909 | Fission product storage |
| Cell 911 | Transfer pipes and demisters |

Even after the plant wash out, radiation levels in the shielded cells were found to range from a few tens of mGy to 1 Gy. These high levels of radiation precluded any human access to cells 903 and 904 and only limited working time in cells 902, 905, 908 and 909. In addition, cells 903, 904 and 905 were completely blind, i.e. there were no windows or manipulators and hence it was necessary to design special equipment, called ATENA, to remotely dismantle these cells. The ATENA machine consisted of a carrier and an electrically actuated tele-manipulator. The carrier comprised a containment housing, a transfer carriage and a support arm which was a remotely controlled, multi-jointed arm which could be retracted into a steel hood. The hood worked both as a confinement and as a biological shield for the operators. The tip of the telescopic arm could be equipped with a cutting tool or with a MA23 M or RD 500 type remote manipulator.

### 15.2.2.2 Decommissioning Work

The planned sequence of operations for decommissioning was as follows:

- Dismantle unshielded α-cells and glove boxes associated with the fourth cycle (separation, concentration and recovery).
- Dismantle some shielded cells such as cell 911 and filtration for the installation of the ATENA remote dismantling machine.
- Replace the biological shield that covered cells 904 and 905 and opening up access to 904 and 905 for the ATENA machine with specially designed equipment.
- Dismantle three blind cells (cells 903, 904 and 905) using ATENA.
- Dismantle various storage cells (liquid waste stored in cell 907, fission products stored in cells 908, 909 and in the extension building).
- Carry out general decontamination of the building.
- Cutt off the ATENA and dispose of it as waste.

The progress of work can be summarised as follows:

| | |
|---|---|
| **1985–1990** | • Dismantling of α-contaminated cells and unshielded glove boxes. This work was carried out by direct manual access |
| | • Dismantling of storage cells containing wastes, fission products etc. This work was carried out by direct manual access with biological shielding |
| **1990–1994** | • Dismantling and removal of reprocessing equipment from cells |
| | • Dismantling of high radiation cells requiring use of remotely operated equipment |
| **1995** | • Video inspection, mapping and sampling |
| **1996–2001** | • General decontamination of building and monitoring. |

From January 1990 until February 1993, the ATENA machine with a manipulator arm MA23 was used for dismantling operations in hot cells 903, 904 and 905. Initial dismantling operations with the ATENA machine using hydraulic shears were carried out in cell 905. However, it soon became clear that the task was too demanding for the MA23 manipulator and the shears were replaced by a circular saw which was lighter and more manoeuvrable. Once cell 905 had been dismantled, cell 904 was then dismantled. This operation was carried out with a circular saw. In order to allow the introduction of the polyarticulated arm of ATENA in cell 903, it became necessary to make a large opening (dimensions were 1.2 m × 4.5 m) in the wall between cells 903 and 904. However, the level of radiation at that point of the partition wall was too high for human operation. So the concrete was cut remotely using a diamond-tipped disc saw mounted on the ATENA machine without the MA23. The cutting disc was cooled by liquid nitrogen, thereby eliminating the generation of liquid waste.

A special treatment cell, called the workshop cell, for conditioning the waste from high activity cells was built at the northern side of cell 905. It was made of a concrete wall and stainless steel modular panels. After the conditioning process, the workshop cell was decontaminated and dismantled.

After the completion of dismantling operations, a programme to decontaminate the walls of the high activity cells (903, 904 and 905) was carried out. The estimated dose rates were as follows:

- Cell 905   0.05 mGy.h$^{-1}$
- Cell 904   0.2–2 mGy.h$^{-1}$
- Cell 903   10–20 mGy.h$^{-1}$

These conditions did not allow direct human access. The initial technique selected and tested in cell 905 was shot blasting, operated semi-remotely. Concrete was removed to a depth of 4 mm and shot was recycled in order to limit the amount of solid waste.

There were a number of cells such as cells 901 and 902 (fuel receipt and cutting) and fission product storage cells 920, 908 and 909 where radiation levels were comparatively low and so direct human access was possible, although for a limited period of time. Cells 908 and 909 each contained a 15 m$^3$ tank and associated pipe work. During shutdown operations, the fission product solutions were removed and the tanks were aggressively rinsed. This reduced the radiation level to 0.25 mGy.h$^{-1}$ with hot spots up to 100 mGy h$^{-1}$. The tanks were dismantled by linear-shaped explosive charges and the process was completed with traditional cutting. Other dismantling operations that were carried out by direct access of workers were: cell 906 (cells and glove boxes), cell 952 (extraction of U and Pu), cell 907 (solvent washing) and cell 911 (pipe work and demisters).

To avoid dispersion of contamination, modular workshops were developed to carry out these operations. The workshops were built with stainless steel panels of standard dimensions. The smooth panel surfaces were easy to decontaminate and reuse.

### 15.2.2.3 Lessons Learnt

From the decommissioning of AT-1, the following experiences were gained:

- The ageing of plant equipment and systems such as the ventilation system and the electrical installations and subsequent maintenance and replacement should be taken into consideration before reliance is placed on their availability. Failure of systems and equipment had resulted in several unplanned stoppages and emergency maintenance during the dismantling operations.
- The shot blasting technique was not very thoroughly investigated. It produced significant quantities of secondary waste and the need to achieve increasingly lower levels of activity required frequent changing of the shot.
- Manual, semi-remote and remote operations were dictated by radiation levels which were determined by radiation surveys. Such surveys were difficult at remote inaccessible places. Computer modelling with input from measured quantities would improve the modelling output and thereby reduce uncertainty.
- The dismantlers' task would have been much easier if the cells had been kept clean, and waste and redundant equipment had not been stored in them. The operations prior to plant closure should be properly conducted and should include an inventory, a schedule of condition and a radiological mapping.

## 15.2.3 KRB-A in Germany

### 15.2.3.1 Operational Period

The nuclear power plant unit A at Gundremmingen (KRB-A) is a Boiling Water Reactor (BWR) which is located in Bavaria on the river Danube between Stuttgart and Munich. It was constructed during the period 1962–1966 as the first commercial nuclear power plant in Germany with an electrical output of 250 MWe. It was a dual cycle BWR of the General Electric type. Beside the primary steam (1000 Mg.h$^{-1}$), there was a secondary steam (360 Mg.h$^{-1}$), available for load regulation of the reactor. Each of the recirculation loops was equipped with a large recirculation pump and a steam generator. The dismantling experience gained at KRB-A was transferable to other BWR and PWR activities. The plant was operated from 1966 to January 1977 when it was shutdown due to a short-circuit in the grid which caused substantial damage to the plant. It was estimated that the repair and backfitting measures would require an outage for

**Fig. 15.1** *Nuclear power plants at the Gunmremmingen site*

several years and would require an enormous amount of money and hence its subsequent operation could not be justified on economic grounds. In January 1980, the owners (RWE Energie AG (75%) and Bayernwerk AG (now E.ON AG) (25%)) decided to decommission the unit. The planning for decommissioning started in 1980, actual dismantling operations began in 1983 and were completed in 2000. Since 1984 two modern BWRs have been in operation on the site, each with an electric output of 1344 MWe each. Figure 15.1 shows the nuclear site at Gunmremmingen with the small domed building of unit A on the left. Figure 15.2 shows the layout of the BWR plant.

### 15.2.3.2 Decommissioning Work

The whole decommissioning project was divided into various distinct phases, starting with the systems with low contamination and ending with the most highly activated reactor vessel. This approach of tackling progressively higher hazards facilitated the acquisition of experience in a step-by-step fashion. It also helped collection of data for radiation protection purposes, for dismantling and decontamination techniques and for waste management and cost estimation.

The key dates for the KRB-A plant were

| | |
|---|---|
| 1966 | start of operation |
| 1977 | cessation of operation |
| 1980 | decommissioning planning |
| 1983–1992 | decommissioning phase 1: removal of components and systems in the turbine house; phase 2: decommissioning of primary circuit |
| 1992–2004 | phase 3: dismantling of RPV, internals and biological shield |
| 2005 | reactor removal completed |

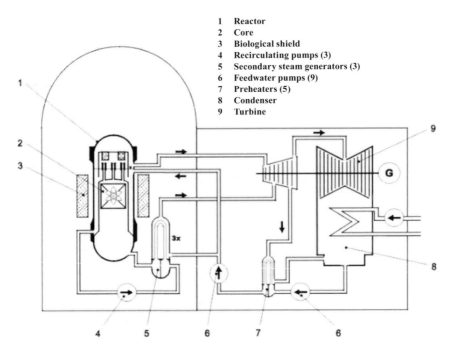

1   Reactor
2   Core
3   Biological shield
4   Recirculating pumps (3)
5   Secondary steam generators (3)
6   Feedwater pumps (9)
7   Preheaters (5)
8   Condenser
9   Turbine

**Fig. 15.2** *KRB-A nuclear power plant (BWR)*

Decommissioning started in 1983 with the removal of components and systems in the turbine house (phase 1), which generated approximately 4500 metric tonne of material, mainly steel and concrete. The total activity was 40 GBq, with surface contamination of up to 1000 Bq.cm$^{-2}$. The collective dose was only 1man-Sv. During phase 2 of decommissioning, which consisted of removing the primary circuit and associated systems, a collective dose of 1.4 man-Sv was received when the surface contamination level was in the range 1000–50,000 Bq.cm$^{-2}$. Phase 3 which began in 1992 and was completed in 2003 included dismantling the RPV, its internals and the biological shield. A collective dose for this phase was estimated to be about 1.5 man-Sv, but because of the use of underwater cutting and remote handling techniques, the actual dose was less than the estimated amount.

Mechanical and thermal cutting techniques were assessed and used, depending on cutting tasks, to acquire experience and to identify best possible methods for dismantling of contaminated components. The techniques used included ice-sawing, a modified plasma torch for thick-walled pipes as well as conventional ones such as circular saws and bandsaws.

For the dismantling of heat exchangers such as shutdown coolers, clean-up coolers and the secondary steam generators, thermal cutting was considered to be unsuitable as it would generate a large amount of radioactive aerosol and the operators would be excessively exposed to radiation. The prevalent dose rate was up to 10.3 mSv.h$^{-1}$. Moreover, the application of saws to cut through a tube bundle was not feasible as non-fixed single tubes would vibrate and clamp the band or blade of a sawing machine. The solution to this cutting problem was found to be an ice-sawing technique which was developed at KRB-A. The idea was to fill up the heat exchanger with water on the secondary side and to freeze the whole component down to about −20°C by blowing cold air through the primary side before cutting it with a saw. This technique minimised workers' dose, helped anchor single tubes in the tube bundle and reduced airborne contaminants. It also helped to cool the cutting blade.

For dismantling activated components of the RPV and reactor internals, various thermal cutting techniques were considered and tested on inactive materials of similar composition and dimensions. These included oxy-acetylene, oxy-propane, powder oxy-acetylene, powder oxy-propane flames and plasma arc techniques. The other consideration was that there were very high dose rates from activated components and hence dismantling needed to be carried out remotely. Remote underwater cutting was eventually selected because

- there are no frictional forces between the cutting tool and the component.
- the cutting tool can be small and manoeuvrable.
- the cutting speed can be higher.

The RPV was cut into segments using an oxy-acetylene torch which was guided around the vessel on a rail. This thermal cutting technique was used in the low activated upper and lower areas of the RPV and these segments were then cut into pieces in air in a tented enclosure. The highly activated central part of the RPV was removed as one large piece (70 tonne) into the fuel storage pool for further segmentation using plasma arc cutting.

In order to maximise recycling, an effective decontamination and electropolishing technique for dis-assembled steel parts was developed. It involved dipping a part in phosphoric acid and passing an electric current of 6000 A (max) for about 4 h. The arrangement is shown in Figure 15.3.

The secondary waste from this decontamination technique was minimised and a special procedure was developed to regenerate the phosphoric acid. After regeneration, the phosphoric acid could be reused. The principle was that the dissolved iron could be precipitated as iron oxalate by adding oxalic acid to the phosphoric acid bath. The iron oxalate was then converted to iron oxide by thermolytic conversion as the final waste product for storage. This thermolytic process was developed using a heated cone propeller and re-cleanable filter device. The phosphoric acid from the precipitation bath could be reused.

For cutting the concrete bioshield a range of techniques was evaluated on a specially constructed mock-up. The technique chosen was diamond cable sawing.

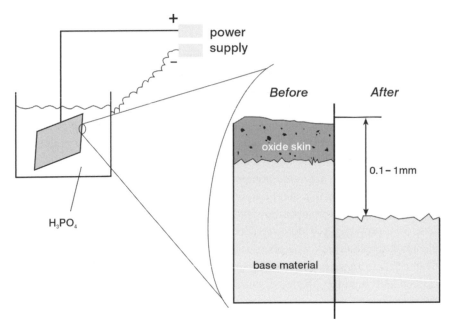

**Fig. 15.3** *Decontamination and electro-polishing technique (Courtesy of the EC pilot project on KRB-A decommissioning)*

### 15.2.3.3 Lessons Learnt

The key lessons that can be learnt from KRB-A decommissioning are

- There is no single dismantling tool, however versatile, that can be used to decommission the entire nuclear power plant. Different tools and techniques are required for different dismantling problems. It was demonstrated that standard tools could be used with some modification for specific uses, such as underwater.
- Standard mechanical tools such as milling and sawing tools, grinders etc. are easy to use and require little personnel training. However, their use is mostly limited to linear movements, which could be problematic in limited space, and they have rather slow cutting speeds.
- Thermal cutting techniques were shown to be both flexible and reliable. However, they produce radioactive aerosol when in air and contaminate water when underwater cutting is done. Both of these issues were satisfactorily controlled at KRB-A.
- The decommissioning cost is highly dependent on the need for minimisation of wastes (primary and secondary). Minimisation of waste is of particular importance in Germany as there is no final disposal facility or centralised storage facility. One way of reducing the volume of waste is by decontaminating components for

free release. The decontamination of components of defined shape and structure, where surface measurements can be made, is by mechanical and chemical means.

- For materials where surface measurements are unsuitable, such as electrical cables, mechanical procedures have been proven to be effective. The cables were segmented and the inner copper core had been mechanically separated from insulation. The contamination was in the insulating material and the copper was mostly clean. The portion of waste was only 30% of the total volume.

The overall experience of KRB-A decommissioning shows that decommissioning of a nuclear power plant can be carried out

- without deferring for a long period of time (safe storage provision).
- within estimated cost and time.
- with low doses to the workers and with little or no environmental impact.
- with a small amount of radioactive waste to dispose of.

# 15.3  KGR Greifswald, Germany

## 15.3.1 Background

At the Greifswald site (KGR), there is a total of eight reactor units of Russian design pressurised water reactor, WWER 440 type. Units 1–4 are of model V 230 and units 5–8 of the more recent model V 213. The reactors were configured on a double unit basis, i.e. two reactors were arranged in one reactor hall with certain mechanical equipment and secondary systems in common. On the other hand, there was only one turbine hall for all the reactors. There were also three plants for treatment and storage of liquid radioactive waste. The solid radioactive waste was stored in concrete pits.

After the reunification of Germany, the four operating units (units 1–4) at Greifswald were shut down, the trial operation of unit 5 and all construction work for units 6–8 was suspended. In 1990 a decision was taken to decommission units 1–4, followed by the same decision for unit 5 in 1991. The Energiewerke Nord (EWN) was given the responsibility of decommissioning eight reactor units at Greifswald, the world's largest decommissioning project [3].

An overall technical and economic assessment was made. On the basis of this assessment covering financial and radiological factors, a decision was taken to undertake an immediate dismantling. This decision was also influenced by unemployment situation in the area. On technical grounds, it was estimated that altogether 1,800,000 tonne of material would be generated from the Greifswald site decommissioning work. Figure 15.4 gives an overview of the expected material categorisation.

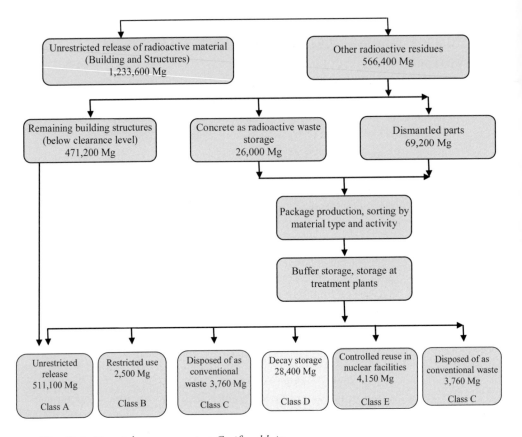

**Fig. 15.4** *Material management at Greifswald site*

The various classes of material that may arise are identified as follows:

Class A    free release
Class B    restricted reuse
Class C    disposal as conventional waste
Class D    storage for decay
Class E    reuse in nuclear facilities
Class F    disposal as radioactive waste

Due to the lack of disposal facilities in Germany, the Interim Storage North (ISN) complex was constructed on site as an integrated treatment and storage facility for radioactive waste and dismantled material. The ISN comprised eight halls, a loading corridor and a treatment/conditioning area; the total storage volume of the ISN being 200,000 m³. Storage hall 8 would house spent fuel in CASTOR casks, whereas halls 6 and 7 would be used for large components from primary circuits, awaiting further treatment. Halls 1–5 would be used as interim and buffer stores for all kinds of packages.

## 15.3.2 Progress and Achievement

The key KGR project dates are

| | |
|---|---|
| 1973–1990 | units 1–4 operated |
| 1989 | trial operation of unit 5 |
| 1990 | shutdown of units 1–5 and construction work halted for units 6–8 |
| 1995 | decommissioning licence issued to EWN |
| 1998 | start of ISN operation |
| 1999–2002 | demonstration of remote dismantling of unit 5 |
| 2004–2007 | dismantling of reactor and internals on units 1–4 |
| 2012 | demolition of building structure |

The basic principles adopted were as follows:

- progression from lower contamination to higher contamination and finally to activated plant parts
- commencement of dismantling in unit 5 and the turbine hall, followed by units 1–4
- use of 'off the shelf' equipment as far as possible
- removal of as large as possible components and parts for storage and/or further treatment in the ISN
- dismantling on a room basis, not on a system basis

As units 6–8 were only being constructed, they were uncontaminated. Consequently, the equipment from these units was to be transported into the steam generator room of unit 5 and was to be cut as model dismantling. The tested tools and equipment would then be used for dismantling units 1–4. Unit 5 reactor components would not be cut immediately. Instead, the individual components would be transported as one part to the interim storage on site. After a decay of 40–70 years they will be cut without remote techniques.

Preparatory to dismantling, measures were taken to reduce dose rates. First, parts of the primary loops were decontaminated electrolytically, and secondly, hot spots were removed by high pressure water jets or by mechanical means. However, before dismantling work began, asbestos (used in thermal insulation of pipe-work) was removed in a safe manner.

The strategy for dealing with the RPV vessel and the internals of units 1–4 was remote dismantling and storage in the ISN. Dismantling work was carried out in the steam generator room which is situated around the RPV. Here cutting (dry and wet), packaging and transfer areas were installed. The complete system was designed to be mobile and was first installed in unit 5 for inactive testing before installation and commissioning in unit 2. Inactive testing started in mid-1999 and was completed by the end of 2002. The techniques that were applied are summarised in Table 15.1. Figure 15.5 shows the bandsaw cutting the core basket and Figure 15.6 shows the strategy that was followed in cutting and packaging reactor components.

Following the successful dismantling of unit 5 RPV, the following strategy was followed:

- The RPVs of units 1 and 2 without the internal will be transported and stored as one piece in the ISN.
- The highly activated RPV internals of units 1 and 2 will be cut in the wet cutting area in unit 2. The dry cutting area will be used for components with lower levels of activity and internal parts.
- The RPVs of units 3 and 4 will be transported with their internals to the ISN.

The use of casks is an essential part of waste management. The highly activated reactor components will be dismantled and then loaded, transported and stored in casks in the newly built ISN for at least 40 years for decay. For the storage in the ISN, the dose rate of the casks must be below the limit of 0.1 mSv.h$^{-1}$ at a distance of 2 m. Moreover, a licence for the transport of casks on road and rail is required. Figure 15.7 shows the transport of the RPV of unit 8 to unit 5.

### 15.3.3 Lessons Learnt

After initial difficulties caused by large reduction in plant personnel combined with West German nuclear laws and regulations and, even more so, the application of the West German market economy in the erstwhile East German centralised economy for the plant, EWN succeeded in drawing up a compromise socio-economic decommissioning strategy for the Greifswald site. There

**Table 15.1** *Cutting techniques for remote dismantling*

| Cutting | Components | Technique |
|---------|-----------|-----------|
| Dry | Reactor pressure vessel<br>Upper part of protection tube unit<br>Upper part of reactor cavity | Bandsaw<br>Disc cutter<br>Plasma arc |
| Wet (pool) | Core basket<br>Lower part of protection tube unit<br>Lower part of reactor cavity<br>Cavity bottom | Bandsaw<br>CAMC*<br>Plasma arc<br>Fretsaw |

\*    Contact Arc Metal Cutting

**Fig. 15.5** *Cutting of core basket by band-saw*

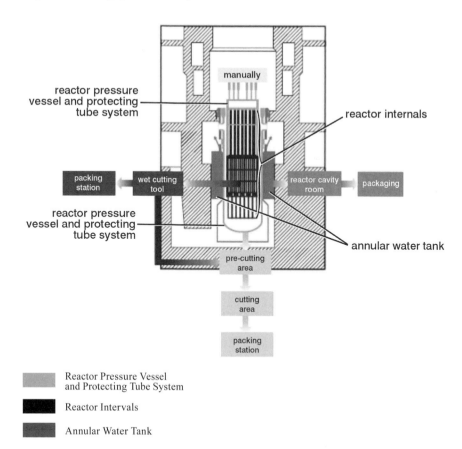

**Fig. 15.6** *Removal, cutting and packaging of reactor components*

**Fig. 15.7** *Transport of unit 8 RPV to unit 5 by road*

were no major technical problems as such in decommissioning of the Russian WWER-type reactors. But the sheer size of the project and the ensuing waste management task was vast. The following lessons can be learnt from this project.

- Decommissioning of a nuclear facility or a suite of nuclear facilities is not so much of a technical problem, but a challenge to programme/project management and logistics, once the legal and socio-economic boundary has been established.
- Multiple licensing requirements in Germany may cause logistic problems, a single licence structure may remove some of the project management problems.
- Overall planning of the decommissioning programme is vital, particularly when several projects need to be co-ordinated for cost-effective operation.
- The use of tools and equipment on inactive components offers experience, which can then be effectively applied on active components with confidence and with a desirable outcome, i.e. low doses for workers, faster completion time etc.

## 15.4 Calder Hall in the UK

### 15.4.1 Background

The Calder Hall power station, within the Sellafield nuclear site in West Cumbria, UK, contained four Magnox-type nuclear reactors generating $4 \times 180$ MW (thermal) with a net electricity export of 138 MWe to the national grid.

This power station was the first commercial nuclear power station in the world and was officially opened in October 1956 by Her Majesty Queen Elizabeth II. It operated for 47 years, well beyond its original design lifetime, and was taken out of service in March 2003. During its operational lifetime, Calder Hall station supplied strategic nuclear materials to the MoD as well as supplying electricity to the national grid. The reactors were approximately 21.5 m high and 11.3 m in diameter. The four reactors were serviced by eight steam turbines (each of 30 MW capacity), four in each of two separate turbine halls. There are also four hyperbolic concrete cooling towers each approximately 90 m high.

Due to the availability of essential facilities on the main Sellafield site, this Magnox station did not have some of those essential facilities. For example, there was no fuel cooling pond since spent fuel was transferred directly to a Magnox reprocessing plant on the main Sellafield site. There was no active effluent treatment plant as liquid radioactive effluent was transferred to Sellafield Effluent Treatment Plant (SETP) and there was no ILW vault as all ILWs were transferred to the Sellafield miscellaneous $\beta/\gamma$-store.

In April, 2005 the ownership of the site was transferred to the NDA and the British Nuclear Group (BNG) (now Sellafield Limited) became the Site Licence Company (SLC). The decommissioning of the site is now in progress under the NDA decommissioning strategy and financial discipline. So the plans and procedures that are being applied to the decommissioning of this plant are typical of those to be applied to all the plants in the UK under the NDA ownership.

## 15.4.2 Decommissioning Strategy and Preparation

Under the NDA strategy, the scope, schedule and costs of work needed to decommission a site are reviewed annually during the Life-Cycle Base Line (LCBL) review process and the Near Term Work Plan (NTWP) is produced. The initial LCBL is submitted by the SLC to the NDA for scrutiny and approval. The cost is also estimated and the total discounted cost (NPV) for complete decommissioning of the plant right up to delicensing is estimated to be 1032.5 million pounds sterling [4]. Alongside the financial management, regulatory requirements need to be met by the SLC.

Immediately after the cessation of operation of the plant in March 2003, the BNG carried out a number of assessments to establish the safest and most cost-effective method to decommission the station and the LCBL was produced, primarily for use by the NDA. Key milestones in decommissioning the site, taken from the 2004 LCBL, are shown in Table 15.2.

The LCBL study shows the following waste arising:

Total liquid discharge from $\beta/\gamma$-emitting radionuclides
$= 4$ to $5 \times 10^9$ Bq

**Table 15.2** *LCBL for Calder Hall (from 2004 report)*

| Year | Milestones |
| --- | --- |
| 2006 | Cooling tower demolition |
| 2008 | Completion of defuelling |
| 2020 | Start of C&M – all fuel removed from site, all plants, facilities and buildings other than reactor building demolished. |
| 2105 | End of C&M – commencement of demolition of reactor buildings |
| 2112 | Demolition of four reactor buildings completed |
| 2117 | Final site clearance and potential delicensing. |

α-emitting radionuclides are negligible
Solid LLW arising by 2020 = 9000 m³
Solid LLW from 2020 to site delicensing = 62,000 m³
Total ILW to site delicence = 7200 m³
Hazardous wastes containing asbestos material in the period
    2005 to 2010 = 5200 m³

In August 2004, the BNG/SL submitted the Environmental Impact Assessment for Decommissioning (EIAD) (ES) under the EIADR99 (see Section 6.3.10) to the HSE/ND. Following public consultation and after consultation with other regulatory bodies, in July 2005 the HSE granted consent for the decommissioning process. Conditions were attached to the consent, primarily related to production and maintenance of an Environmental Management Plan (EMP) to prevent, reduce and, if possible, offset any significant adverse environmental effects of the decommissioning work. The EMP is to be issued annually or at intervals agreed with the HSE. The first EMP was issued by the SL in September 2005 and the second one in September 2006.

The EMP provides details of environmental effects covering location, geology, hydrogeology, ecology, noise, vibration, archeology and cultural heritage, human habitation etc. as well as environmental measures that need to be taken during the C&M phase of the plant. Further details for all the phases of the decommissioning project are presented in the ES. The EMP is divided into three phases

- C&MP (Care and Maintenance Preparations)
- C&M period
- Final site clearance

The C&MP is the first phase and it has been estimated that it will take about 15 years for Calder Hall power station. During this phase most of the radioactive and non-radioactive plants and buildings (other than the reactor buildings) are to be removed. The C&M is the second phase during which reactor buildings and other enclosed facilities are managed, monitored and maintained. This period should last from 2020 to 2105 during which no significant dismantling will be carried out. The last phase, final site clearance, will last for about 12 years. During this phase the remaining structures including the reactor buildings will be dismantled and demolished. The site will be cleared leading to delicensing by around 2117.

### 15.4.3 Progress

The SL strategy is to demolish the cooling towers at the earliest stage. A preliminary safety case was prepared for this work identifying methods and techniques to be used, hazards involved in the work etc. which were then put together in the PCmSR for approval by the HSE/ND. The cooling towers are to be brought down by a process known as 'explosive demolition' using a number of charges at strategic points. Figure 15.8 shows the cooling towers which will be demolished. But before they are demolished their inside sections are to be dismantled and cleared, as shown in Figure 15.9.

Insulation containing asbestos needs to be removed under stringent safety conditions using specialist personnel working in tented areas subject to airlocks and a negative air pressure system. All work will be carried out in strict accordance with the control of Asbestos at Work Regulation 2002. The tents will fully enclose the work areas and the entire volume will be smoke tested to ensure integrity before asbestos removal begins.

### 15.4.4 Lessons Learnt

The decommissioning of Calder Hall power station is at a very early stage. Consequently all the financial and administrative disciplines to be applied under the NDA owned sites are evidenced in decommissioning this plant. The lessons that can be learnt from this plant even at this early stage are

- Production of LCBL by the SLC for the NDA scrutiny and approval is useful not only to obtain an overview of the whole process but also to identify constraints, limits and risks.
- Production of the ES for regulatory consent from the HSE/ND and other regulatory bodies is essential and can highlight issues which need to be addressed throughout the lifespan of this project.

**Fig. 15.8** *Calder Hall cooling towers which are to be demolished (Courtesy BNG)*

**Fig. 15.9** *Dismantling and clearance of internals of a Calder cooling tower (Courtesy BNG)*

## REFERENCES

[1] EU website, http://www.eu-decom.be.
[2] Nokhamzon J.G., A Garcia A., Nolin D. and Salou J.J., AT1 Decommissioning Feedback Experience, CEA/DEN/DPA/JGN 01-507, 2001.
[3] http://ec-cnd.net/eudecom/EWN-Dismantling ReactorPressureVessel.pdf.
[4] http://www.nda.gov.uk.

# 16

# RADIOACTIVE WASTE CLASSIFICATION AND INVENTORY

## 16.1 Introduction

The management of radioactive material and radioactive waste is an essential follow-up operation in nuclear decommissioning. Various types of radioactive material or radioactive waste in various physical and chemical forms arise from any nuclear decommissioning activity and they pose different levels of risk. Such materials or wastes need to be categorised properly in order to manage them safely, securely and in an environmentally acceptable way for handling, storage, transportation and eventual disposal. But, first, it is essential to have a clear understanding of the meaning of radioactive material and radioactive waste.

## 16.2 Radioactive Substances and Radioactive Waste

Under the Radioactive Substances Act 1993 [1], a radioactive substance is defined as any material whose activity concentration is higher than that specified in Schedule 1 of the Act. The elements specified in Schedule 1 are given in Table 16.1. A radioactive substance may be categorised as a 'radioactive material' or a 'radioactive waste'. A radioactive material, despite its radioactivity, may have some positive use; if not, it may be considered as a radioactive waste. Irradiated nuclear fuel, if destined for reprocessing or some other use, would be considered as a radioactive material, not waste. On the other hand, if no such use is envisaged, then it is a waste. So the label of 'material' or 'waste' against a substance is based on the anticipated use or value envisaged by the owner of that material. However, there are some substances such as fission products, activation products, contaminated tools etc. for which no possible use can be foreseen within the safety requirements and they are obviously categorised as wastes. Once categorised, it is not easy to change the category and the implications of such categorisation are also quite profound. For example, a radioactive material cannot be disposed of, or a radioactive waste cannot be reused unless it is cleared by the relevant regulatory body.

The minimum levels of activity shown in Schedule 1 can also be considered as the clearance level. The 'clearance level' of waste is that level which

Table 16.1 *Radioactive substances as per RSA93 [1]*

| Elements | $Bq.g^{-1}$ | | |
|---|---|---|---|
| | Solid | Liquid | Gas or Vapour |
| Actinium | 0.37 | 7.40E-2* | 2.59E-6 |
| Lead | 0.74 | 3.70E-3 | 1.11E-4 |
| Polonium | 0.37 | 2.59E-2 | 2.22E-4 |
| Protactinium | 0.37 | 3.33E-2 | 1.11E-6 |
| Radium | 0.37 | 3.70E-4 | 3.70E-5 |
| Radon | – | – | 3.70E-2 |
| Thorium | 2.59 | 3.70E-2 | 2.22E-5 |
| Uranium | 11.1 | 0.74 | 7.40E-5 |

* 1.0E-2 signifies $1 \times 10^{-2}$

represents an insignificant hazard to humans and the environment and hence it can be released unconditionally into the environment. The details regarding exemption and clearance levels are given in Section 17.3.

## 16.3 Classification of Radioactive Waste

Radioactive wastes can be classified in many different ways, according to: source, physical state (i.e. solid, liquid or gaseous), levels of radioactivity, half-lives, final disposal route and, of course, radiotoxicity. The two most significant and widely used parameters which are taken into account for waste classification are (i) the half-lives of the radionuclides, and (ii) activity concentrations. The half-life consideration leads to separation into short-lived or long-lived nuclides. Such consideration is consistent with the disposal route: long-lived wastes require long-term isolation from humans and the environment, preferably in a geological structure; short-lived wastes do not require such long-term isolation and the disposal route is much less stringent. Classification based on half-lives of the radionuclides leads to the IAEA waste categorisation [2, 3], whereas a classification based on the concentration levels of the wastes is used in the UK (HLW, ILW, LLW and VLLW) [4]. There are, of course, some variations in these two types of classification. For example, French waste classification is a hybrid of the two [5] and the EC recommendation [6] is a slight modification of the IAEA classification.

Another factor which should be taken into consideration when dealing with radioactive content and associated hazards is the chemotoxicity of the material. Chemotoxicity arises from stable heavy metals resulting from the decay of heavy radionuclides or from degradable organic materials. Such wastes are currently called Radioactive Mixed Wastes (RMWs).

## 16.3.1 IAEA Waste Classification

A classification system used by the IAEA is shown in Table 16.2 [2, 3]. The main advantage of this classification is its simplicity and consistency with general practice. The wastes in the different categories are to be disposed of in a different manner and, possibly, in different repositories.

**Table 16.2** *IAEA waste classification*

| Waste class | Characteristics | Disposal options |
|---|---|---|
| 1 Exempt Waste (EW) | Activity levels at or below clearance levels. It is based on annual dose to a member of the public < 10 µSv | No restrictions on radiological grounds |
| 2 Low and Inter-mediate Level Waste (LILW) | Activity levels > clearance levels but thermal power < 2 kW.m$^{-3}$ | |
| 2.1 Short lived waste (LILW–SL) | Mainly SLWs, LLWs conc. < 400 Bq.g$^{-1}$ | Near surface disposal |
| 2.2 Long-lived waste (LILW–LL) | Mainly LLWs | Geological disposal |
| 3 High level waste (HLW) | Thermal power > 2 kW.m$^{-3}$ and mainly LLWs | Geological disposal |

## 16.3.2 UK Waste Classification

In the UK, the wastes are classified mainly on the basis of activity concentrations [4]. Four classes of wastes are specified: HLW, ILW, LLW and VLLW. The UK waste categorisation is shown in Table 16.3. Each class of waste can have a variety of physical and chemical forms.

### 16.3.2.1 High Level Waste or Heat Generating Waste

Waste which is very radioactive and in which temperature may rise significantly as a result of its radioactivity is categorised as a HLW or Heat Generating Waste (HGW). This heat generation characteristic as well as high levels of activity needs to be taken into account in designing storage or disposal facilities for these wastes. Such wastes arise from the reprocessing of irradiated nuclear fuels as concentrated aqueous residues (in nitric acid) at the primary stages of the separation of U and Pu. This aqueous solution is vitrified into passive safe solid form and contained in 150 litre stainless steel containers at Sellafield in the UK. Plant items from the vitrification process which become contaminated with vitrified HLW are themselves categorised as HLW. Irradiated fuel is usually classified as HLW if it is to be disposed of without reprocessing. Over 95% of all activity arising from a nuclear power plant falls into this category. It is envisaged that after vitrification, HLW will be stored for a minimum of 50 years in accordance with government policy. There is no disposal route for HLW in the UK.

### 16.3.2.2 Intermediate Level Waste

Wastes whose activity levels are low enough to require heating to be taken into account in storage or disposal facilities but whose activity levels exceed the upper boundaries of LLW are categorised as ILWs. The ILWs consist principally of metals such as fuel cladding and reactor components, graphite from reactor cores and sludges from radioactive effluent treatment plants, as well as some wastes from medical and industrial use. ILW requires radiation shielding to comply with radiological safety requirements. These wastes are immobilised in cement grout in 500 litre stainless steel drums or for larger items in concrete boxes.

No disposal route is currently available for ILW in the UK and at the moment these wastes are stored at sites where they are produced.

### 16.3.2.3 Low Level Waste

Wastes containing radioactive materials whose activity exceed those of VLLW but not exceeding 4 GBq.te$^{-1}$ (4000 Bq.g$^{-1}$) of $\alpha$-activity or 12 GBq.te$^{-1}$ (12,000 Bq.g$^{-1}$) of $\beta/\gamma$-activity are categorised as LLWs. The LLWs arise from nuclear industry, research and hospital facilities and may include general rubbish such

**Table 16.3** *Waste classification in the UK*

| Waste class | Characteristics | Disposal options |
|---|---|---|
| Very Low Level Waste (VLLW) | Activity concentration between 0.4 to 4 Bq g$^{-1}$ b/g or 40 kBq of b/g per single item of waste | Dustbin disposal, disposal at landfill sites or incineration |
| Low Level Waste (LLW) | Activity concentration higher than 4 Bq g$^{-1}$ b/g but less than 4 kBq g$^{-1}$ of a or 12 kBq g$^{-1}$ of b/g | Shallow surface disposal |
| Intermediate Level Waste (ILW) | Activity concentration higher than LLW but less than when heat generation is anticipated | No defined disposal route at the moment |
| High Level Waste (HLW) or HGW | Wastes of very high concentration or of heat generating capacity | No disposal route, min storage period is 50 years |

as used paper towels, discarded protective clothing, laboratory equipment etc. Building materials and larger items of plant and equipment generated from the decommissioning of a nuclear facility produce LLW. These wastes do not normally require radiation shielding.

Since 1959, most of the solid LLWs have been disposed of at the LLWR near Drigg in Cumbria, which was managed by BNFL. The management of the LLWR was awarded on 31 March 2008 by the NDA to UK Nuclear Waste Management (UKNWM) Ltd – a consortium comprising URS-Washington Division, Studsvik, Areva and Serco Assurance for an initial period of five years. The UKAEA also operates a LLW disposal facility at Dounreay in Caithness. Normally all suitable LLW are compacted or super-compacted before final disposal.

### 16.3.2.4 Very Low Level Waste

These wastes have very low levels of activity such that each 0.1 $m^3$ of material must have less than 400 kBq of $\beta/\gamma$-activity (4 Bq.g$^{-1}$) or 40 kBq per article unless the activity is due to C-14 or tritium (H-3) in which case the limits are a factor of ten greater. Such waste may be disposed of by various means such as domestic refuse at landfill sites or by authorised incineration. The method depends on the nature and quantity of the material.

With special precautions and within certain limits, solid wastes which are too radioactive (for example, a single item exceeding 40 kBq) for 'dustbin disposal' may be disposed of at suitable landfill sites. Demolition wastes and other high volume wastes with low specific activity can be authorised for burial at landfill sites. It should be noted that activity levels below 0.4 Bq.g$^{-1}$, as shown in Schedule 1 of RSA93, are cleared for unrestricted release.

Incineration can be used for radioactive wastes which are obnoxious or toxic. An authorisation for such disposal takes into account the quantity and nature of activity. Wastes which are disposed of as VLLW are not recorded in the UK waste inventory.

## 16.3.3 French Waste Classification

Radioactive wastes in France are classified as follows [5]:

**Type A**: These are LLWs with short half-lives ($t_{1/2}$<30 years). The long-lived α-concentration is limited to 3.7 kBq.g$^{-1}$. These wastes are generated from normal plant operations, routine maintenance work, refuelling operations etc. and consist of protective clothes and shoes, disposable handkerchiefs etc.

**Type B**: These are LLWs or ILWs with half-lives ranging from short ($t_{1/2}$< 30 years) to long ($t_{1/2}$> 30 years), but with low thermal power.

**Type C**: These are HLWs with long half-lives and significant thermal

**Table 16.4** *French waste categories and their disposal routes*

| Category | Half life | |
|---|---|---|
| | Short ($t_{1/2} < 30y$) | Long ($t_{1/2} > 30y$) |
| FA and TFA | Surface disposal (under investigation) (possible site is Centre de l'Aube) | |
| Type A | Surface disposal (Centre de l'Aube) | Long term interim storage (under invest.) |
| Type B | | Geological disposal |
| Type C | (under investigation: Dec 30, 1991 Law) | |

power. Fission products from reprocessing operations and untreated spent fuel fall into this category.

In addition, France categorises two other waste streams.

**FA** (faible activité): These are wastes containing long-lived radionuclides in very low concentrations. They may arise from uranium mining.

**TFA** (très faible activité): These are VLLWs (activity concentration < 100 Bq.g$^{-1}$). They may arise from dismantling operations during decommissioning.

These types of wastes and their disposal routes are shown in Table 16.4.

In France the FA and TFA wastes (short-lived as well as long-lived wastes), which may be categorised jointly as VLLWs, are not released into the environment under the free release criteria, although this is permitted under the EU regulations. This is a precautionary approach adopted in France. These wastes are disposed of at a surface disposal facility. A surface disposal facility has been constructed near the village of Morvilliers which is in the vicinity of the Centre de l'Aube. The disposal facility, called the Centre de Morvilliers, has a capacity of 650,000 m³ and started operation in October 2003.

Type A wastes are disposed of at a surface repository following encapsulation by cementation. The repository at Centre de l'Aube is used for type A waste. The wastes, encapsulated in drums, are placed in a layer in the concrete compartment and the gaps between the drums are back-filled with cement grout in preparation for the next layer. The drums are placed remotely under strict quality control and record keeping procedures. Layer upon layer of encapsulated drums are placed until the top of the compartment is reached. The top layer

**Fig. 16.1** *Centre de l'Aube for type A wastes (Photograph courtesy of* Les Films Roger Leenhardt*)*

is then grouted. Finally, a moveable roof is placed on top for protection against rain-water (Figure 16.1). Rows and rows of such compartments have been built. Underneath such rows, there is a water collection system and monitoring system. Collected water is treated and then discharged to the nearby brook.

For some of the type B (long-lived) and type C wastes, there are no disposal routes at the moment. They are to be placed in deep geological repositories.

### 16.3.4  EC Recommendation on Waste Classification

Recognising the fact that radioactive waste classifications among the member states of the EU are diverse and that there is a need for harmonisation in order to facilitate communication as well as cross-boundary movement of wastes in a single market economy, the EC under the Euratom treaty produced a recommendation on waste classification [6] for all member states to adopt. Although this recommendation was produced in October 1999, the EC granted an adaptation period until 1 January 2002, after which this classification system should be used for providing information concerning solid radioactive waste to the public, the national and international institutions and non-governmental organisations. It would not, however, replace technical criteria, where required, for specific safety considerations such as licensing of facilities or other purposes.

This classification system is based on the IAEA classification scheme with some changes to take into account the views and practical experiences of European national experts. First of all, the specification of 2 kW.m$^{-3}$ as the limiting heat generation rate for LILW in the IAEA classification was removed, as there was no foundation for such a specific value. In fact, when a heat generation rate is to be taken into account it can only be related to site-specific safety analysis. The proposed classification is summarised below.

### 16.3.4.1 Transition Radioactive Waste

This is the type of waste, mainly from medical sources, which would decay within the temporary storage period and which can then be removed from regulatory control under the provisions of the clearance criteria.

### 16.3.4.2 Low and Intermediate Level Waste

In LILW, the concentration of radionuclides is such that the generation of thermal power during the disposal is sufficiently low. The thermal power level is to be determined on a site-specific basis following safety assessments.

- Short-lived LILW (LILW-SL): This category includes waste with nuclide half-lives less than about 30 years with a restriction on long-lived α-radionuclide concentration to 4000 $Bq.g^{-1}$ in individual waste packages and to an overall average of 400 $Bq.g^{-1}$ in the total waste volume.
- Long-lived LILW (LILW-LL): This category of waste consists of long-lived radionuclides and α-emitters whose concentration exceeds the limits for short-lived waste.

### 16.3.4.3 High Level Waste

In this type of waste, the concentration of radionuclides is such that heat generation is of concern during storage and disposal. This type of waste would arise mainly from treatment and conditioning of spent nuclear fuel.

## 16.4  Waste Inventory in the UK

The DEFRA of the UK government in association with the UK Nirex Limited (now merged with NDA) periodically (about once every five years) publishes a comprehensive inventory of waste in the UK. The latest such report was published in October 2005. This section draws materials from that publication.

There are 37 sites of major producers of radioactive waste in England, Scotland and Wales. The sites under the NDA management undergoing decommissioning are shown in Figure 16.2. Other nuclear sites such as the AGRs and PWR power plants, although producing wastes, are not shown in the diagram, as they are outside NDA's remit. There are no major waste producers in Northern Ireland. In addition, there are a large number of small waste producers: hospitals, educational and research establishments, industrial producers etc. They are not shown here. About 86% of all radioactive wastes are produced in England, nearly 10% in Scotland and the remaining 4% in Wales.

It should be noted that not all radioactive materials produced in the UK are classified as wastes. Examples are depleted and enriched U, Pu and spent nuclear fuel. Enriched U and Pu are used to produce a new nuclear fuel (MOX fuel) and depleted U is used to produce hard munition shells. The NDA is now evaluating various options from technical and commercial point of view for

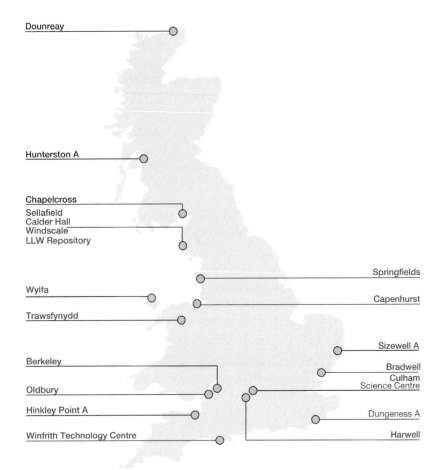

**Fig. 16.2** *Decommissioning sites under NDA management producing radioactive wastes (Courtesy of the NDA)*

long-term management of these materials. If these materials are found to have net negative value, i.e. they are considered to be a liability, they may well be categorised as waste.

## 16.4.1 Waste Types

As mentioned above, wastes can be categorised according to source, physical state, levels of radioactivity, half-lives and radiotoxicity. For the purposes of decommissioning, waste arising can be differentiated as operational waste or decommissioning waste. **Operational waste** arises from normal day-to-day operations of a plant or a facility from its start-up to final shutdown. Wastes such as contaminated materials, redundant equipment as well as those arising from defuelling of nuclear reactors and POCO operations are considered as operational wastes. **Decommissioning waste** arise from the decommission-

**Table 16.5** *Waste from all sources at 1 April 2004 and estimated future arising*

|  | HLW* | ILW** | LLW*** | Total |
|---|---|---|---|---|
| Volume (m³) | 1340 | 217 000 | 2 060 000 | 2 270 000 |
| Mass (te) | 3600 | 250 000 | 2 800 000 | 3 100 000 |

| | |
|---|---|
| * | These are the conditioned HLW |
| ** | ILW includes 11,600 m³ (11,000 te) of waste that are expected to become LLW as a result of decontamination and decay |
| *** | Nearly half (947,000 m³) of LLW is from contaminated soil from site clean-up |

ing of the plant. It consists of dismantled plant items and equipment, pipes, process vessels, building materials and rubbles, contaminated soil from land remediation etc.

There is an element of prediction in the estimation of waste arising presented here. For example, current operational plants may have their operational lives extended resulting in an increase in both the operational and decommissioning wastes. New plants, which are not currently foreseen, may also be constructed in the near future. All these aspects introduce a large amount of uncertainty into the estimation of waste volumes which may be required for planning purposes of a repository. For the purposes of maintaining some degree of reality in the waste volume estimates, wastes which may arise from extended planned activities of the existing plants have been included here as future waste arising and no new nuclear facility has been assumed.

## 16.4.2 Waste Volume

The total volume of wastes in the UK arising from all sources at 1 April 2004 and all future arising from HLW, ILW and LLW is given in Table 16.5. The

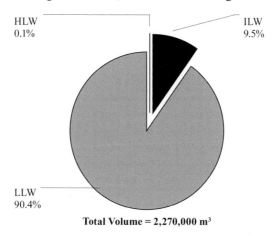

**Fig. 16.3** *Distribution of waste from all sources by volume*

total estimate of wastes from all major sources in the UK (British Nuclear Group (BNG), British Energy (BE), UKAEA, MoD, GE Healthcare (formerly known as Amersham International plc) and Urenco as well as from small users is included in Table 16.5. The total volume is 2,270,000 m³, of which 90.4% (2,060,000 m³) is LLW, 9.5% (217,000 m³) is ILW and less than 0.1% (1340 m³) is HLW. The volumetric distribution of these wastes is shown in Figure 16.3.

Most of the LLW in packaged form are sent to the Low Level Waste Repository (LLWR) near Drigg in Cumbria. The site has a remaining capacity of about 40,000 waste packages (800,000 m³) before it is expected to close around the middle of this century. If all the LLW in the inventory is packaged, it will require a volume of 2,520,000 m³ at the disposal site. So there is a shortfall of about 1,700,000 m³ for LLW disposal.

The existing wastes, as at 1 April 2004, from all sources are shown in Table 16.6. The table shows the breakdown of wastes in terms of conditioned and yet to be conditioned waste categories. The total volume is 105,290 m³. HLW and ILW are accumulated in stores on site as there are no disposal routes for these wastes. Liquid HLW is treated by evaporation and then vitrified in stainless steel canisters. The ILW is normally immobilised in a cement mixture and placed in stainless steel or concrete containers. Some ILW is immobilised in

**Table 16.6** *Volumes, masses and package numbers at 1 April 2004*

| Waste category | | Volume (m³) | Mass (te) | No. of packages |
|---|---|---|---|---|
| HLW | Conditioned | 456 | 1 200 | 3 037 |
| | Not conditioned | 1 430* | 2 100 | – |
| | Total | 1 890 | 3 400 | 3 037 |
| ILW | Conditioned | 16 400 | 32 000 | 31 028 |
| | Not conditioned | 66 100 | 68 000 | 529 |
| | Total | 82 500** | 100 000 | 31 557 |
| LLW | Conditioned | 1 870 | 1 400 | 123 |
| | Not conditioned | 19 000 | 30 000 | – |
| | Total | 20 900*** | 31 000 | 123 |
| All wastes | Total | 105 290 | 134 400 | |

* When this waste stream is treated, there will a reduction of 545 m³ of volume and an increase of 260 te of mass to this waste stream
** Anticipated ILW arising in future: volume = 134 000 m³ and mass = 150 000 te
*** Anticipated LLW arising in future: volume = 2 030 000 m³ and mass = 2 800 000 te

polymer and placed in mild steel containers. These mild steel containers are going to be overpacked before disposal. Most of the LLW is routinely sent to LLWR where it is packaged and disposed of in the vault. The LLW arising at Dounreay is kept in temporary storage for disposal pending approval of a storage facility on site.

## 16.4.3 Material Composition of Waste

The masses of various categories of wastes existing on 1 April 2004 and likely future arising are shown in Table 16.5. These masses comprise metals, organic and inorganic substances, soil, concrete, building rubbles etc. Table 16.7 gives the composition of these wastes.

HLW arises as a concentrated nitric acid solution containing fission products from the first stage of the reprocessing of spent nuclear fuel. Conditioning of HLW by the vitrification process would contain glass and ceramic products. The mass of conditioned HLW at 1 April 2004 was about 1200 te. When all the HLW is conditioned, a total of 3600 te of wastes would arise.

The ILW and LLW are composed of a variety of materials. Figures 16.4 and 16.5 show the composition of unconditioned ILW and LLW.

**Table 16.7** *Material composition (major contents) of wastes*

| Material[*] | Mass (te) | | |
|---|---|---|---|
| | **HLW** | **ILW** | **LLW** |
| **Metals** | | | |
| Stainless steel | 150 | 26 000 | 67 000 |
| Other steel | 0 | 35 000 | 350 000 |
| Magnox | 0 | 6 900 | 200 |
| Aluminium | 0 | 980 | 5 700 |
| Other metals | 0 | 4 300 | 31 000 |
| **Organics** | | | |
| Cellulose | 0 | 1 500 | 83 000 |
| Plastics | 0 | 4 400 | 28 000 |
| Other organics | 0 | 1 200 | 4 800 |
| **Inorganics** | | | |
| Concrete, rubble etc. | 0 | 56 000 | 550 000 |
| Graphite | 0 | 59 000 | 34 000 |
| Glass and ceramics | 2 600 | 1 100 | 4 400 |
| Sludges, flocs etc. | 0 | 34 000 | 2 800 |
| Other inorganics | 860 | 3 400 | 3 300 |
| Contaminated soil | 0 | 2 900 | 1 600 000 |
| Total | 3 600 | 250 000 | 2 800 000 |

\* Other minor items of material have been omitted

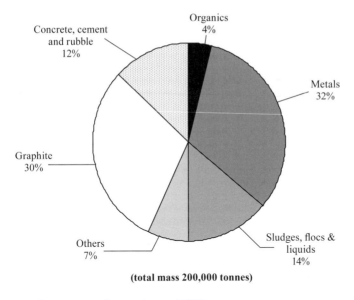

**Fig. 16.4** *Composition of unconditioned ILW*

## 16.4.4 Radioactive Content of Waste

The radioactivity of various categories of wastes on four specified dates is given in Table 16.8. It should be noted that the total activity of a particular waste category at a particular time is estimated on the basis of specific activity of the waste category multiplied by the total volume of waste of that category. Activity decays with time, as per the half-life of the radionuclide, and so the category of waste under consideration would contribute a lower level of activ-

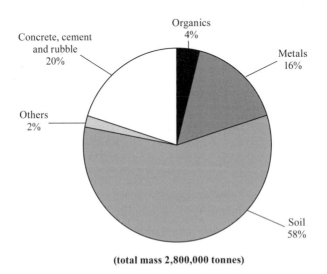

**Fig. 16.5** *Composition of unconditioned LLW*

**Table 16.8** *Activity of all wastes at different times*

| Waste category | Total activity | | | |
| --- | --- | --- | --- | --- |
| | At 1 April 2004 | At 1 April 2005 | At 1 April 2100 | At 1 April 2150 |
| HLW | 7.5E+19 | 4.2E+19 | 1.3E+19 | 4.4E+18 |
| ILW | 4.5E+18 | 1.8E+18 | 8.3E+17 | 5.1E+17 |
| LLW | 2.1E+13 | 2.5E+14 | 2.8E+14 | 3.1E+14 |
| **Total** | **8.0E+19** | **4.4E+19** | **1.4E+19** | **4.9E+18** |

ity at a future date. However, as decommissioning work progresses, new waste streams within that category would increase the amount of activity. Figure 16.6 shows the variation in activity of various categories of wastes as a function of time. At any specific date, the relative contributions of various categories are worth examining, as shown in figure 16.7 based on activity at 1 April 2004.

It should be noted that although HLW offers only 0.1% by volume (see Figure 16.3) of all radioactive wastes, it constitutes 94% of the total activity. On the other hand, LLW offering 90.4% by volume constitutes only 0.0003%

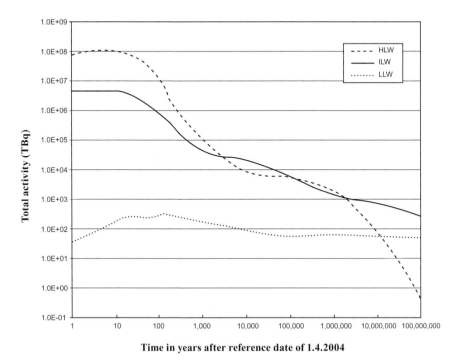

**Fig. 16.6** *Variation of total activities of HLW, ILW and LLW with time*

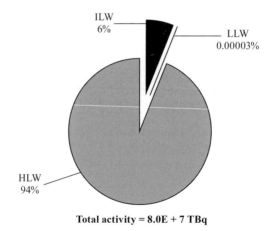

**Total activity = 8.0E + 7 TBq**

**Fig. 16.7** *Proportion of activity by waste type at 1 April 2004*

of the total activity. Figure 16.6 shows the actual levels of activity of various categories of waste in time. The activities of waste arise from contributions from various types of radionuclides. Figures 16.8–16.10 show the time distributions of various radionuclides in the HLW, ILW and LLW categories.

It should be noted that equilibrium activities of short-lived daughter products of Cs-137 (Ba-137m) and of Sr-90 (Y-90) have been included in their

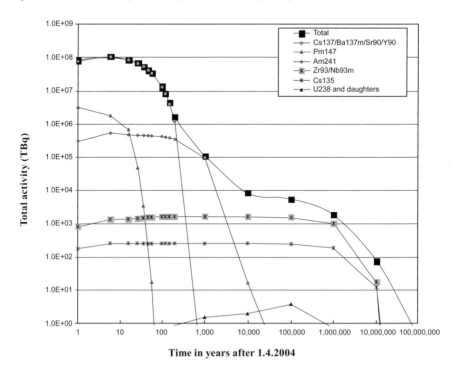

**Fig. 16.8** *Contributions from various radionuclides in HLW*

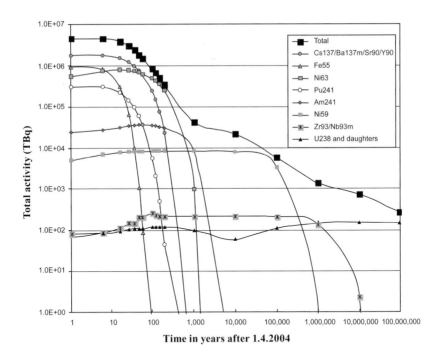

**Fig. 16.9** *Contributions from various radionuclides in ILW*

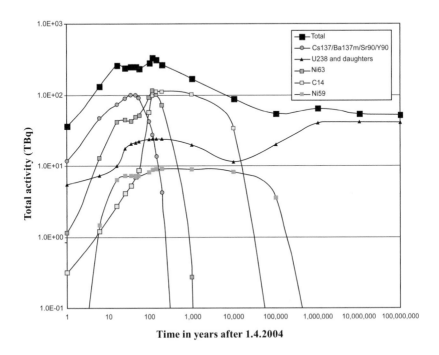

**Fig. 16.10** *Contributions from various radionuclides in LLW*

parent nuclides. Although initially Pm-147 ($t_{1/2}$ = 2.62 y) contributes most to the total activity, over a longer time scale Zr-93 ($t_{1/2}$ = 1.5 × 10$^6$ y) becomes most significant.

### 16.4.5 Waste Substitution

When irradiated fuel is reprocessed HLWs, ILWs and LLWs are produced. The contracts for reprocessing of fuel between the contractor (erstwhile BNFL in the UK) and overseas customers stipulate that various types of wastes be returned to the country of origin. However, the BNFL proposed, and the UK government approved, an arrangement for the substitution of HLW for the generated LLW in their document entitled 'Review of Radioactive Waste Management Policy – Final Conclusions' [8]. In 2004, the government extended this approval to ILW substitution [9]. This waste substitution policy allows that an amount of HLW from the UK stock be added to the overseas HLW that is going to be sent back to oversees customers to compensate for a radiologically equivalent amount of ILW and LLW that is retained in the UK. Thus the overseas customers will receive in concentrated form (HLW) the same amount of activity that is contained in the original spent fuel. That leaves each party to have the same amount of waste in radiological terms, but the volume of waste that is to be shipped and consequently the costs of shipment are significantly reduced. This substitution policy is somewhat controversial. The effect of such a policy is that the HLW stock in the UK is going to be somewhat reduced at the expense of increase in ILW and LLW inventory.

## 16.5 Partitioning and Transmutation

Partitioning and Transmutation (P&T) is a technology under development for reducing the inventory of long-lived high level radioactive wastes [10]. Transmutation is the process of changing one nuclide to another as a result of neutron interactions (capture and emission) with a target nucleus. It is mostly effected by the bombardment of neutrons on target atoms in a nuclear reactor, or, more recently, in a particle accelerator. The aim of this process is to transmute longer-lived nuclides into shorter lived or stable nuclides. As a precursor to transmutation, it is necessary to chemically separate out some of the nuclides from the waste containing long-lived radionuclides; this is known as partitioning. This chemical separation avoids the likelihood of unwanted interaction of neutrons with materials that could produce other long-lived radionuclides and also reduces the length of time the long-lived radionuclides is required to be irradiated.

The P&T process may be successfully applied to long-lived fission products such as Tc-99, I-129 as well as to some minor actinides such as Am-241 and Np-237. The key parameters in this reaction are the neutron capture cross-section by the target atoms, the applied neutron fluence rate and its energy spectrum. It should be noted out that this technology may only be applied, when developed, on HLW (solid or liquid) before conditioning, i.e. before vitrification or encapsulation. Vitrified HLW cannot be used and that is the reason why partitioning is required for this technique.

Research and Development (R&D) programmes on the irradiation facility, called Accelerator Driven System (ADS), are now being pursued at CEA of France, JAERI of Japan, LANL of America and CERN in Europe, which combine a high intensity particle accelerator with spallation targets and a sub-critical core of nuclear fuel. An energetic beam of particles (usually protons) is produced by a particle accelerator, which will then be allowed to impinge on spallation target made of a heavy metal such as Pb or Pb–Bi. The target will be surrounded by a blanket of assemblies containing chemically separated (partitioned) waste and then a core of fissile material which will be operated at sub-critical levels. The sub-critical core is flexible in operation: it can be operated at either thermal or fast neutron spectrum. The waste will experience neutron fluence of the chosen energy spectrum in excess of that available from a small reactor, thus expediting the transmutation process.

## 16.6  Waste Inventory Produced by the CoRWM

The Committee on Radioactive Waste Management (CoRWM) was set up by the UK government in November 2003 as an independent body in response to public comments made in the consultation process following the publication of 'Managing Radioactive Waste Safely' [4]. The committee was then asked to review the options for long-term management of the UK's higher level solid radioactive wastes and make recommendations to the government by July 2006. The higher levels of waste include ILW, HLW and other wastes which are not acceptable in the LLWR at Drigg. The CoRWM had produced the 'Recommendations for the Long Term Management of Radioactive Waste' [11] with 15 recommendations (see Section 20.5). In order to identify the extent of the problem and define disposal options, CoRWM produced an inventory of radioactive wastes and materials projected up to year 2120 (see Table 16.9). In this estimation, contributions from 10 new AP1000 nuclear reactors which could use up the UK stockpile of MOX fuel have been assumed.

The CoRWM estimates that the volume of waste that will have to be managed (to 2120) approximates 478,000 $m^3$ and the activity is 78 million terabecquerels ($78 \times 10^{18}$ Bq) when these wastes are treated and conditioned. The ILW makes up approximately 74% of the volume and uranium is about 16% and the other categories make up around 10% of volume. While the combined volume of HLW and spent fuel is less than 2% of total volume, they constitute 92% of the total radioactivity.

**Table 16.9** *Radioactive waste inventory in the UK to 2120*

| Material | Packaged volume (m$^3$) | Volume (%) | Activity (Bq) | Activity (%) |
|---|---|---|---|---|
| HLW | 1290 | < 0.3 | 3.8E+19 | 50 |
| ILW | 353 000 | 73.9 | 2.4E+18 | 3 |
| LLW (non-Drigg)* | 37 200 | 7.8 | <1.0E+14 | <0.001 |
| Pu (separated)** | 3 270 | 0.7 | 4.0E+18 | 5 |
| U*** | 74 950 | 15.7 | 3.0E+15 | <0.01 |
| Spent fuel | 8 150 | 1.7 | 3.3E+19 | 42 |
| Total | 477 860 | 100 | 7.7E+19 | 100 |

\* LLWs that cannot be disposed of at the LLWR due to radioactive content, (such as α-emitters) or physical/chemical properties that do not meet the site waste acceptance criteria
\*\* Pu extracted from the irradiated nuclear fuels by reprocessing
\*\*\* U extracted from the irradiated nuclear fuels as well as arising from processing of raw U in the form of highly enriched, low enriched and depleted U

## Revision Questions

1. How is 'radioactive substance' defined in the RSA93 of the UK? What are the categories of radioactive substance? Describe them and explain their significance.

2. On what basis are radioactive wastes normally classified? Describe the two most widely used waste classification schemes.

3. Describe the IAEA waste classification scheme and the associated disposal options.

4. What is the prime consideration in the waste classification scheme in the UK? Describe the waste categories with numerical values and specify the disposal routes.

5. Define the EU waste categories. Compare and contrast these categories with those of the IAEA scheme.

6. What are the radioactive waste types that may arise from nuclear activities? Give a short description of their sources.

7. What is the waste substitution policy, as practised in the UK reprocessing industry? Describe this policy and highlight its significance.

8. Explain the principle and the proposed practice of transmutation of long-lived HLW. Why is partitioning necessary before transmutation?

9. Write a short note on the functions and responsibilities of CoRWM.

## REFERENCES

[1] UK Government, Radioactive Substances Act 1993, The Stationary Office.

[2] International Atomic Energy Agency, Classification of radioactive waste, A safety guide, IAEA Safety Series No. 111-G-1.1, Vienna, Austria, 1994.

[3] International Atomic Energy Agency, Radioactive Waste Management, Status and Trends, IAEA/WMDB/ST/1, Vienna, Austria, August 2001.

[4] UK Government, DEFRA and the Devolved Administrations for Scotland, Wales and Northern Ireland, Managing Radioactive Waste Safely – Proposals for Developing a Policy for Managing Solid Radioactive Waste in the UK, September 2001.

[5] Rahman A., Nuclear Waste Management in France, *Nuclear Energy, Journal of the British Nuclear Energy Society*, 2001,40(6) pp. 391–396.

[6] European Commission, Commission Recommendation of 15 September 1999 on a Classification System for Solid Radioactive Waste, 1999/669/EC (Euratom), L 265/37 of 13 October 1999.

[7] UK Government, DEFRA, The 2004 UK Radioactive Waste Inventory, Main Report, DEFRA/RAS/05.002, (Nirex Report N/090), October 2005.

[8] UK Government, Review of Radioactive Waste Management Policy – Final Conclusions, Cm 2919, July 1995.

[9] UK Government, Department of Trade and Industry, Intermediate Level Radioactive Waste Substitution, December 2004, http://www.dti.gov.uk/files/file30058.pdf.

[10] UK Nirex Limited, Applicability of Partitioning and Transmutation to UK Wastes, Technical Note, 2002.

[11] Committee on Radioactive Waste Management, Recommendations for Long Term Management of Radioactive Waste, July 2006, http://www.corwm.org.uk.

# 17

# MANAGEMENT OF RADIOACTIVE WASTE

## 17.1 Introduction

Radioactive wastes are generated during the operational phase of a nuclear facility as well as during decommissioning which encompasses decontamination, dismantling and land remediation. The safe management of radioactive wastes at all stages is an essential regulatory requirement. The management here covers the whole sequence of operations starting with the generation of waste and ending with its final disposal. Various types of wastes are produced at various stages of the life of the plant. Figure 17.1 shows the various activities in the management of waste, whether it is generated during the operation or during decommissioning. Major activities such as treatment and conditioning; storage and transportation; and disposal are considered in full detail in separate chapters.

This chapter deals primarily with the national and international regulatory requirements and safety standards associated with the management of radioactive waste. As waste disposal or dispersal has implications beyond the site or even beyond the country, many of these regulations are internationally driven. The regulatory issues associated with the transportation of radioactive wastes by land, sea and air are not covered here, although waste management, in principle, includes all of these issues including transportation. The transportation regulations covering transport packaging, testing, safety requirements etc. are considered in Chapter 19.

## 17.2 General Principles of Waste Management

The general principles for the management of radioactive wastes can be described under the headings of: (i) waste minimisation; (ii) sustainable development; and (iii) the polluter pays principle.

### 17.2.1 Waste Minimisation

UK government policy [1] as well as the regulatory requirement [2] states that the production of radioactive waste either during the operation and/or during decommissioning should be avoided as far as possible. Where the production

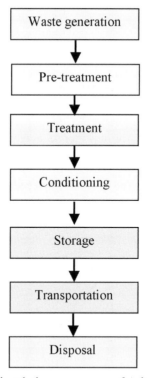

**Fig. 17.1** *Activities associated with the management of radioactive waste*

of waste is unavoidable, it should be minimised. This requirement ensures improved plant design and better operational practice. One example where the waste minimisation technique can be applied is that if heavy metals are used in areas where there are significant neutron fluxes, activation products would be generated, this could possibly be avoided by using non-metals. The essential properties of metals such as their structural strength, thermal conductivity etc. will have to be taken into account when non-metals are considered as their replacements. But within these constraints, there may be occasions when non-metals could be used. Another example is that the surfaces of building structures, glove boxes etc. may be made of non-absorbent or low absorbent materials so that during decommissioning, simple skimming of the surface would decontaminate the structure. Once wastes have been generated, they need to be appropriately identified, treated and conditioned. These steps are necessary to minimise the volume of the waste, which can then be safely disposed of at appropriate times and in appropriate ways.

## 17.2.2 Sustainable Development

Sustainable development is a rather complex concept embracing the environmental issues and socio-economic priorities of the present generation and the perceived priorities of future generations. The present generation may like to benefit financially and achieve higher standards of living now from the nuclear practices, but it must be done without harming the environment or leaving a legacy of technical mismanagement for the future generations. These objectives are reconciled within the 'sustainable development' principle. Briefly this principle can be described as 'development that meets the needs of the present generation without compromising the ability of future generations to meet their own needs' [3]. This principle may be incorporated by utilising best possible scientific and technical standards and knowledge in implementing present activities. The assessment of risks now and in the future from all anticipated causes needs to be addressed. Where there are uncertainties, precautionary, conservative assessment as against best-estimate assessment may be undertaken to view the outcome of the operation.

### 17.2.3 'Polluter Pays' Principle

The 'polluter pays' principle is, to some extent, an offshoot of sustainable development.

The producers and owners of radioactive waste are responsible for bearing the costs of managing the wastes. These responsibilities are not limited to bearing the costs of managing and disposing of waste, but also include the cost of the R&D undertaken by themselves and by the regulatory bodies [1].

## 17.3 Regulatory Issues of Waste Management

It is important to note that the safety principles and practices that apply during the operational phase also apply to the management of radioactive waste during the decommissioning phase. This is due to the fact that the UK regulators consider decommissioning to be an integral and essential part of the operational activity and so all the initial safety requirements and licence conditions that were applied under the operational licence remain valid during decommissioning. The regulatory aspects of decommissioning have been described in detail in Chapter 6.

Regulation of radioactive substances with regard to use, accumulation, disposal or reuse is carried out in the UK under the provisions of RSA93 [4] and the European BSS [5]. The European BSS was promulgated in 1996, later than the RSA93 which was promulgated in 1993. However, there is a large measure of agreement between these sets of provisions. But, in places where they differ, the general principle is that the more restrictive clause is adhered to, as the less restrictive clause would then automatically be satisfied. The RSA93 requires that the use or keeping of radioactive material is subject to notification and registration with the regulatory body. After the use of a radioactive material, the material may be considered as radioactive waste.

The waste may be disposed of, but the disposal (in whatever form) requires authorisation from the regulatory body. The definition for disposal includes both the emplacement of solid waste in a disposal site and the dispersion of effluents (liquid or gaseous substances) in the environment.

## 17.3.1 The Joint Convention

In the field of radioactive waste management, an international legal instrument under the title 'The Joint Convention on the Safety of Spent Fuel Management and on the Safety of Radioactive Waste Management' (Joint Convention) [6] came into force on 18 June 2001. Contractual parties are legally bound to meet the obligations of the convention. The UK is a signatory to this convention.

The Joint Convention applies to spent fuel and radioactive waste resulting from civilian nuclear activities and to spent fuel and radioactive waste from military or defence programmes when such substances are transferred permanently to and managed within exclusively civilian programmes. The convention also applies to planned and controlled releases into the environment of liquid or gaseous radioactive materials from regulated nuclear facilities.

The obligations laid down under the Joint Convention on the contracting parties are based to a large extent on the principles contained in the International BSSs [7]. They include, in particular, the obligation to establish and maintain a legislative and regulatory framework to govern the safety of spent fuel and radioactive waste management and the obligation to ensure that individuals, society and the environment are adequately protected from radiological and other hazards. These objectives can be achieved by appropriate provisions of siting, design and construction of facilities and by making provisions for ensuring the safety both during operation and following closure.

There are 44 articles in this convention. When a contracting party complies with these articles, that party will automatically comply with all other relevant international treaties and conventions. The major elements of the convention are:

**Article 4: General Safety Requirements**
- To take appropriate steps to ensure that at all stages of spent fuel management, individuals, society and the environment are adequately protected against radiological hazards. This can be done by
    - ensuring criticality and residual heat removal during spent fuel management are adequately addressed
    - ensuring generation of radioactive waste associated with spent fuel management is kept to the minimum practicable

- providing effective protection to the individuals, society and the environment by applying nationally and internationally endorsed safety criteria and standards

### Article 5: Existing Facilities

- To take appropriate steps to review the safety of any spent fuel management facility and to ensure that all necessary practicable improvements are made to upgrade the safety.

### Article 6: Siting of Proposed Facilities

- To take appropriate steps to ensure that procedures are established and implemented for a proposed spent fuel management facility by
  - evaluating all relevant site-related factors likely to affect the safety of such a facility during its operating lifetime
  - evaluating the likely safety impact on individuals, society and the environment
  - making safety information available to members of the public
  - consulting other contracting parties in the vicinity and giving them general data to evaluate the safety impacts of the facility on their territory

### Article 8: Assessment of Safety of Facilities

- To take appropriate steps, before construction of a spent fuel management facility, that a systematic safety assessment and an environmental assessment are carried out.

### Article 19: Legislative and Regulatory Framework

- To establish and maintain a legislative and regulatory framework to govern the safety of spent fuel and radioactive waste management. The legislative and regulatory framework should cover
  - the establishment of national safety requirements and regulations for radiation safety
  - a system of licensing of spent fuel and radioactive waste management activities
  - a system of institutional control, regulatory inspection and documentation and reporting

### Article 21: Responsibility of the Licence Holder

- To make sure that the prime responsibility for the safety of spent fuel or radioactive waste management lies with the holder of the relevant licence.

### Article 22: Human and Financial Resources

- To ensure adequate numbers of qualified staff are available for safety related activities.
- To make provisions for adequate financial resources to support the safety of facilities for spent fuel and radioactive waste management during their operating lifetime and for decommissioning.

### Article 24: Operational Radiation Protection

- To ensure that radiation exposure of the workers and the public is kept as low as reasonably achievable, economic and social factors being taken into account.
- To ensure that no individual shall be exposed, in normal situations, to doses which exceed national prescriptions for dose limitation.
- To ensure that discharges shall be limited to keep exposures to radiation as low as reasonably achievable and no individual shall exceed national dose limits.

### Article 26: Decommissioning

- In order for the decommissioning work to be carried out safely, qualified staff and adequate financial resources are to be made available.
    - The provisions of Article 25 dealing with emergency preparedness are applied.
    - Records of information important to decommissioning are kept.

### Article 27: Trans-boundary Movement

- To ensure that trans-boundary movement of radioactive waste and spent fuel are undertaken in a manner consistent with the provisions of this convention and other binding international instruments. This is achieved by the following steps:
    - The contracting party originating spent fuel or radioactive waste shall take steps to ensure that trans-boundary movement is authorised and takes place only with the prior notification and consent of the State for which it is destined.
    - Trans-boundary movement through States of transit shall be subject to international obligations.

It should be noted that this Joint Convention on the management of spent fuel and radioactive wastes can be viewed as a legal framework which complements the 1989 Basel Convention on the control of trans-boundary movements of hazardous wastes and their disposal [8]. The Basel Convention was developed in 1989 under the umbrella of the United Nations Environment Programme (UNEP) to address the issues of hazardous wastes from industrialised countries being dumped into developing countries. It may be noted that a radioactive substance below the clearance level is not covered by the Joint Convention and hence Article 27 does not apply in matters of trans-boundary movement. However, if the cleared substance is still regarded as hazardous due to its chemical or biological toxicity, it will be covered by the Basel Convention.

In the UK, the issue of authorisations for the import and export of radioactive waste into and out of England and Wales is dealt with the EA under the provisions of the Council Directive 92/3/Euratom [9]. However, the UK government maintains that radioactive wastes should not generally be imported to or exported from the UK unless in exceptional circumstances [1]. The exceptional circumstances may include situations when small users such as hospitals situated in EU Member States produce such small quantities that the provision of their own specialised facilities would be impracticable or developing countries which cannot reasonably be expected to acquire suitable disposal facilities.

## 17.3.2 The OSPAR Convention

The OSPAR Convention 1992 [10] came into force on 25 March 1998 and replaced the 1972 Oslo Convention (prevention of marine pollution by dumping from ships and aircraft) and the Paris Convention 1974 (prevention of marine pollution from land-based sources). The original Oslo and Paris Conventions were administered by the Oslo and Paris Commissions and these also ceased to exist when the OSPAR Commission (OSPARCOM) was created to administer the new convention. The contracting parties to the 1992 OSPAR Convention are: Belgium, Denmark, the European Community, Finland, France, Germany, Iceland, Ireland, Luxembourg, the Netherlands, Norway, Portugal, Spain, Sweden, Switzerland and the United Kingdom of Great Britain and Northern Ireland. The contracting parties are required to prevent and, where possible, eliminate pollution of the marine environment (previous conventions merely required a reduction in pollution). The new convention places particular emphasis on the use of the 'polluter pays principle' and the 'precautionary approach'. It also places particular emphasis on preventing pollution from diffuse sources and, to this end, a list of substances contributing to diffuse pollution has been identified. But its most significant element is that actions must be taken to ensure that discharges, emissions and losses of radioactive substances are reduced by the year 2020 to levels where the additional concentrations in the marine environment above historic levels, resulting from such discharges, emissions and losses, are close to zero.

At the OSPAR Commission and the ministerial meeting at Sintra, Lisbon on 20–24 July 1998, the contracting parties committed themselves to end all opt-out clauses on the sea dumping of nuclear wastes and to ban dumping of all steel offshore installations to sea.

In order to comply with the commitments of the OSPAR Convention, the UK Government produced a strategy document for radioactive discharges 2001–2020 [11] which gives detailed action plans for the implementation of the convention in the UK.

### 17.3.3  Waste Disposal: Article 37 of the Euratom Treaty

The disposal of radioactive waste in any form is subject to a stringent control under the Article 37 of the Euratom treaty [12]. Specifically this Article 37 requires:

'Each Member State is to provide the commission with such general data relating to any plan for the disposal of radioactive waste in whatever form as will make it possible to determine whether the implementation of such plan is liable to result in the radioactive contamination of the water, soil or airspace of another Member State'.

The implementation of this requirement is that before any disposal of radioactive waste is authorised by the competent authority of a Member State, the commission must be provided with the general data regarding such a plan and commission's opinion must be sought.

It is thus necessary to determine which types of operation may result in the disposal of wastes (in solid, liquid or gaseous forms) within the meaning of Article 37 of the treaty. The OSPAR Commission recommends that the following operations should be included:

    (1)   operation of nuclear reactor
    (2)   reprocessing of irradiated nuclear fuel
    (3)   mining, milling, conversion of uranium and thorium
    (4)   enrichment of uranium
    (5)   fuel fabrication
    (6)   storage of irradiated fuel
    (7)   processing and storage of radioactive waste
    (8)   dismantling of nuclear reactors and reprocessing plants
    (9)   placement at surface or underground of radioactive waste

In addition to the above-mentioned operations, if there are modifications to the plant requiring discharge limits to be increased or increased accident consequences as a result of increased radioactive inventory or changes in storage arrangements, OSPARCOM needs to be provided with sufficient details.

#### 17.3.3.1  General Data Requirement

The general data requirement for items (1)–(7) above is specified in Annex 1 of [12] which specifies

- **The site and its surroundings**: Geographical, topographical and geological features of the site and the region should be provided.

The location of the installation in relation to other installations and the location of the site with regard to other Member States giving the distances from frontiers and nearest conurbations together with their population should be provided.

- **Seismology**: Information on the degree of seismic activity in the region (probable maximum seismic activity and designed seismic resistance of the installation) should be provided.
- **Hydrology**: If the installation is close to a water body, potential contamination pathways to another Member State should be given. This should include a brief description of tributaries, estuary, water abstraction, underground water table with levels and flows, flood plains etc. It should also include average, maximum and minimum water flows, direction and force of currents, tides, circulations patterns etc.
- **Meteorology**: Local meteorological conditions such as: average wind speeds and directions, rainfall, atmospheric dispersion conditions, average duration of temperature inversions etc. should be given.
- **Natural resources and foodstuffs**: Soil characteristics and ecological features of the region should be given. A brief description of the principal food resources such as crops, fishing, hunting and for discharges into the sea, data on fishing in territorial and extra-territorial waters should also be provided.
- **Details of the installation**: Brief description of the installation including type, purpose, site layout plan and safety provisions should be given. A description of the ventilation, filtration and airborne discharge systems, in normal and accident conditions, is to be provided. Descriptions of liquid waste treatment and solid waste treatment facilities as well as storage capacities are to be given.

The requirements for operations listed under items 8 and 9, respectively, are given in Annexes 2 and 3 of [12]. The general provisions are similar to those specified above except that in dismantling operations, unplanned releases, emergency provisions and environmental monitoring provisions should be provided. For radioactive repositories, radiological impact assessment post-closure is required.

### 17.3.3.2 Application of Article 37

The UK, being a part of the EU, is bound by the requirement of this Article. The government bodies in charge of making a submission under this Article are the DEFRA in England and the Scottish Government in Scotland. They decide which operations require submissions. The regulatory bodies (EA in England and Wales and SEPA in Scotland, working together with DEFRA) must agree on the scope of submission and when it should be made. In England, the operator in consultation with the EA prepares a submission which normally takes 6–18 months, depending on the complexity of the plant in question. After the submission by the UK government, OSPARCOM takes up to six months to consider the submission, during which time it consults Article 37 experts. It then decides whether or not the plan is liable to lead to contamination of other Member States and then publishes its opinion.

To prepare a single reactor decommissioning submission, takes between 12 and 24 months including six months of assessment time by OSPARCOM. For the operator, it may take six person-months costing about £50,000 to prepare the submission.

## 17.4 Exemption and Clearance Levels

In order to optimise regulatory functions and effective implementation regulatory duties, the regulatory bodies establish levels, called **clearance levels**, below which disposal, recycling or reuse of materials would not be subject to regulatory controls. Alongside this clearance level, there is another term, the **exemption level,** which is used to exempt sources from the regulatory requirements of notification and registration. Whereas the exemption level of a particular radionuclide allows the owner/operator to be relieved of the responsibilities of notification and registration, it does not allow the owner/operator to dispose of the material at the end of its use without authorisation. If the material to be disposed of is below or equal to the clearance level, the owner/operator may apply for authorisation from the regulatory body for uncontrolled release in the environment. The European BSS has also introduced the term **exclusion** of radiation sources. The significance of these terms are explained and, where possible, quantified below.

### 17.4.1 Exclusion of Sources

A source giving an exposure should be excluded from regulatory controls if it is not amenable to control. Such sources are natural radioactive sources. An example is the cosmic radiation at ground level. Human beings are continuously exposed to it, albeit at very low levels. By setting limits of exposure from this source, the regulatory body may set a condition which is difficult to monitor and impossible to implement. In any case, the risk from such low levels of exposure is so small that it can be considered to be trivial. Consequently it is sensible to put the source outside the regulatory regime. In other words, it is excluded from regulatory control. However, there may be situations when the exclusion of cosmic radiation may be rescinded, for instance when its levels become high enough to confer significant risks to people, that may arise when people make frequent high altitude flights. Other examples of exclusion are primordial long-lived radionuclides such as K-40, Rb-87 etc. which exist in the earth's crust at varying but extremely low concentrations. The exposures from these sources are so small that they may be considered totally insignificant.

There is some controversy whether the same principle of exclusion can be applied to artificial, man-made radionuclides when exposures are at very low levels. The IAEA considers that such exclusions would be unwise and the best approach to tackling such radionuclides with very low levels of exposures would be through specifying exemption levels.

## 17.4.2 Exemption Levels

Practices and sources within a practice may be exempted from the regulatory controls such as notification and registration for use and accumulation provided that the following basic criteria are met [7]:

- The radiological risks to individuals caused by the exempted practice are sufficiently low as to be of no regulatory concern.
- The collective radiological impact of the exempted practice is sufficiently low.
- The exempted practice is inherently without radiological significance, with no appreciable likelihood of scenarios that could lead to a failure to meet the above criteria.

The above generic criteria need to be met along with the following specific radiological criteria in order that a practice or a source may be exempted:

- The effective dose incurred by a member of the critical group is less than or equal to $10\ \mu Sv.y^{-1}$.
- The equivalent dose to the skin is limited to 50 mSv. $y^{-1}$.
- The collective effective dose commitment from one year of operation is no more than 1 man.Sv.

The effective dose of $10\ \mu Sv.y^{-1}$ to an individual is in fact a rounded value. If the dose is in the range $3$–$30\ \mu Sv.y^{-1}$, it is rounded to $10\ \mu Sv.y^{-1}$ as the geometric mean. The general principle is that if the value lies between $3 \times 10^x$ and $3 \times 10^{x+1}$, the rounded value is $10^{x+1}$.

The exemption levels for a total of 300 radionuclides with actual and potential uses have been calculated and presented in the European BSS [5] and the International BSS [7]. In calculating these levels, all possible physical forms of these radionuclides and three basic scenarios, e.g. normal use, accidental exposure and the disposal were considered. Four exposure pathways were considered and these were: inhalation, ingestion, external $\gamma$-radiation and $\beta/\gamma$-skin contamination. For accidental exposure calculation, the most conservative scenario was considered with a limiting effective dose of 1 mSv with a frequency of occurrence that is lower than once in every 100 years ($10^{-2}\ y^{-1}$). Thus the requirement of $10\ \mu Sv.y^{-1}$ is fulfilled in this probabilistic calculation.

**Table 17.1** *Exemption levels for some radionuclides*

| Radionuclide | Concentration (Bq.g$^{-1}$) | Quantity (Bq) |
|---|---|---|
| Tritiated compounds | $10^6$ | $10^9$ |
| C-14 | $10^4$ | $10^7$ |
| K-40 | $10^2$ | $10^6$ |
| Fe-55 | $10^4$ | $10^6$ |
| Co-60 | $10^1$ | $10^5$ |
| Ni-63 | $10^5$ | $10^8$ |
| Zn-65 | $10^1$ | $10^6$ |
| Sr-90 | $10^2$ | $10^4$ |
| Ru-105 | $10^1$ | $10^6$ |
| I-131 | $10^2$ | $10^6$ |
| Cs-137 | $10^1$ | $10^4$ |
| Pb-210 | $10^1$ | $10^4$ |
| Po-208 | $10^1$ | $10^4$ |
| Ac-225 | $10^1$ | $10^4$ |
| U-235 | $10^1$ | $10^4$ |
| Pu-239 | $1$ | $10^4$ |
| Pu-241 | $10^2$ | $10^5$ |
| Am-241 | $1$ | $10^4$ |

The skin dose of 50 mSv.y$^{-1}$ is also taken into account in order to avoid deterministic effects.

The exemption levels for some of the radioactive substances are given in Table 17.1. As long as the total activity of a given radionuclide present on the premises at any time or the activity concentration does not exceed the value specified, the practice or the source may be exempt from notification and registration for use.

In addition to a practice or a source, some radiation generating equipment such as cathode ray tube may also be exempt provided that

- They do not cause under normal operating conditions an ambient dose equivalent or directional dose equivalent in excess of 1 μSv.h$^{-1}$ at a distance of 0.1 m from the accessible surface.
- The maximum energy of the radiation produced is no greater than 5 keV.

In the UK, exemption orders are produced by the authorising body (DEFRA) under the RSA93 in order to allow use of radioactive substances by minor users where there is a clear benefit from their use whilst ensuring protection of the environment and the public. At present there are 18 exemption orders and they are listed under the following categories [13]:

(1) exemption orders relating to natural radioactivity
(2) exemption orders relating to products containing radioactivity

(3)  exemption orders relating to specific types of undertaking
(4)  exemption orders relating to the transit of radioactivity
(5)  exemption orders relating to substances of low activity

The concepts of exemption and clearance are shown diagrammatically in Figure 17.2.

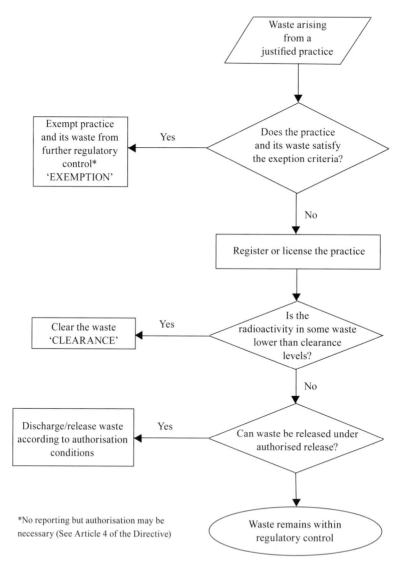

**Fig. 17.2**  *Concepts of exemption and clearance*

## 17.4.3 Clearance Levels

Clearance is a related concept to exemption. Whereas exemption levels for radiation sources define levels which do not enter into the regulatory regime, i.e. exempted from the requirements of notification and registration, clearance levels define levels which are released from regulatory controls after consideration of radiological consequences of release of those substances to the environment. A clearance level is specified by the regulatory body on a case-by-case basis. A clearance may be made either under the general clearance level when no restrictions are placed on its destination or under the specific clearance level when its destination or particular use may be specified. Obviously, the general clearance level must be lower than the specific clearance level, as it entails no restriction.

### 17.4.3.1 General Clearance Levels

When a material is cleared under the general clearance level, it can be recycled, reused or disposed of without any further restriction. The term recycling means the use of the material as a raw material for the manufacture of a new material or a new product. The scrap metal can be recycled to manufacture metal products whereas building rubble can be recycled to make building materials. During the manufacturing process, uncontaminated materials are usually mixed with recycled material with the result that the specific activity of the recycled material is reduced further. However, while specifying the clearance level, this process of dilution must not be taken into account. The general clearance may also be given for reuse of material such as equipment, tools, gears etc. Once cleared, these items may end up in public use and hence great care needs to be applied. The disposal of cleared materials is also without restriction. Such materials can be disposed in landfill sites, old quarries etc.

The radiological assessments to specify the general clearance levels should cover consequences to the workers as well as to members of the public from significant pathways such as inhalation, ingestion, external $\gamma$-radiation and $\beta/\gamma$-skin dose arising from normal use of material during the process of clearance, accidental exposures and disposal. The radiological criteria which are applied here are similar to those used for the exemption level. However, whereas in the estimation of exemption levels, rounding approximations in exposure level and probabilistic considerations are given, the clearance level estimation is primarily based on the effective dose of 10 $\mu$Sv.y$^{-1}$. Additional criteria of 50 mSv.y$^{-1}$ of skin dose and the collective effective dose of 1 man. Sv are also applied. Finally, a check is made that the estimated clearance level is never higher than the exemption level. If it is, then the lower value is the accepted clearance level.

The derivation of clearance levels based on each of the stated pathways to each of the population group is given below [14]. The final accepted value is the the one which gives the lowest value.

## Inhalation

The inhalation dose is calculated by using the following equation:

$$H_{inh} = h_{inh}.t_e.f_d.C_c.V.e^{-\lambda t_1}\frac{1-e^{-\lambda t_2}}{\lambda t_2} \tag{17.1}$$

where

$H_{inh}$ is the committed effective dose from inhalation per year of unit activity concentration of cleared material $(Sv.y^{-1})/(Bq.g^{-1})$

$h_{inh}$ is the dose coefficient for inhalation $(Sv.Bq^{-1})$

$t_e$ is the exposure time in hours per year $(h.y^{-1})$

$f_d$ is the dilution factor for activity concentration in air

$C_c$ is the concentration of dust of material under consideration in air $(kg.m^{-3})$

$V$ is the breathing rate $(m^3.h^{-1})$

$\lambda$ is the decay constant $(y^{-1})$ of the cleared material

$t_1$ is the decay time before the start of scenario (y)

$t_2$ is the decay time during the scenario (y)

Using this equation, $H_{inh}$ can be calculated for a specified scenario. When the dose limit of 10 $\mu Sv.y^{-1}$ is divided by this quantity, $H_{inh}$, one obtains the activity concentration in terms of $Bq.g^{-1}$. As stated above, the lowest value from any of the pathways is the final clearance level.

---

### Example 17.1

Calculate the committed effective dose arising from inhalation per year by a worker of unit concentration of Co-60 dust.

Given that $h_{inh} = 7.1 \times 10^{-9}$ $Sv.Bq^{-1}$ (for 5 $\mu$m AMAD particle and with moderate absorption rate: from Table II-III of [7]

$t_e = 1800$ h.y$^{-1}$ $\qquad$ $f_d = 1$

$C_c = 10^{-6}$ kg m$^{-3}$ $\qquad$ $V = 1.2$ m$^3$ h$^{-1}$

$t_{1/2} = 5.27$ y $\qquad$ $t_1 = 0$ i.e. no decay before the scenario

$t_2 = 0$ i.e. no decay before the scenario

*Solution*

As $t_1 = 0, e^{-\lambda t_1} = 1$

and, as $t_2 = 0, \dfrac{1-e^{-\lambda t_2}}{\lambda t_2} = 1$

$H_{inh} = 7.1 \times 10^{-9}$ (Sv.Bq$^{-1}$) $\times 1800$ (h.y$^{-1}$) $\times 10^{-6}$ (kg.m$^{-3}$) $\times 1.2$ (m$^3$.h$^{-1}$)

$\qquad = 1.5 \times 10^{-11}$ (Sv.y$^{-1}$)/(Bq.kg$^{-1}$)

$\qquad = 1.5 \times 10^{-2}$ ($\mu$Sv.y$^{-1}$)/(Bq.g$^{-1}$)

## Ingestion

Both workers and the general public may ingest contaminated material. For workers, it is mainly via hand-to-mouth contact, whereas for the public it is via the food chain. The dose arising from ingestion can be calculated by the following equation:

$$H_{ing} = h_{ing} \cdot q \cdot f_d \cdot f_c \cdot e^{-\lambda t_1} \cdot \frac{1 - e^{-\lambda t_2}}{\lambda t_2} \tag{17.2}$$

where $H_{ing}$ is the committed effective dose from ingestion per year of unit activity concentration of cleared material $(Sv.y^{-1})/(Bq.g^{-1})$
  $h_{ing}$ is the dose coefficient for ingestion $(Sv.Bq^{-1})$
  $q$ is the ingested quantity per year $(kg.y^{-1})$
  $f_d$ is the dilution factor
  $f_c$ is the concentration factor for activity in the ingested material
  $\lambda$ is the decay constant $(y^{-1})$ of the cleared material
  $t_1$ is the decay time before the start of scenario (y)
  $t_2$ is the decay time during the scenario (y)

---

### Example 17.2

Calculate the committed effective dose from ingestion (e.g. via hand-to-mouth) per year of Co-60 by a worker.

Given that $h_{ing} = 3.4 \times 10^{-9}$ $Sv.Bq^{-1}$

$$q = 20 \times 10^{-3} \text{ kg.y}^{-1}$$
$$t_{\frac{1}{2}} = 5.27 \text{ y}$$
$$t_1 = 0, \text{ i.e. no decay before the scenario}$$
$$t_2 = 0, \text{ i.e. no decay during the scenario}$$

*Solution*

As $t_1 = 0$, $e^{-\lambda t_1} = 1$

$f_c$ and $f_d$ are assumed to be 1

and, as $t_2 = 0$, $\dfrac{1 - e^{-\lambda t_2}}{\lambda t_2} = 1$

$$\begin{aligned} H_{ing} &= 3.4 \times 10^{-9} \text{ (Sv.Bq}^{-1}) \times 20 \times 10^{-3} \text{ (kg.y}^{-1}) \\ &= 6.8 \times 10^{-11} \text{ (Sv.y}^{-1})/(Bq.kg^{-1}) \\ &= 6.8 \times 10^{-1} \text{ (}\mu Sv.y^{-1})/(Bq.g^{-1}) \end{aligned}$$

---

## External Irradiation

External irradiation from cleared materials can arise from a number of scenarios. For example, a worker working on the disposal of cleared waste, i.e. a landfill worker may be exposed to external radiation from disposed material or a person may live in a house built using cleared building rubble. The dose from external irradiation is calculated as

$$H_{ext} = h_{ext} \cdot t_e \cdot f_d \cdot e^{-\lambda t_1} \cdot \frac{1-e^{-\lambda t_2}}{\lambda t_2} \qquad (17.3)$$

where $H_{ext}$ is the effective dose per year from external irradiation per unit activity concentration in the cleared material $(Sv.y^{-1})/(Bq.g^{-1})$

$h_{ext}$ is the effective dose rate per unit activity concentration $(Sv.h^{-1})/(Bq.g^{-1})$

$t_e$ is the exposure time in a year $(h\ y^{-1})$

$f_d$ is the dilution factor

$\lambda$ is the decay constant $(y^{-1})$

$t_1$ is the decay time before the scenario $(y)$

$t_2$ is the decay time during the scenario $(y)$

The quantity $h_{ext}$ needs to be evaluated taking into account the geometry of the source, intervening shielding material, the separation distance of the dose point from the source and the exposure geometry. The Microshield code can be used to evaluate this quantity.

## Skin Contamination

Skin contamination by dust containing radionuclides can only occur to a significant extent at workplaces with dusty environments. The effective dose can be calculated by

$$H_{skin} = h_{skin} \cdot w_{skin} \cdot f_{skin} \cdot t_e \cdot L_{dust} \cdot f_d \cdot p \cdot \exp(-\lambda t_1) \frac{1-e^{-\lambda t_2}}{\lambda t_2} \qquad (17.4)$$

where $H_{skin}$ is the annual effective dose to an individual from skin contamination with $\beta/\gamma$-emitters per unit activity concentration in cleared material $(Svy^{-1})/(Bq.g^{-1})$

$h_{skin}$ is the sum of skin dose coefficients for $\beta$-emitters (4 mg.cm$^{-2}$ skin density) and for $\gamma$-emitters per surface specific unit activity $(Sv.h^{-1})/(Bq.cm^{-2})$

## Example 17.3

Calculate the effective dose per year from external irradiation arising from unit activity of Co-60.

*Solution*

The scenario is that a person lives for a total of 7000 hours per year in a house built using cleared building rubble. The fraction of cleared material constitutes only 2% of building material. The built room size is $3 \times 4\,m^2$ and height is 2.5 m with floor, walls and ceiling of 20 cm thickness. Using Microshield, doses are calculated at the middle of the room at a height of 1 m with a rotational geometry. A decay period of 100 days before the use of the cleared material in the building construction and a further decay of 365 days before occupancy are assumed. So the quantities used are

$$h_{ext} = 7.9 \times 10^{-7} \left(Sv.h^{-1}\right)/\left(Bq.g^{-1}\right)$$

$$t_e = 7000\ h\ y^{-1}$$

$$f_d = 0.02$$

$$t_{1/2} = 5.27\ y$$

$$t_1 = 100\ d = 0.27\ y$$

$$t_2 = 365\ d = 1\ y$$

$$\lambda = \frac{0.693}{5.27} = 0.1315\ y^{-1}$$

Using equation (17.3),

$$H_{ext} = 7.9 \times 10^{-7} \left(\frac{Sv}{h}\right) / \left(\frac{Bq}{g}\right) \times 7000 \left(\frac{h}{y}\right) \times 0.02 \times 0.9651 \times 0.937$$

$$= 10^2\ (\mu Sv.y^{-1})/(Bq.g^{-1})$$

$w_{skin}$ is the skin weighting factor (ICRP 60)
$f_{skin}$ is the fraction of body surface which is contaminated
$t_e$ is the exposure time in a year (h.y$^{-1}$)
$L_{dust}$ is the thickness of the layer of dust on the skin (cm)
$f_d$ is the dilution factor
$\rho$ is the density of the surface layer (g.cm$^{-3}$)
$\lambda$ is the decay constant (y$^{-1}$)
$t_1$ decay time before the start of the scenario (y)
$t_2$ decay time during the scenario (y).

## Example 17.4

Calculate the effective dose per year from skin contamination to a worker from unit activity of Co-60.

*Solution*

The scenario is that a worker works for the whole year (1800 h.y$^{-1}$) in a dusty environment contaminated with Co-60 dust. During this time, his forearms and hands making up to 10% of his body surface are covered with a layer of dust of 100 μm thickness. The density of the surface layer is 1.5 g.cm$^{-3}$. There is no decay time before the scenario or during the scenario. So the quantities used are

$h_{skin} = 1.7 \times 10^{-2} \left( Sv.y^{-1} \right) / \left( Bq.cm^{-2} \right)$ (from Table 5-2 of [13])

$\qquad = 1.94 \left( \mu Sv.h^{-1} \right) / \left( Bq.cm^{-2} \right)$

$w_{skin} = 0.01$

$f_{skin} = 0.1$

$t_e = 1800 \ h.y^{-1}$

$L_{dust} = 0.01 \ cm$

$f_d = 1$

$\rho = 1.5 \ g.cm^{-3}$

$t_1 = 0 \ d$

$t_2 = 0 \ d$

Using equation (17.4),

$H_{skin} = 1.94 \left( \dfrac{\mu Sv}{h} \right) / \left( \dfrac{Bq}{cm^2} \right) \times 0.01 \times 0.1 \times 1800 \left( h.y^{-1} \right) \times 0.01 \left( cm \right) \times 1.5 \left( g.cm^{-3} \right)$

$\qquad = 5.2 \times 10^{-2} \left( \mu Sv.y^{-1} \right) / \left( Bq.g^{-1} \right)$

To estimate the accepted concentration of cleared material, all the above mentioned four exposure pathways giving individual values of committed effective dose per year from unit activity concentration of cleared material e.g. $H_{inh}$, $H_{ing}$, $H_{ext}$ and $H_{skin}$ are evaluated. The parameter with the highest effective dose is used to estimate the concentration of the cleared material. Let us call this $H_{max}$. The concentration of cleared material, $C_c$, is then calculated as

$$C_c = \frac{10 \ \mu Sv.y^{-1}}{H_{max} \left( \mu Sv.y^{-1} \right) / \left( Bq.g^{-1} \right)} \qquad (17.5)$$

Now considering the results of four examples above, it is seen that the maximum dose arises from external irradiation to a person living in the house constructed of rubble containing 2% of Co-60 material. Let us call this $H_{max}$. The concentration of cleared material, $C_c$ is then calculated as

$$C_c = \frac{10 \, \mu Sv.y^{-1}}{H_{max}} \tag{17.6}$$

$$= \frac{10 \mu Sv.y^{-1}}{10^2 \, (\mu Sv.y^{-1})/(Bq.g^{-1})} = 0.1 \, Bq.g^{-1}$$

Following the same procedure, the concentration of other radionuclides can be calculated.

As mentioned before, unconditional clearance of radioactive materials to allow recycling, re-use or disposal had been estimated by the Department of Environment, Transport and Regions (DETR) in the UK and was reported in [15]. The DETR was subsequently reorganised as the DEFRA. The whole spectrum of man-made radionuclides had been divided into three groups and the corresponding clearance levels for solid materials have been specified. Table 17.2 shows these values.

**Table 17.2** *Clearance levels of solid radioactive materials*

| Group name | Clearance level (Bq.g$^{-1}$) | Group description |
|:---:|:---:|:---|
| Group I | 0.1 | Strong γ-emitters, e.g. Co-60, Cs-134, Cs-137 etc. |
| Group II | 1 | α-emitters, most β-emitters and γ-emitters, e.g. Pu-239, Am-241, Ru-106, I-125 etc. |
| Group III | 10 | Low β-emitters, e.g. H-3, C-14, S-35, Ni-63 etc. |

It should be noted out that the estimated clearance level for a certain radionuclide depends on the estimated value of $H_{max}$, which is the highest of $H_{inh}$, $H_{ing}$, $H_{ext}$ and $H_{skin}$. The estimated values for these parameters $H_{inh}$, $H_{ing}$, $H_{ext}$ and $H_{skin}$ depend on how the accident scenarios are constructed. Different countries may have different accident scenarios and consequently different $H_{max}$ values would arise. The lower the $H_{max}$ value, the higher is the $C_c$ value. Table 17.3 gives the values of general clearance levels of selected radionuclides in some of the EU countries.

### 17.4.3.2 Mixed Nuclides

In practical situations, a radioactive waste may contain a number of individual radionuclides. To clear such a waste, the following summation formula of

**Table 17.3** *General levels of clearance (Bq.g⁻¹) of some radionuclides in some EU countries*

| Country | H–3 | Co–60 | Sr–90 | Cs–137 | Pu-239 |
|---------|------|--------|--------|---------|---------|
| Belgium | 1.0E+02 | 1.0E–01 | 1.0E+00 | 1.0E+00 | 1.0E–01 |
| Germany | 1.0E+03 | 1.0E–01 | 2.0E+00 | 5.0E–01 | 4.0E–02 |
| Greece | 1.0E+03 | 1.0E–01 | 1.0E+00 | 1.0E+00 | 1.0E–01 |
| France | 0.0E+00 | 0.0E+00 | 0.0E+00 | 0.0E+00 | 0.0E+00 |
| Ireland | 0.0E+00 | 0.0E+00 | 0.0E+00 | 0.0E+00 | 0.0E+00 |
| Netherlands | 1.0E+06 | 1.0E+00 | 1.0E+02 | 1.0E+01 | 1.0E+00 |
| UK [13] | 1.0E+01 | 1.0E–01 | 1.0E–01 | 1.0E+00 | 1.0E+00 |
| EU (RP122 Part 1)[12] | 1.0E+02 | 1.0E–01 | 1.0E+00 | 1.0E+00 | 1.0E–01 |

the ratio of concentrations of individual nuclides to their clearance levels is applied:

$$\sum_{i=1}^{n} \frac{c_i}{c_{ci}} \leq 1.0 \qquad (17.7)$$

where $c_i$ is the concentration of radionuclide, $i$ (Bq.g⁻¹); $c_{ci}$ is the clearance level of radionuclide, $i$ (Bq.g⁻¹); and $n$ is the number of radionuclides in the mixture.

## 17.5   Environmental Discharge

Current dose limits incorporated in UK legislation (IRR99) are based on the 1990 recommendations of the ICRP in their publication 60 [16]. This gives a dose limit for the public as 1 mSv per year. (This may be compared with the average background radiation dose of 2.6 mSv per person per year in the UK).

Since members of the public may be exposed to radiation from more than one source of radiation, the UK operates a dose target of 0.5 mSv.y⁻¹ with respect to radioactive discharges from any nuclear site, irrespective of the size of that site or the number or type of nuclear installations in it. The discharge limits contained in authorisations reflect the totality of operations on site. These discharge limits are expressed in terms of activity. Following ICRP Publication 60, NRPB (now the RPD of the HPA) stated in 1993 that it considered there was a need for constraints to assist in the optimisation of new facilities. It recommended that the constraint on dose to members of the public arising from discharges from a single source should not exceed 0.3 mSv.y⁻¹, although it recognised that in some cases this might not be achievable, in which case the operator should demonstrate that the doses are ALARP. The dose limits and dose constraints are given in Table 17.4.

**Table 17.4** *Dose limits and dose constraints*

| Type of person | Dose limit | Dose constraint |
|---|---|---|
| An individual member of the public (critical group) | 1 mSv.y$^{-1}$ | 0.3 mSv.y$^{-1}$(source related) 0.5 mSv.y$^{-1}$ (site-related) |

# Revision Questions

1. What steps are applied in the management of radioactive waste from the generation to disposal? Show these steps diagrammatically.

2. What is the UK government's policy with regard to limiting generation of unnecessary wastes and transferring responsibility of its management to the polluter? Briefly describe these principles.

3. To what extent is RSA93 applicable to a licensed as well as non-licensed nuclear site? Clearly specify the lines of duty under this legislation applicable to these two classes of nuclear facilities.

4. What is the 'Joint Convention'? Briefly describe the main obligations on the contracting party under this convention.

5. Describe how does the Basel Convention of 1989 complement the Joint Convention in matters of hazard reduction from nuclear and chemical substances?

6. What is the 1992 OSPAR Convention? Describe its main features and state why 2020 is a landmark under this convention.

7. Describe the main obligations imposed by Article 37 of the Euratom treaty on the Member States of the EU.

8. Explain clearly with numerical examples the concepts and significance of the following terms:
    (i) exclusion level
    (ii) exemption level
    (iii) clearance level

What are the clearance criteria in terms of radiological dose?

## REFERENCES

[1] UK Government, Review of Radioactive Waste Management Policy – Final Conclusions, Cm 2919, July 1995.

[2] UK Health and Safety Executive, Management of Radioactive Materials and Radioactive Waste on Nuclear Licensed Sites, 13 March 2001.

[3] UK Government, Sustainable Development: the UK Strategy, Cm 2426, January 1994.

[4] UK Government, Radioactive Substances Act 1993, The Stationary Office.

[5] European Union, Council Directive 96/29/Euratom of 13 May 1996, laying down Basic Safety Standards for the Protection of the Health of Workers and the General Public Against the Dangers from Ionising Radiation, *Official Journal of the European Union*, L159, 29 June 1996.

[6] International Atomic Energy Agency, The Joint Convention on the Safety of Spent Fuel Management and on the Safety of Radioactive Waste Management, INFCIRC/546, IAEA, Vienna, Austria, 1997

[7] International Atomic Energy Agency, International Basic Safety Standards for Protection Against Ionising Radiation and for the Safety of Radiation Sources, FAO, IAEA, ILO, OECD/NEA, PAHO and WHO, IAEA Safety Series No. 115, Vienna, Austria, 1996.

[8] Basel Convention on the Control of Transboundary Movements of Hazardous Wastes and their Disposal, Final Act, 1989, http://www.basel.int/text/con-e-rev.doc.

[9] European Union, Council Directive 92/3/Euratom of 3 February 1992 on the Supervision and Control of Shipments of Waste between Member States and into and out of the European Community, *Official Journal of the European Union* L 35, of 12 February 1992, http://ec.europa.eu/energy/nuclear/radioprotection/doc/legislation/94c22402_en.pdf.

[10] OSPAR Commission, The Convention on the Protection of the Marine Environment of the North-East Atlantic (OSPAR Convention), 1998, http://www.ospar.org/eng/html/welcome.html.

[11] UK Government, UK Strategy for Radioactive Discharges 2001–2020, Department for Environment, Food and Rural Affairs, July 2002.

[12] European Commission, Recommendation of 6 December 1999 on the Application of Article 37 of the Euratom Treaty, 1999/829/Euratom.

[13] UK Department of the Environment, Food and Rural Affairs, Exemption Orders under the Radioactive Substances Act 1993, December 2002, http://www.defra.gov.uk/environment/radioactivity/discharge/rsact/exemption.htm.

[14] European Commission, Radiation Protection No. 122, Practical Use of the Concepts of Clearance and Exemption – Part I: Guidance on General Clearance Levels for Practices, 2000.

[15] UK Department of the Environment, Transport and Regions (DETR), Derivation of UK Unconditional Clearance Levels for Solid Radioactively Contaminated Materials, DETR Report No. DETR/RAS/98.004, April 1999.

[16] International Commission on Radiological Protection, 1990 Recommendations of the International Commission on Radiological Protection, ICRP Publication 60, *Annals of the ICRP*, Vol. 21, No. 1–3, 1991.

[17] Radiation Protection Division, Ionising Radiation Exposure of the UK Population: 2005 Review, HPA-RPD-001, 2005, http://www.hpa.org.uk.

# 18

# TREATMENT AND CONDITIONING OF RADIOACTIVE WASTE

## 18.1 Introduction

Wastes of various types and various levels of activity arise from the decommissioning of a nuclear facility. These wastes could be solid, liquid, sludge or gaseous; they could be inorganic, organic or mixed. Solid wastes could comprise items such as pipes, tubes, metallic components, worn-out and damaged equipment, cables, insulation material, concrete and rubble. Liquid wastes mainly arise from decontamination work and from water purification systems. Gaseous waste may arise during decontamination and dismantling operations.

The wastes arising from a decommissioning operation can be broadly categorised as primary wastes and secondary wastes. The primary wastes are those which are the contaminated materials of the original construction materials used in the nuclear facility. The amount of primary waste would obviously be less than the amount of material used in the original construction. These wastes are also called the raw wastes. The secondary wastes, on the other hand, are those which arise due to operations such as dismantling and decontamination of contaminated facilities and the processing of the primary wastes. These wastes also include items used to dismantle the facility such as handling/cutting tools, waste sentencing and disposal/storage facilities, dismantling machines and support facilities, ventilation and filtration systems etc. Both of these waste types may be further segregated into the categories described in Chapter 16.

The treatment of waste is carried out so that waste can be managed safely and economically [1]. There are three basic treatment objectives: volume reduction, removal of activity from the waste and change of composition. The objective for conditioning is, on the other hand, to convert waste to a solid form so that it can be packaged suitable for handling, transport, storage and eventual disposal. Conditioned wastes may be packaged in containers and, if necessary, in over-packs such that they are isolated from the environment for a long period of time. The whole process of treatment and conditioning of waste is shown diagrammatically in Figure 18.1 [2].

The HLW could be the spent fuel itself when there is no intention to reprocess it, or HLW can arise in liquid form from the reprocessing of spent fuel. The ILW and LLW can arise from various decommissioning operations as

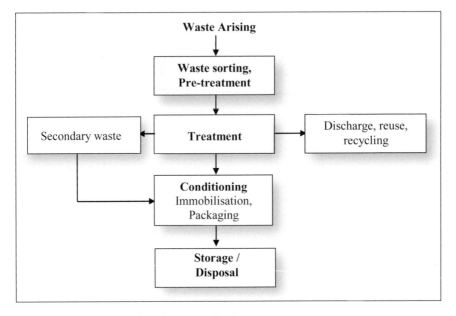

**Fig. 18.1** *Treatment and conditioning of radioactive waste*

well as from activities associated with industrial use of radiation and from medical practices. All of these wastes need to be prepared for isolation from human beings for a period commensurate with the levels of hazards they pose. The need for compatibility between the conditioned waste and the waste container, and between waste container and the future disposal facility stresses the importance of a systematic approach whereby all the problems of waste arising, treatment, conditioning and disposal are all addressed in a comprehensive way [1].

## 18.2 Operations Preceding Treatment and Conditioning

### 18.2.1 Segregation at Source

Separation and segregation of various types of waste at source by activity contents, physical states, chemical composition etc. are important for subsequent steps of radioactive waste management. The steps may include

- Segregation of waste according to half-lives: the long- and short-lived wastes need to be separated as they dictate the disposal routes.
- Waste may be separated into waste categories: small volumes of HLW are best kept separated from larger volumes of LLW.
- Waste containing organics and complexing agents may be separated out.

- Solid, liquid and aqueous wastes are segregated as they require different treatment and conditioning methods.
- Biodegradable materials, biotoxic and chemotoxic materials should be segregated as they need separate treatment processes.

## 18.2.2 Decontamination

Decontamination produces a variety of radioactive wastes. Decontamination of activated structural material such as steel, pipes, equipment as well as surface contaminated objects can be effective in reducing activity levels in those materials, but they produce secondary wastes which need to be handled properly. Efficient decontamination steps do now exist, but they can give rise to additional problems in waste handling. For example, chemical complexing agents or organic solvents used to decontaminate metals need to be separated out as they may cause corrosion of the metal containers.

## 18.2.3 Size Reduction

Some large pieces of equipment, such as glove boxes, heat exchangers or reactor pressure vessels which may be contaminated throughout, may require reduction in size before they are removed and conditioned.

## 18.2.4 On-site Storage and Transportation

Wastes cannot always be treated and conditioned immediately, either because of the non-availability of a treatment facility or for programme scheduling reasons. In such cases, a temporary on-site storage facility should be available. Waste may also need to be transported to the temporary storage facility in flasks, drums, shielded containers etc. The safety aspects of such activities should be taken into account.

## 18.3  Treatment of Waste

A prerequisite of most treatment processes is to sort wastes into classes which are, as far as possible, homogeneous. Solid, liquid and gaseous wastes are treated in different ways. Treatment of liquid wastes generally concentrates on separating as much of the radioactive components as possible from the waste volumes into a liquor form of concentrated activity. The treated effluent can then be released into the environment and the concentrated liquor would undergo conditioning prior to disposal. The treatment process obviously depends on the physical state of the waste. For example, compaction can be carried out on loose solid radioactive waste, but not on incompressible liquid waste. Some of the treatment and conditioning practices in various countries are shown in Table 18.1.

Table **18.1** *Current practices for treatment and conditioning of LLW and ILW*

| Country | Treatment Process | Conditioning Process |
|---------|-------------------|----------------------|
| Belgium | Compaction Incineration Evaporation | Bituminisation Vitrification |
| France | Compaction Incineration Evaporation Ion exchange | Bituminisation Cementation polymerisation |
| Germany | Compaction Incineration Evaporation | Cementation Polymerisation |
| U K | Compaction Incineration Evaporation | Bituminisation Cementation Polymerisation |
| U S A | Compaction Incineration | Bituminisation Cementation Polymerisation |

## 18.3.1 Compaction

Compaction is normally carried out on solid wastes, but can be applied to sludge, slurries etc. The main aim is to reduce volume. It is normally carried out on steel drums containing raw waste using drum compactors which are hydraulically rammed onto the top of the drums (in-drum compaction) or the whole drums may be crushed to form 'pucks'. The reduction in size depends on the nature of waste and the compaction force used. Originally, only modest compaction forces of less than 1 MN (mega-newton) were used to minimise volumes in storage and transportation or to increase packing density in a waste container. If the process uses high compaction forces ($\geq$ 10 MN), the process is termed super-compaction. The waste to be compacted must be chosen careful-ly. For example, compressed gases and explosive materials must be excluded from the compaction process.

There are basically two types of compactor: one with a horizontal piston and the other with a vertical piston. They may be either mobile or stationary. Low force compaction is used for in-drum compaction of materials such as ion exchange resin at pressures of about 0.20 MN or for general compaction for economic storage and transportation of waste. As pre-compacted wastes burn less efficiently than loose wastes, compaction for incineration purposes is not desirable. A low force compactor is described in Section 18.3.1.1. In the nuclear industry, the force is often specified in terms of tonne-force*. For super compaction, the compaction force is greater than 1000 tonne-force (10 MN). This amount of force that is applied to a waste package before its disposal is in the super-compaction range so that the package experiences a comparable pressure under disposal conditions in deep underground repositories. These geological forces were estimated to be approximately 300 bar for the Konrad mine in Germany.

### 18.3.1.1 Low Force In-drum Compactor

A schematic diagram of a low-force in-drum compactor is shown in Figure 18.2 [3]. Wastes are filled in a drum which is placed on the working platform of the compactor. The vertical shaft having an applied force of about 0.2 MN compresses the waste whose volume is reduced due to extraction of moisture, collapse of cavities etc. The process of filling and compacting is repeated until the desired filling level is reached. The compacted drum may be further treated before conditioning.

### 18.3.1.2 Super-compactor with Horizontal Piston

A mobile super-compactor with a horizontal piston used in Germany, called FAKIR, is shown in Figure 18.3 [3]. It can easily be transported in a container to a place where waste may arise. The working pressure is slightly above 300 bar and the maximum area for compaction is 1 m × 0.54 m. For this area, a force of about 16 MN is required to achieve 300 bar pressure.

The main parts of the system are

- pre-compactor
- ram with hydraulic device
- jib crane and grab
- exhaust device for charger and exit

The height of the plant with jib crane is about 2. 8 m and with the pre-compactor it is about 4 m. The mass without the pre-compactor is about 47 Mg and with the pre-compactor, it is about 52 Mg. It is made of steel and is painted throughout to help subsequent decontamination. Beneath the cylinder, there is a device for collecting fluid that may be released during compaction. The

---

* One tonne-force (1 Mg-force) is $10^3$ kg-force. One kg-force is defined as the force exerted by 1 kg of material under the gravitational acceleration of g (9.81 m.s$^{-2}$). So, 1 kg-force = 1 kg × 9.81 m.s$^{-2}$ = 9.81 N $\cong$ 10 N. So 1 tonne-force = $10^3$ × 10 N = 10 kN. Thus 100 tonne-force = 1 MN. 1 MN exerting a pressure on 1 m$^2$ area is equal to $10^6$ N.m$^{-2}$ = 10 bar (as 1 bar = $10^5$ N.m$^{-2}$). (1 N.m$^{-2}$ is the SI unit of pressure, called the pascal, Pa.)

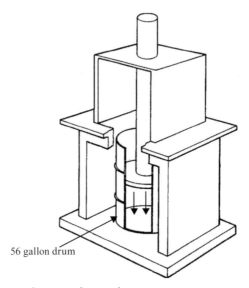

56 gallon drum

**Fig. 18.2** *Schematic diagram of an in-drum compactor*

pre-compactor, with a mass of about 5 Mg, is positioned above the charge opening on the working platform. Aerosols may be produced during compaction. They are removed by mobile filtering devices at the charging position, the exit and above the collecting device for liquids. Although the maximum area for compaction is 1 m × 0.54 m, larger diameter drums can be reduced to the correct diameter using the pre-compactor. During compaction all information regarding waste and the operational features are collected and recorded.

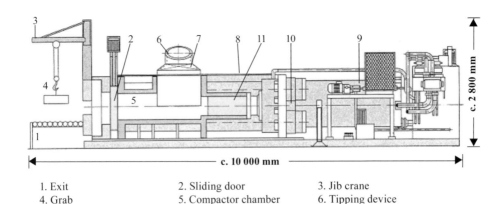

1. Exit 2. Sliding door 3. Jib crane
4. Grab 5. Compactor chamber 6. Tipping device
7. Charger 8. Working platform 9. Hydraulic system and control
10. Hydraulic cylinder 11. Ram

**Fig. 18.3** *Super-compactor with a horizontal piston*

The possible modes of operation of this super-compactor are
- Compaction of wastes that have been delivered in 180 litre or 200 litre drums.
- Compaction of wastes contained in transport drums which are tipped out into the compactor and compacted into special cartridges.
- Compaction of wastes that have been delivered in bags.

Loose debris is compacted by tipping the contents of the transport drum into the super-compaction channel using a hydraulic drum grabber. Before charging, a cartridge is put into the compaction channel via the slide and the lid of the cartridge is placed directly in front of the ram. The waste is then filled into the channel and compacted into the cartridge which is finally sealed with a lid. The final product is a steel-mantled pellet.

Wastes delivered in bags are compacted as described above. All pellets are measured automatically. Collated data contains the mass and height of each pellet and the dose rate (measured manually). The pellets are then marked and packed into a drum or into a container with the help of a grab or a crane.

Following several campaigns, indicative data about the super-compactor can be obtained. Depending on the type of waste that is super-compacted, typical volume reduction factors can be derived (see Table 18.2).

**Table 18.2** *Results of several campaigns of the super-compactor*

|  | Campaign A | Campaign B | Campaign C |
|---|---|---|---|
| Volume of raw waste | 20 m$^3$ | 15 m$^3$ | 14 m$^3$ |
| Mass (wet) | 3310 kg | 6500 kg | 2510 kg |
| Volume after compaction | 4 m$^3$ | 6.3 m$^3$ | 3.2 m$^3$ |
| Mass after compaction (dry) | 3145 kg | 6100 kg | 2380 kg |
| Volume reduction factor | 5 | 2.4 | 4.4 |

### 18.3.1.3 Auxiliary Equipment

During the process of compaction, free liquids may be forced out of the waste. Pellets (cartridges) containing waste that have discharged free liquids have to be dried in order to fulfil special requirements for interim or final storage. A German plant for drying pellets, called PETRA, is described here [3]. The main components of this system are
- a drying and heating chamber
- a condensation unit operating at low pressure
- an energy supply and data processing unit

- a cooling device
- a drum weighing device

Up to eight drums can be placed into the heating chamber. Its dimension is 3.4 m (length) × 1.8 m (width) × 2.5 m (height). The electrical heating is carried out at the back wall of the facility within an air channel. Movement of air is carried out by two ventilators. Maximum operating temperature is about 300°C. An 80 mm thick layer of mineral wool is placed between the leaves of the heating chamber wall. The heating chamber is constructed of galvanised carbon steel with a zinc layer.

Drying is carried out at low pressures of about 20–50 mbar absolute. Vaporised water is condensed in the condensation chamber with a total condensation surface of about 1.5 m² and finally released to the water collecting device approximately every 10 min. The dimensions of the cooling device are 1.0 m × 1.0 m × 1.5 m. A glycol–water mixture at about 5°C is used as the coolant. Before and after drying, drums are weighed with a drum scale. This and other process related data are stored in a data processing unit, so that they can be used to obtain approval for interim or long-term storage.

The process of drying is stopped when the following criteria are fulfilled:

- inner pressure of drums: < 50 mbar
- temperature at drum outer surface > 130°C
- amount of condensate per 200 litre drum < 100 ml.h$^{-1}$

In order to ensure that drying is complete, the drums are left for another two to three hours before being removed from the PETRA facility. Although this system is used after super-compaction, drying systems may also be implemented before super-compaction.

### 18.3.1.4 Use of Super-compaction

Super-compaction as a means of volume reduction of waste had been in use in various countries for well over 20 years. Table 18.3 shows the status of super-compaction in some of these countries [2].

## 18.3.2 Incineration

The incineration process is carried out under a stringent regulatory regime which controls the release of activity and chemical compounds into the environment. Incineration plants were usually designed to cover a specific type of waste, although recent developments tend to aim towards a wider range of wastes. In incineration, sufficient air is provided to burn all the waste and is favoured for high volume LLW. Incineration is carried out on both solid and liquid wastes. However, the proportion of non-combustible waste in one feed is somewhat restricted. Often it is not allowed to exceed 20% by weight. The main advantage of incineration is the elimination of all organic and combustible material within the waste, thus resulting in improvement in fire safety.

**Table 18.3** *Status of super-compaction in some countries*

| Country | Facility | Pressure (MN) | Comment |
|---------|----------|---------------|---------|
| Belgium | Mol | 20 | |
| France | La Hague / Cogema | 25 | ILW |
| | EdF / Bugey | 20 | Mobile |
| Germany | Karlsruhe | 15 and 20 | |
| Netherlands | COVRA | 15 | |
| UK | UKAEA Dounreay | 20 | |
| | BNFL Sellafield | 20 | |
| USA | INEEL | 20 | |
| | Hanford WRAP | 20 | |

Also, as there is no biological activity left in the waste, bio-degradation would not be a concern in the subsequent disposal considerations. The disadvantage of this process is that some chemicals may cause corrosion in the disposal containers. From that point of view, the amount of chlorinated plastics, e.g. PVC should be restricted in the incineration process so as to minimise corrosion by chlorination. Sulphur containing feeds such as ion exchange resins and rubbers should also be limited to minimise $H_2SO_4$ formation and corrosion.

Liquid wastes with a high content of organic material can be incinerated in specially designed plants or in plants adapted from solid waste incineration. Organic liquid wastes arise during decontamination process such as removal of coatings, paint etc. by organic solvents. Contaminated oils may arise in pumps, transformers etc. The heating requirements for organic wastes are normally up to 40,000 $kJ.kg^{-1}$ and throughputs are in the range 10–40 $kg.h^{-1}$.

As incineration facilities normally work within so-called campaigns, and as wastes are delivered from different nuclear facilities to the incineration plant, the likelihood of cross-contamination of wastes, particularly from $\alpha$-bearing wastes of trans-uranic compounds, is very real. Great care is taken to avoid this. Discharge of $\alpha$-bearing off-gas is highly restricted.

The residue from the incineration process can be compacted or super-compacted, thus allowing further reduction in volume. Incineration is used in many countries such as: Belgium, Canada, France, the Netherlands, Germany, the UK and the USA.

A related process to incineration, known as pyrolysis, where air is controlled in order to prevent vigorous burning, is used to limit the generation of corrosive products and contain ashes within the incinerator. This process is described in Section 18.3.3.3.

### 18.3.2.1 Incineration of Solid Waste

Many techniques for the incineration of solid wastes are used throughout the world such as shaft kiln furnaces with stationary or movable grates and rotary kilns. In Europe, mainly shaft kiln furnaces are used. Therefore, a short description of shaft kiln furnaces is given below, including details about sorting the wastes before incineration and off-gas cleaning.

### 18.3.2.2 Shaft Kiln with a Tippable Grate

Incineration in a shaft kiln with a tippable grate is carried out in the primary chamber and all ashes from the incineration are collected in the ash box, as shown in Figure 18.4 [3].

The initial temperature of the furnace is 200–300°C and it is heated to a temperature of about 800°C. When the incinerator reaches 800°C, the charging of burnable waste is initiated. Batches of about 20 kg are fed into the incinerator at intervals of about 5–10 min. Charging is carried out via a sluice and the off-gases of the sluice are fed back into the primary chamber. The waste starts to burn. After a few charges the furnace temperature exceeds 800°C and the primary oil burner is shut off. The temperature of the primary chamber is then kept constant at 800–1000°C by regulating the incineration air.

The off-gas leaves the primary chamber and enters the secondary chamber. In order to obtain a complete incineration, the off-gas passes the secondary oil burner, where air is added and a continuous flame is burning. The off-gas leaves the furnace and is cooled down to about 200°C by passing through a water-cooled heat-exchanger. The off-gas is then cleaned out in the bag filter

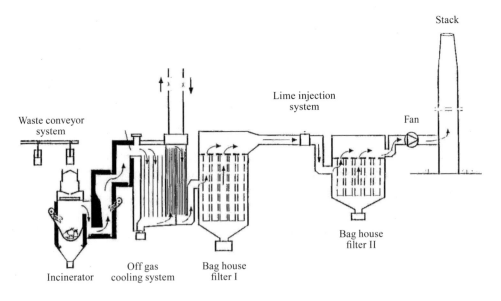

**Fig. 18.4** *Incineration plant: shaft kiln with tippable grate*

which contains solid particles. A large fraction of low activity contained in the off-gas is separated out with these particles.

Cleaning of conventional hazardous compounds such as dioxin and $SO_2$ from the off-gas is carried out by the injection of lime and charcoal into the off-gas tube. Dust particles containing dioxin compounds etc. are filtered out in the second bag filter. Before being released to the environment via the stack, off-gases are continuously sampled and monitored.

In a single campaign about 5–6 Mg of waste can be incinerated producing 500 kg of ash and about 50 kg of dust from the first bag filter. The masses that arise from filtering in the second bag filter during cooling are negligible.

The majority of the activity is contained within the ashes and the dust filtered out in the first bag filter. The activity content of the waste delivered for incineration is normally very low, leading to a dose rate at the surface of the bags of the order of few $\mu Sv.h^{-1}$. Control of the activity flow and the activity content of the products are carried out by a $\gamma$-scanning of the ash-drums, $\gamma$-, $\beta$- and $\alpha$-measurements of samples of the products in a laboratory and by continuous dose-rate measurements during operation.

After an operational period of several days, the ashes are removed from the primary chamber by hydraulically lowering and tipping the bottom of the chamber. Ashes fall down into the ash collector and, after allowing time for cooling, they are loaded into the ash drums with the help of long-handled tools. Repeated analyses of ash samples have shown the remaining combustible material in the ashes to be approximately 1% by weight.

### 18.3.2.3 Pyrolysis

Pyrolysis is similar to incineration but is based on the thermal decomposition of materials, mainly organic, under an inert or oxygen-deficient atmosphere to destroy waste and to convert it into an inorganic residue. The operating temperature is in the range 500–550°C which is significantly lower than that used for conventional incineration. The generated pyrolysis gas is burnt in a simple combustion chamber and then treated in a flue gas cleaning section. At the operating temperatures, corrosive species such as phosphoric oxides form stable inorganic phosphates and are thus of little concern in further considerations. At lower temperatures and reduced oxygen levels, volatile species such as Ru and Cs are largely retained within the pyrolysis chamber.

## 18.3.3 Compaction and Incineration

Treatment of solid wastes using compaction and incineration may have the following modes:

- no compaction, no incineration
- no-compaction, incineration

- pre-compaction, no incineration
- pre-compaction and incineration

In addition there may also be
- super-compaction and incineration including super-compaction of ashes

Compaction and incineration may be seen as complementary but the actual choice depends on technical and economic factors. Pre-compaction of combustible wastes prior to incineration leads to lower transport volumes and costs, but it requires more staff and technical equipment because of the additional handling requirements. As pre-compacted wastes may show less advantageous incineration features than loose or non-compacted wastes, shredding of the pre-compacted wastes before incineration may become necessary.

### 18.3.4 Evaporation Facilities

Evaporation of radioactive liquid wastes is carried out in order to reduce liquid content. The solid content of liquid wastes is increased due to evaporation to about 25% by weight. This concentrate is either solidified with a flux material (such as cementation) or converted into a solid waste product by subsequent drying. The plant is not mobile and hence the liquid wastes have to be transported for treatment.

Chemical compounds contained in liquid wastes are subjected to various kinds of pre-treatment such as filtration, decantation, neutralisation, precipitation or, in the case of strong acids, dilution. The last option of pre-treatment is, in comparison to the other methods, much easier to perform, but requires a longer treatment time or evaporation for a longer timescale.

A typical evaporation facility is shown in Figure 18.5 [3]. Throughputs are in the range 1–10 $m^3.h^{-1}$. Evaporation does not lead directly to a final product, it requires further steps such as solidification, drying etc. The decontamination factor for evaporation is in the range $10^4$–$10^5$. In most cases, the evaporated water can be released to the environment via a normal purification plant.

A special type of the evaporator is the so-called thin-film evaporator (Figure 18.6). This type of evaporator is used in order to solidify liquid concentrates from other evaporation facilities, active ion exchange resins (LLW and ILW) mixed with sludge and spent oil and organic solvents in very small quantities.

Concentrates from evaporation facilities are normally pumped into storage vessels having volumes of 7 $m^3$. From there the concentrates are pumped into the head of the vertically arranged evaporator which is equipped with movable strippers at eight different heights. The concentrates are evenly spread onto the top of the inner surface of the evaporator. Movement from the top to the bottom is carried out by gravity with the help of the rotating strippers, which are forced against the surface due to the rotation. At the bottom of the evaporator

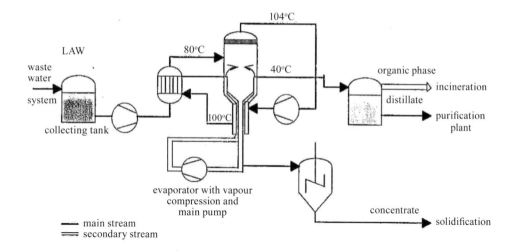

**Fig. 18.5** *Evaporation plant for LLW*

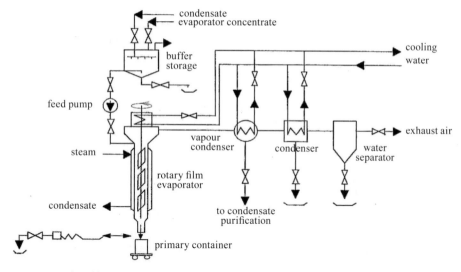

**Fig. 18.6** *Thin film evaporator*

a highly viscous dehydrated product is drained into a waste container (drum). The quality of the product can be varied from liquid-aqueous to powder-like solids by the following actions:

- The feed rate can be adjusted.
- The partial pressure of water vapour can be reduced, the water/water vapour equilibrium can be shifted by maintaining an air input (up to 20 m³.h⁻¹ bypass of air).

- Low pressure operation can be used by applying 2 kPa to the head of the evaporator. The specific activity in the condensate is in the range $10^{-3}$–$10^{-4}$ of the specific activity of the liquid evaporator concentrate.

## 18.3.5 Drying Facilities

Liquid wastes and concentrates are converted into solid state by means of mobile or stationary drying facilities. Drying is defined as lowering of the water content of the waste far beyond the range that can be reached by de-watering (mechanical) techniques. As a type of pre-treatment, it may follow de-watering processes such as evaporation. To avoid the spread of contaminants, drying is often carried out at very low pressures (vacuum) and by means of heat. It may be regarded as a type of pre-treatment as well as a type of final treatment.

Radioactive wastes may be dried as loose or bulk material that has already been packaged into a waste container. Here heating of the total waste package is required. Evaporation of water vapour is facilitated by operating at very low pressure. An example of a device used to dry loose or bulk materials is a 'drum dryer'.

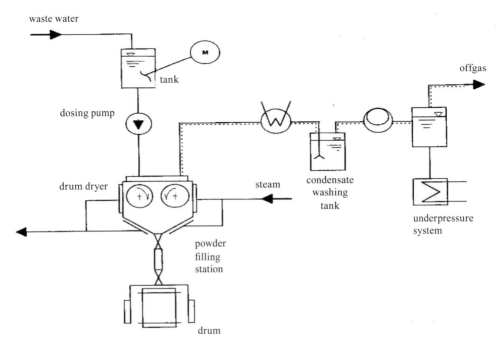

**Fig. 18.7** *Schematic diagram of a drum dryer system*

### 18.3.5.1 Drum Dryer

Drum dryers are used for the continuous drying of evaporation concentrates to powder-like interim products. A very thin layer of concentrate is spread onto the surface of a drum heated by vapour, as shown in Figure 18.7 [3]. It sticks to the surface of the rotating drum and is finally removed by a scraper. It then falls down into a waste drum and may be subjected to further conditioning techniques such as immobilisation.

## 18.4 Conditioning of Waste

The objective of conditioning is to convert the waste to a solid form with decreased solubility and improved mechanical stability. Various conditioning processes are currently available (see Table 18.4). In the selection of the process, it is essential to take account of the compatibility of the waste with the matrix material and between the matrix material and the future disposal environment.

**Table 18.4** *Currently available conditioning process for radioactive waste*

| Waste type | Conditioning process |
|------------|---------------------|
| LLW | Cementation, bituminisation, polymerisation |
| ILW | Cementation, bituminisation, polymerisation, calcination |
| HLW | Vitrification |
| Irradiated fuel | Encapsulation |

### 18.4.1 Bituminisation

Bituminisation is a process of conditioning of radioactive wastes by using bitumen. The bitumen is a mixture of hydrocarbons and other substances which occur naturally or are obtained from distillation of coal or petroleum. They are a component of asphalt and tar which are dark, heavy, viscid substances. The basic principles can be summarised as follows: the liquid or solid radioactive wastes are mixed with molten bitumen at a temperature of 110–230°C. Water and other volatile constituents in the heated molten bitumen evaporate. The remaining water-free product is then packaged into a suitable container ready for disposal.

Types of wastes that are suitable for incorporation into bitumen are: chemical sludge, ion exchange resins, reagents and concentrate salt solutions, organic solvents, incinerator ashes, plastic waste and other solid waste. Bituminisation is normally applied to wastes that will be disposed of at surface or near-surface facilities, so the aspect of leachability of the product is extremely important. It should be noted, however, that it is no longer used extensively because of specific developments in the final repository requirements.

## 18.4.2 Cementation

Solid materials which are not compactable and which require uneconomic major efforts to segment for compaction, or which cannot be treated further because of high dose levels, are encapsulated in waste containers by pouring cement or grout over them. Typical wastes are: fuel cladding, highly contaminated surfaces, contaminated equipment and cables, loose building materials, etc.

A variety of LLW, ILW and α-contaminated wastes are suitable for incorporation into cement matrices. There are certain advantages to this conditioning process.

- It requires relatively simple process plant operating at low temperatures.
- It has a high density which provides considerable self-shielding.
- It is highly alkaline and has properties which lower the solubility of radionuclides.
- It is a low-cost option.

But the main disadvantage is that it increases the volume of the waste and consequent increase in disposal cost.

Prior to cementation, liquid wastes containing chemicals arising from decontamination processes such as electro-polishing or acid treatment may be subjected to neutralisation or precipitation or, in the case of strong acids, they may be subjected to dilution. The nature of the pre-treatment depends on the particular situation and the effectiveness of the pre-treatment process also has to be considered.

### 18.4.2.1 Magnox Encapsulation Plant

The Magnox Encapsulation Plant (MEP) [4] at Sellafield has been designed to process intermediate-level solid radioactive waste, packaging it into a convenient form for efficient and simple handling, transport, storage and eventual disposal. The waste is made up primarily of the cladding or 'swarf' from fuel elements that have been used in Magnox nuclear power stations. Magnox stations were the first generation of nuclear power stations in the UK, deriving their name from the fuel they use, natural uranium metal enclosed in a magnesium alloy can. Reprocessing the fuel recovers valuable uranium and plutonium for re-use as fuel.

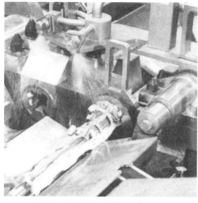 

(A)   Fuel handling plant            (B)   'Swarf' from Magnox
                                            fuel elements

**Fig. 18.8** *Fuel handling plant and magnox 'swarf'*

In order to be able to reprocess the Magnox fuel rod, its outer cladding must first be stripped off. By using specially designed remote control equipment, the cladding is peeled off into small pieces a few centimetres in length (Figure 18.8). Before the MEP was built, Magnox swarf was stored under water in purpose-built bays at Sellafield. Transport, storage and eventual disposal of the waste are better managed if a more manageable, consistent solid waste form can be produced. The plant has therefore been designed to encapsulate the waste within a cement grout matrix in stainless steel drums.

The encapsulation process takes place inside a heavily shielded concrete cell which ensures maximum protection for the workers from the radioactive swarf. All the in-cell equipment had been designed to operate remotely or automatically and to be as maintenance-free as possible, although items can be removed remotely should large-scale maintenance be necessary.

The cell has several process positions and 'posting-in' ports which allow the various items which are required to be loaded into it at the appropriate positions. Plant personnel view and supervise operations by way of lead-glass viewing windows and a number of television cameras are placed at strategic points within the cell. Television monitors are positioned outside the cell near the operating positions, giving the plant operators a choice of in-cell views (see Figure 18.9).

The waste drums are moved around within the cell by the In-Cell Drum Transporter (ICDT), which is in effect an advanced computer-controlled crane. A control system tracks all drum movements and ensures efficient use of the in-cell equipment. There are nine separate drum movements required in the cell, each of which has been assigned a priority. The most time-consuming operation is given the highest priority, so that a backlog of drums does not build

**Fig. 18.9** *Operations within the cell are supervised through viewing ports and TV cameras*

up. The transporter picks the drums up by means of four remotely controlled 'fingers' which engage underneath the top flange of the drum.

There are two sets of lifting fingers: the first set transfers the drum before it has been decontaminated. The second set handles the drum after decontamination. This ensures that clean, decontaminated drums arc not touched by contaminated lifting equipment. Other drum transfers are carried out by special trolleys called 'transfer bogies' which manoeuvre drums across the cell between the transporter set-down positions and the process/posting positions.

The cell also has a power manipulator (a telescopic remote control arm) which can reach some parts of the cell not accessible by the ICDT. Essentially for maintenance work, the manipulator will remain in a parked position under normal operation.

Before the encapsulation process can take place, the plant must have the raw materials necessary. These are

- The waste itself: Magnox swarf is transported from the fuel handling plant in a swarf container which is inside a swarf flask. The waste is covered with water at all times to prevent very small particles of uranium from drying out, as they could cause a fire to start.
- Initial grout to encapsulate the waste: A mixture of ordinary Portland cement, blast furnace slag and water is used. The cement and blast furnace slag are delivered by road and loaded into silos. From there, they can be conveniently mixed in the correct proportions with water. Extensive testing of different grout mix designs is carried out to ensure that the quantities used are correct. The mix

is computer-controlled for consistency and accuracy. Slightly more concrete than required is mixed for each batch to allow samples to be taken and tested, and to stop air from entering the system as this could cause the concrete to splash. The grout is mixed to tight QA assurance parameters.

- Capping grout: This is a second type of grout, more fluid than that used for the initial filling. This second grout is a self-levelling mixture of cement, pulverised fuel ash and water. The mix, like that of the initial grout, is computer-controlled and is the result of considerable testing to ensure that the correct quantities and concrete properties are obtained. The grout is pumped into the drum to form a seal over the encapsulated waste.

- Drums: High integrity stainless steel drums, specially designed by the erstwhile BNFL, are manufactured off-site to extremely high standards. The drum itself is made from 316L stainless steel and the lid is constructed from 304L stainless steel, both proven materials in radioactive environments. Every drum is checked before use to ensure that it conforms to the rigorous standards required. A unique number is etched onto the outside of each drum to allow thorough checks to be carried out including the exact material composition, manufacturer's name, date of manufacture, even the drum's eventual position in the on-site store. This detailed accountability means that the plant operators have a bank of information about every drum in the cell.

### 18.4.2.2 Cementation Process

A drum is lowered through the cell roof onto a waiting transporter bogie and moved to a position at the base of the swarf tipping chute. On the cell roof, a swarf flask is positioned at the tipping machine. An operator at the tipping machine can remotely open the flask door to lower the container holding the Magnox swarf and water. The container is tipped over to allow the swarf and water to flow down the tipping chute and into the drum. The chute is then washed down to remove any traces of uranium and swarf. The empty container is refilled with water before being hoisted back into the swarf flask.

The drum containing the swarf and water is then transferred to a position underneath another posting station where an anti-flotation plate can be lowered into the cell and fitted inside the drum over the top of the swarf. The anti-flotation plate is a disc made from a steel mesh which stops the light swarf from floating in the cement grout and reduces the amount of grout that splashes during the filling. The drum, containing swarf, water and an anti-flotation plate, is then moved to the heart of the process – the de-water and grout station where, in an inert atmosphere, water is removed from the drum. The inert atmosphere replaces the water and by excluding oxygen prevents uranium fires.

Once most of the water has been removed, the drum is vibrated whilst the initial grout is pumped in through a hole in the anti-flotation plate. The vibration of the drum ensures that the swarf is evenly distributed throughout the concrete grout. The grout is pumped through a pipe from outside the cell. To clean the pipe after grout transfer, a non-metallic device (known as a 'pig') is sent through the pipe to remove any traces of grout. The 'pig' falls into the drum and is encapsulated along with the waste. The drum is then transferred to the capping and curing area where it is left to allow the cement grout to set.

During the initial grouting, it is possible that some small particles of swarf may float to the surface and be exposed to the atmosphere. Capping grout is added to counter this. The capping grout is a more fluid self-levelling concrete which covers any exposed swarf. There are ten stations which are able to perform the capping operation, thus preventing a backlog of drums building up. Once again the pipe is cleaned by the use of a 'pig' when the operation is complete.

The capping grout is left to cure for up to 24 h before the drum is transferred to the lidding station. Here, a lid, complete with 12 specially made bolts, is posted into the cell and fitted to the drum. The lid is bolted on by 12 automatic nut runners which tighten the bolts to a pre-determined torque and in a set order so that stresses do not build up. The lid contains a specially designed filter which allows the small amounts of hydrogen generated during the storage of the drum to escape, without the release of any activity. It is important that the top of the drum and the bottom of the lid make an airtight seal.

Remotely operated tools are installed in the cell to ensure satisfactory lid seals. After lidding, the drum is transferred to a decontamination chamber, where it is sprayed with high pressure water to remove any surface contamination. The drum is then monitored. Its surfaces are automatically wiped and swab samples are remotely withdrawn from the cell for monitoring.

Once monitored and confirmed as clean, the ICDT uses its clean set of fingers to place the drum in a special container, known as a stillage. The stillages, which each hold four drums, are posted into the cell to collect the encapsulated waste drums. Once full, the stillage is moved into a transfer area where it can be picked up and lowered through a shield door into the transfer tunnel that connects the plant to the encapsulated product store.

### 18.4.2.3 Encapsulation Product Store

The Encapsulation Product Store (EPS) provides interim storage for ILW pending disposal in a final repository. It has a single storage vault with a capacity of some 12,500 drums. This capacity can be extended if necessary and the store has been designed with this in mind. Drums of encapsulated waste arrive four at a time in purpose-built stillages. The stillages are placed in channels by the computer-controlled store charge machine, and may be stacked up to 16 high. With 196 storage channels in all, arranged in a 14 × 14 horizontal array, this allows for the storage of over 3000 stillages.

The store charge machine makes it possible to load or retrieve any stillage from any part of the store whilst its computer system carefully tracks and records all drum movements. The store is cooled by a chilled air recirculation system to remove any heat given off by the drums.

A particular feature of the store is the single drum export facility. This allows drums to be removed in a suitable way for eventual disposal in an underground repository. This facility can also double as a way of taking drums into the cell.

### 18.4.3 Vitrification

Vitrification is a technique whereby hazardous waste is mixed with glass-forming chemicals at high temperatures to form molten glass which then solidifies, immobilising the waste. Vitrification is specifically applied to HLWs, although recently France and the USA have been trying to extend the technology to LLWs and ILWs. Aqueous effluents arising from the reprocessing of spent fuel containing fission products (when the fissile and fertile U and Pu have been separated out) is mixed with borosilicate glass at a temperature of about 1100°C. The mixed product is poured into a stainless steel canister for temporary cooling and storage.

This vitrification process may be attained in one step where the effluent is directly injected into the molten borosilicate glass or in two steps where the first step consists of evaporation and calcination of effluent and then the second step is the incorporation of calcinated residue in the borosilicate melt. The one-step vitrification process was applied in Marcoule, France in 1967. The plant was operated in batches, with the operations of calcination and vitrification being carried out in the same equipment and consequently throughput was low. The plant ceased operation in 1972.

The two-step process is an evolution from batch processing to continuous processing where calcination and vitrification stages are carried out in separate compartments and the melt is then poured into a canister pre-heated by the induction method. This vitrification process developed in France is known as AVM (Atelier de vitrification de Marcoule) [4]. A commercial-scale plant came into operation at Marcoule in 1978 to vitrify aqueous products from reprocessing of Gas Cooled Reactor (GCR) fuels. The containment glass is mainly borosilicate glass with the following composition: 45% of $SiO_2$, 14% of $B_2O_3$, 10% of $Na_2O$, 5% of $Al_2O_3$ and 12% of fission products and actinides; the remainder being CaO, ZnO, $Fe_2O_3$, $Li_2O$, $ZrO_2$, NiO and $P_2O_5$. This glass immobilises radioactive waste in its structure and shows strong thermal and chemical stability. This structure was tested for chemical, physical, mechanical and radiation stability. In standard test conditions, the leach rate of aqueous medium is less than $10^{-5}$ per year of leached fraction in the saturated condition.

[351]

This vitrification process has been adapted at Sellafield for vitrification of aqueous HLW and is known as the Windscale vitrification process (see Section 18.4.3.1)

An alternative to solidification of aqueous HLW is the Synroc process which involves transforming the wastes and mixture into a synthetic rock, similar in structure and composition to natural minerals. Section 18.4.3.2 describes the process in greater detail.

### 18.4.3.1 Windscale Vitrification Plant

Highly Active Liquor (HAL) arrives at the Windscale Vitrification Plant (WVP) [5] from the HAL storage facility, and is stored in buffer tanks. Before the HAL can be vitrified, the nitric acid needs to be removed and the HAL made somewhat alkaline. This is carried out in the calcination process. The resulting friable solid is then mixed with glass frit before being melted together in a crucible. The molten glass is then poured into a product container. The whole process of vitrification at WVP is shown in Figure 18.10.

For practical reasons, HAL is transferred from the HAL storage facility to the WVP buffer storage in batches. This allows the plant the flexibility of operating for several days without receiving any new liquor.

For the calcination process to work properly, the volumetric flowrate of the HAL feed must be kept absolutely constant. This is achieved using a pair of motor-driven Constant Volume Feeders (CVFs). Each CVF contains a rotor which revolves at a constant speed taking feed from a container which is maintained at a static level by an overflow mechanism. The rotor meters a steady flow of liquor to the upper end of the calciner.

Conversion of HAL to a solid is achieved by evaporating the nitric acid in which the fission products and actinides are dissolved, a process known

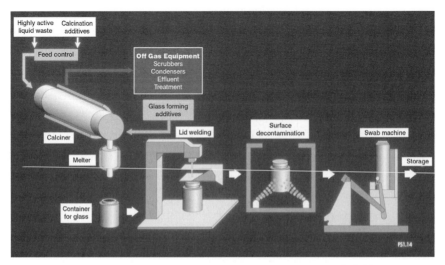

**Fig. 18.10** *Schematic diagram of vitrification process at WVP at Sellafield*

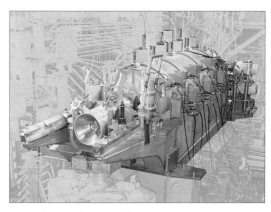

**Fig. 18.11** *Calciner in the vitrification cell*

as calcination. Sugar solution is added to the liquor to help produce a friable calcine rather than a solid lump. The evaporation is carried out in a rotating tube which is at a slight incline to allow the liquor to flow from one end to the other. The heat is applied by four furnace shells containing electrical resistance elements which are evenly spaced around the calciner tube (see Figure 18.11).

Glass granules are passed down the frit feed pipe to be mixed with the calcine in the lower end of the calciner. The mixture then falls through the grating of the calciner tube into the melter crucible positioned underneath.

Once filled with a mixture of calcine and glass frit, the crucible (see Figure 18.12) is heated by induction coils until the glass melts. A lead-glass seal is used to maintain a hermetic seal between the melter crucible and the lower end of the calciner.

**Fig. 18.12** *Crucible to melt calcine and glass frits together*

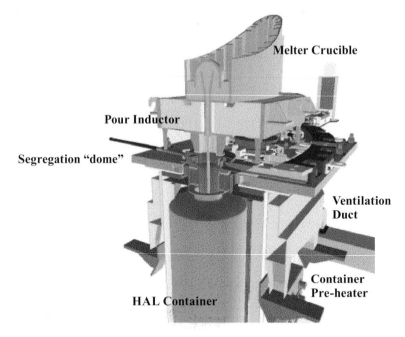

**Fig. 18.13** *Pre-heater to the HAL container*

Beneath the melter, a new product container is raised up to the base of the crucible. The pre-heater acts to reduce the thermal shock of hot molten glass falling into a cold container. A bellows unit and gas extraction system prevent off-gases escaping (see Figure 18.13).

The off-gas is drawn off via an annular duct to prevent contamination entering the cell whilst a bellows unit ensures a seal with the neck of the container.

Each product container from the pour cell is decontaminated whilst it traverses to beneath the control cell inlet trap door. After decontamination the container is raised into the control cell for its final monitoring. Before any container can be exported, it must be thoroughly checked for surface contamination. An industrial robot swabs the container while it rotates on its stand. The swab is then monitored. The swabbed container is exported via a hoist well in the ceiling of the control cell after being weighed. It is then raised directly into the export flask.

### 18.4.3.2 Synroc Process

Synroc, an abbreviation of Synthetic Rock, is an advanced ceramic material comprising geochemically stable natural minerals which can bind all the elements of HLW in their crystalline structures and thus immobilise them. It was invented by late Professor Ted Ringwood of the Australian National University in 1978 from his study of natural structures of U and Th compounds. Subsequent research and development on Synroc has been carried out at the Australian

Nuclear Science and Technology Organisation (ANSTO) Research Laboratories at Lucas Heights, NSW as well as at the Australian National University.

Synroc can be tailored to immobilise various forms and components of HLW [6]. The original form, Synroc-C, was intended to immobilise liquid HLW arising from the reprocessing of spent fuel. The main minerals in Synroc-C are hollandite ($BaAl_2Ti_6O_{16}$), zirconolite ($CaZrTi_2O_7$) and perovskite ($CaTiO_3$). The zirconolite and perovskite components are the major hosts for long-lived actinides such as Pu, though perovskite is principally for Sr and Ba. Hollandite principally immobilises Cs, along with K, Rb and Ba. Synroc-C can hold up to 30% HLW by weight. However, as the vitrification process was very well entrenched and technologically advanced by the 1980s, the Synroc-C process did not make much headway.

Over the last few years, however, various forms of Synroc have been developed. Synroc-D contains nepheline, $(Na, K)AlSiO_4$ instead of hollandite, as host for Cs, Rb and Ba. Another variant, Synroc-F, is rich in pyrochlores, $(Ca, Gd, U, Pu, Hf)_2Ti_2O_7$, and was developed for the disposal of unreprocessed spent fuel from light water and Candu reactors. The pyrochlore-rich titanate ceramic was developed, in association with Lawrence Livermore National Laboratory (LLNL) of the USA, particularly for military waste with large content of plutonium compounds. The waste loading could be as much as 50% by weight of $PuO_2$ and/or $UO_2$.

Of late, there has been a move to develop a glass–ceramic composite wasteform and research is ongoing at ANSTO. The principle is that the radioactive wastes are incorporated in the extremely durable crystalline titanate phases such as zirconolite and pyrochlore (which will hold actinides), which are then mixed within a glass matrix. The composite wasteforms are melted at 1200–1400°C to form a glass matrix. Composite wasteform is also the subject of a collaborative research programme with the French Atomic Energy Commission (CEA). The emphasis is on developing a Synroc–glass wasteform using the French cold-crucible melting technology. This process can achieve up to 50% waste loading and may also be used for the French Partitioning and Transmutation (P&T) programmes. Further details about the P&T technique can be found in Section 20.3.

## 18.4.4 Polymerisation

Polymerisation is the technique of immobilising waste concentrates such as resins, sludges, evaporator bottoms, ashes etc. in organic polymers. Several types of polymers are available such as urea formaldehyde, polyethylene, styrene di-vinyl benzene (for evaporator concentrates), epoxy (for ion exchange resins), polyester, PVC, polyurethane and so forth. The use of this process requires a knowledge of the chemical composition of the wastes and adequate understanding of the chemical reactions that are involved.

## 18.5 Characterisation of Conditioned Waste

The waste concentrates to be conditioned must meet the following two requirements:

- Be compatible with the waste matrix and the waste container
- Be compatible with the future disposal environment.

Compatibility of waste with the waste matrix is extremely important for the long-term stability of the conditioned package. If, for example, due to self-heating of certain types of wastes the physical and chemical integrity of the conditioned waste is disturbed, then the isolation of the waste over a long timescale may not be maintained.

The next consideration is the compatibility of the conditioned waste with the future disposal environment. However, as the permanent disposal facility in the UK has not yet been identified, wastes cannot be conditioned permanently. The parameters which are important in characterising a disposal facility are groundwater flow, chemical behaviour, pressure, temperature etc. In fact, because of the non-availability of the disposal facility, one of the government's requirements is to store waste in a retrievable form.

## Revision Questions

1. What types of waste arise as a result of decommissioning a plant? Briefly describe these types.

2. Why is the process of treatment undertaken on radioactive waste? State the objectives of treatment process clearly.

3. Show in a flow diagram the processes of treatment and conditioning of radioactive waste. Briefly describe the various processes.

4. List the various treatment and conditioning processes that are applied to various types of wastes in the UK, USA and France. Briefly describe each of these processes.

5. What differentiates a low force compactor from a super-compactor? Describe the principle and operation of a low force in-drum compactor with a diagram.

6. Why is 'auxiliary equipment' used in the treatment of radioactive waste? Briefly describe the function of PETRA.

7. What is the purpose of an 'incinerator'? Describe the operation of a shaft kiln with a tippable grate incinerator.

8. Why is an 'evaporator' used? Describe the advantages of using an evaporator to treat radioactive wastes.

9. Which conditioning processes are applied to which categories of wastes?

10. Describe the following conditioning processes stating their suitability in application:
    (i) bituminisation
    (ii) cementation
    (iii) Magnox encapsulation process

11. What is the process of vitrification? Briefly describe the operation of the Windscale Vitrification Process (WVP).

12. What is the Synroc process? Describe its principle and applications.

## REFERENCES

[1] International Atomic Energy Agency, Radioactive Waste Management, IAEA Source Book, IAEA, Vienna, Austria, 1992.
[2] International Atomic Energy Agency, Predisposal Management of Organic Radioactive Waste, Technical Reports Series No. 427, IAEA, Vienna, Austria, 2004.
[3] European Commission, 6th Framework Programme Contract No. FI60-CT-2003-509070, EUNDETRAF II, 2002–2006.
[4] Nuclear Energy Agency, The Safety of Nuclear Fuel Cycle, OECD, 2005
[5] Private communication to A. Rahman by K. Mayer of BNFL, 2004.
[6] http://www.uic.com.au/nip21.htm.

# 19

# STORAGE AND TRANSPORTATION OF RADIOACTIVE WASTE

## 19.1  Storage

Storage of radioactive waste means placement of wastes in a way that provides for its temporary containment with the intention of retrieving it at a later date. A wide variety of wastes may be encountered with varying physical, chemical and radiological characteristics. Storage is a temporary measure that can be applied at any stage of the waste management scheme: before treatment, before conditioning or before disposal. Storage of waste is widely practised for a variety of reasons [1] and these are:

- Raw wastes, in solid, liquid or gaseous forms, are collected and maintained to retain flexibility for future treatment and conditioning options.
- As a buffer to optimise specific treatment and conditioning operations.
- To allow for the decay of short-lived radionuclides to a level at which waste can be released from regulatory control (clearance), or authorised for discharge, or recycled and reused.
- To reduce the rate of heat generation of the high level (heat-generating) waste before undertaking predisposal management action.
- To provide long-term storage where there is no suitable disposal facility.

The storage facility may be located within the site generating the waste, such as a nuclear power plant, a hospital, a research laboratory; or it may be at a centralised location. A wide variety of waste types and storage needs are encountered: large or small amounts of radioactive inventory, short or long half-lives of the radionuclides, short- or long-term storage requirements etc. If storage is carried out for a short period of time awaiting a well-defined waste management operation, it may be considered to be interim storage. On the other hand, if storage is undertaken without any further operational plan or while waiting for further development in the management strategy, it is known as long-term interim storage.

All of these types of waste need to be managed according to the prevailing safety standards. The IAEA Safety Standards [2] mentions that adequate measures should be undertaken in order to protect human health and the environment. The IAEA also recommends that the implementation of safety measures should be commensurate with the nature and levels of the anticipated hazard. For example, liquid HLW (arising from the reprocessing of spent fuel) is generally stored in double-walled stainless steel tanks. Various designs of double-walled and cooled stainless steel tanks are in use at Sellafield in the UK. For radiological protection, safety measures should comply with the international Basic Safety Standards (BSSs) [3]. In addition to the radiological hazards of the waste, there may be non-radiological hazards associated with corrosiveness, flammability, explosiveness, toxicity and pathogenicity of the waste. For safety reasons storage requires a surveillance and maintenance programme.

If wastes are to be stored for an extended period of time, the owner/operator needs to comply with safety provisions such as maintenance of the facility, compliance with the package acceptance criteria for storage, record keeping, inventory and material transfer records, surveillance and monitoring etc. Thus a long-term storage facility is regarded by the regulatory body, for all intents and purposes, as an operational nuclear facility, although no active physical operation may take place. All safety provisions and regulatory standards remain in force for such waste storage facilities. (See also Section 20.1 for the distinction between storage and disposal.)

The physical state of the waste, as far as practicable, should comply with the national and international regulations/recommendations [1] such as

- The radioactive material/waste should be immobilised.
- The waste and its container should be physically and chemically stable.
- A multi-barrier approach should be adopted in order to ensure adequate containment.
- The need for active safety provision should be minimised.
- The waste package should be retrievable for treatment and conditioning, if necessary.
- If the waste package is conditioned, it should be in a state acceptable for final disposal.

## 19.2 Safety Aspects of Radioactive Material Transport

The transportation of radioactive material is required to comply with the IAEA regulatory requirements as set out in the IAEA document, 'Regulations for the Safe Transport of Radioactive Material' [4]. The radiological safety requirements that must be complied with in the whole process of transportation are described in the BSSs [3, 5]. To implement the BSSs, a Radiation Protection Programme (RPP) needs to be set up which contains systematic arrangements for radiation protection measures by the consignors, carriers and consignees. This programme is best established by the cooperative effort of all those involved in the transport process.

While compliance with the transport regulations [4] is important, there may be additional requirements for transport by sea and air. For air shipment, the International Civil Aviation Organisation's (ICAO) Technical Instructions for the Safe Transport of Dangerous Goods by air and the International Air Transport Association's (IATA) Dangerous Goods Regulations are relevant. For sea transport, the International Maritime Organisation's (IMO) International Maritime Dangerous Goods Code is quite important.

In the UK, there are separate but consistent sets of regulations for the transport of radioactive materials by road, rail, sea and air. The overall executive role of the competent authority is carried out by the Radioactive Materials Transport Division (RMTD) of the Department for Transport (DfT) [6]. The enforcement of regulations for transport by road, rail, sea and air is carried out in the UK by the DfT, HM Railway Inspectorate, Maritime and Coastguard Agency, and the Civil Aviation Authority, respectively. The IAEA Advisory Material on Transport of Radioactive Material [7] covers the requirements for an RPP. The major elements of this RPP are

- Every consignor, carrier and consignee, when setting up an RPP must
    - account for the nature and extent of the measures to be taken in respect of the magnitude and likelihood of radiation exposures ('graded approach').
    - adopt a structured and systematic approach to the framework of controls to satisfy radiation protection principles.
- Every consignor, carrier and consignee must adhere to the relevant provisions of the IRR99 [8].

The RSA93, which requires registration for the use of radioactive materials and authorisation for the disposal of radioactive wastes, must also be adhered to in the UK [9].

### 19.2.1 Scope of RPPs

The general principle is that the nature and extent of the measures to be used in controlling radiation exposures shall be related to the magnitude and the likelihood of the exposures. For an individual transport worker where it is determined that the dose received

- is unlikely to exceed 1 $mSv.y^{-1}$, no detailed monitoring or assessment of radiation doses is required.

- is likely to be in the range 1–6 mSv.y$^{-1}$, periodic environmental monitoring and assessment of radiation doses are required.
- is likely to be in the range 6–20 mSv.y$^{-1}$, individual monitoring and health supervision are required.

In addition, the radioactive materials need to be sufficiently segregated in regularly occupied areas such that the transport workers do not receive doses in excess of 5 mSv.y$^{-1}$ or a member of the public does not receive a dose in excess of 1 mSv.y$^{-1}$.

# 19.3 Transportation of Radioactive Materials

Transportation of radioactive material is deemed to comprise all operations and conditions associated with and involved in the movement of radioactive material/waste. These include the design, fabrication and maintenance of packaging and the preparation, consigning, handling, storage in transit and final receipt of the package [4]. Transportation of radioactive waste from the production site to the treatment/conditioning site and from the latter to the disposal site are radiologically significant in the overall management of radioactive waste. Transportation may cause radiation exposure not only to the workers but also to the population at large, particularly those in and around the transport route. Hence there is a need for proper treatment and conditioning of waste before they are transported.

Radioactive materials arise in gaseous, liquid or solid forms. Radioactive materials are usually transported in solid form and less frequently in liquid form. The most common mode of transport within a country is by land, either by road or rail. Sea and air transport are also quite prevalent. To carry out transport operations safely and cost effectively and to make regulations clear and transparent, the IAEA in the Transport Regulations document [4] has defined a number of quantities which are given below.

## 19.3.1 Definition of Transport Regulation Quantities

$A_1$ and $A_2$ **Values**: $A_1$ and $A_2$ are the specific limits of radioactivity in Bq for radionuclides in special form and in non-special form respectively, that are allowed in a type A package. These limits have been set for the type A package such that the radiological consequences of severe damage to this type of package during transport would not breach the international BSSs [3]. These limits and multiples/sub-multiples thereof are also used for several other purposes in the transport regulations such as in specifying type B package leakage limits, Low Specific Activity (LSA) and excepted packages contents limits. The values of $A_1$ and $A_2$ are given in Table 19.1.

**Contamination**: Contamination arises when an object has radioactive

**Table 19.1** *A$_1$ and A$_2$ values for some of the important radionuclides*

| Radionuclide | Half-life | A$_1$ (TBq)* | A$_2$ (TBq)* |
|---|---|---|---|
| C-14 | 5730 y | $4 \times 10^1$ | 3.0 |
| Fe-55 | 2.7 y | $4 \times 10^1$ | $4 \times 10^1$ |
| Co-60 | 5.27 y | $4 \times 10^{-1}$ | $4 \times 10^{-1}$ |
| Ni-59 | $7.5 \times 10^4$ y | Unlimited | Unlimited |
| Sr-90 | 29.12 y | $3 \times 10^{-1}$ | $3 \times 10^{-1}$ |
| Mo-93 | 3500 y | $4 \times 10^1$ | $2 \times 10^1$ |
| I-131 | 8.04 d | 3.0 | $7 \times 10^{-1}$ |
| Cs-137 | 30.0 y | 2.0 | $6 \times 10^{-1}$ |
| Ru-106 | 368.2 d | $2 \times 10^{-1}$ | $2 \times 10^{-1}$ |
| Ir-192 | 74.02 d | 1.0 | $6 \times 10^{-1}$ |
| U-238 | $4.47 \times 10^9$ y | Unlimited | Unlimited |
| Pu-239 | 24065 y | $1 \times 10^1$ | $2 \times 10^{-3}$ |
| Am-241 | 432.2 y | $1 \times 10^1$ | $2 \times 10^{-3}$ |

* TBq is tera-becquerel (= $10^{12}$ Bq)

substances on its surfaces or embedded in its surfaces. This means that activated components are not considered as contaminated objects in transport regulations. There are two types of contamination: fixed contamination and non-fixed contamination. Although it is difficult to separate these two quantities completely, a distinction may be made from practical point of view. A contamination on a body which remains in situ during routine conditions of transport may be considered as fixed. Fixed contamination does not give rise to hazards arising from inhalation or ingestion under normal conditions, but may contribute to these pathways in accident situations. Non-fixed contamination gives rise to such hazards even in normal conditions.

Contamination below levels of 0.4 Bq.cm$^{-2}$ for β/γ-emitters and low toxicity α-emitters, or 0.04 Bq.cm$^{-2}$ for all other α-emitters is considered non-contaminated for the purposes of transport regulations.

**Criticality Safety Index (CSI)**: The CSI assigned to a package, over-pack or freight container containing fissile material is a number which is used to provide control over the accumulation of such packages. The CSI is calculated by dividing 50 by a number $N$ which is given a value in order to ensure sub-criticality in the array of consigned packages in all transport conditions, both normal and accident situations. The lower the risk of criticality in a package is, the higher the value of $N$ that is assigned. For example, if $N = 11$, CSI = 50/11 = 4.545. It is always rounded up to the first decimal place, giving CSI = 4.6. Thus, the number of packages containing fissile materials that can be transported in one conveyance is 50/4.6 ≅ 10. Theoretically, $N$ could be given a very high value for sub-critical packages, resulting in the value of CSI approaching zero. However, for any consignment of fissile material not under exclusive use, the CSI must not exceed 50.

**Exclusive use**: The definition of 'exclusive use' in the shipment of a consignment is that a single consignor takes the shipment and has the sole use of the conveyance or the freight and it is directly under the control of the consignor or the consignee. As the consignment is in 'exclusive use', normal regulatory requirements are somewhat relaxed.

**Low dispersible radioactive material**: Low dispersible radioactive material is that which is mainly not dispersible in its intrinsic form, such as in solid, or encapsulated in a high integrity sealed capsule, in which the encapsulated material acts as a non-dispersible solid. Powder or powder-like materials cannot qualify as low dispersible material. The total amount of such material must be such that the radiation level at 3 m from the unshielded surface must not exceed 10 mSv.h$^{-1}$. This type of material is normally transported in type B packaging.

**Low Specific Activity materials**: The LSA materials are those whose specific activities are so low that it is highly unlikely that, during the transport of radioactive material, a sufficient amount of such material can be taken into the body to give rise to a significant radiation hazard. The concentration limit for LSA material is around $10^{-4}$ $A_2$/g. This type of material should not present a radiation hazard during transport greater than that of a type A package.

The LSA materials are U and Th ores and their physical and chemical concentrates, concrete blocks containing distributed activity or other irradiated objects or objects with fixed contamination. There are three types of LSA materials

- **LSA-I materials**: These are low activity materials such as ores of naturally occurring radionuclides, materials having activity concentrations up to 30 times the exemption levels. They are normally shipped in industrial package type 1 (type IP-1) packages.
- **LSA-II materials**: These materials may include reactor operational wastes such as lower activity resins, filter sludge, absorbed liquor etc. They may also include decommissioning wastes such as activated components, liquid waste; or other types such as scintillation vials, hospital waste etc. The activity limit for such substances is $10^{-4}$ $A_2$/g. They are shipped in type IP-2 packages.
- **LSA-III materials**: These materials include concentrated liquids encapsulated in concrete materials, solidified resins and cartridge filters in a suitable matrix. As these materials are solidified and are highly unlikely to be dispersed in

transport conditions, a slightly higher activity concentration limit of $2 \times 10^{-3}\ A_2/g$ is specified for them. Another condition for LSA-III materials (as well as for LSA-II solid materials) is that the activity should essentially be uniformly distributed. If activities are concentrated in a small volume, the risk from a transport accident would be higher and that is not acceptable. Industrial package type 3 (type IP-3), which is roughly the same as type A, is used for these materials. However, there are additional restrictions on the overall amount of activity that can be transported. For example, if a 200 litre drum is used and is filled with LSA-III material of concentration $2 \times 10^{-3}\ A_2/g$ and if the material density is 1 g.cm$^{-3}$, the total amount of activity in the package will be $2 \times 10^{-3}\ A_2/g \times 1$ g.cm$^{-3} \times 200 \times 10^3$ cm$^3$ $= 400\ A_2$. This amount is higher than the $10\ A_2$ allowed by inland waterways and $100\ A_2$ by other modes.

**Package and packaging**: A package is a packaging plus its radioactive content as presented for transportation. For design and compliance purposes, the package may include structural equipment required for handling or securing the package.

**Over-pack**: An over-pack is the outer casing which may contain one or more packages, each of which fully complies with transport requirements. An over-pack does not need to comply with the design, test or approval requirements, as it is the packaging, not the over-pack, which performs the protective function.

**Special form radioactive material**: A special form radioactive material is either non-dispersible solid radioactive material or a sealed capsule containing radioactive material. A sealed capsule is one which can only be opened by destroying the strong metallic capsule. Materials protected this way from the risk of dispersion are termed 'special form'. If a material is in special form, the package limit for type A package changes from $A_2$ to $A_1$ or a multiple thereof, depending on the radionuclide concerned. Any material other than special form is called a non-special form radioactive material. The package limit for a type A package of non-special form material is $A_2$.

**Surface-Contaminated Object (SCO)**: A surface contaminated object is one which has either fixed or non-fixed contaminants on the surface. There are two types of SCO and they are described below. It may, however, be noted that these are contaminated objects, not objects which are activated. An activated object, even though there may be some surface contamination, would be regarded as a LSA object.

SCO-I: A surface contaminated object-1 (SCO-I) is one which may have fixed and non-fixed contamination on the surface. The non-fixed contamination on the accessible surface does not exceed 4 Bq.cm$^{-2}$ for $\beta/\gamma$-emitters and low toxicity $\alpha$-emitters or 0.4 Bq.cm$^{-2}$ for all other $\alpha$-emitters. If the contamination is fixed, then the limit may be enhanced by a factor of 10,000.

In an accident situation, it is anticipated that 20% of the surface is scraped and 20% of fixed contamination from the scraped surface is released. Also, all of the non-fixed contamination is assumed to be released. So, if a SCO-I has a total fixed contamination of 4 GBq distributed uniformly over the surface area and non-fixed contamination of 0.4 MBq, the amount of activity that is likely to be freed is $4 \times 10^9 \times 0.2 \times 0.2 + 0.4 \times 10^6$ Bq $= 160.4$ MBq. Using an $A_2$ value of 0.02 TBq for mixed $\beta/\gamma$-emitting fission products, the released amount would be is $160 \times 10^6/0.02 \times 10^{12} = 8 \times 10^{-3} A_2$. If the intake by a person in the vicinity of the accident is $10^{-2}$ of the released amount, then the total intake would be $8 \times 10^{-5} A_2$. Hence it provides a level of safety equivalent to type A package. Industrial package type 1 (type IP-1) is normally used for such objects. However, if the activity of an SCO is so low that the activity limits for excepted packages are met, then it can be transported in an excepted package.

**SCO-II**: The SCO-II is one where the non-fixed contamination on the accessible surface does not exceed 400 Bq.cm$^{-2}$ for $\beta/\gamma$-emitters and low toxicity $\alpha$-emitters or 40 Bq.cm$^{-2}$ for all other $\alpha$-emitters. If the contamination is fixed, then the limits are 800 kBq for $\beta/\gamma$-emitters and 80 kBq for $\alpha$-emitters, respectively.

The SCO-II is similar to SCO-I but the activity limits are 100 times higher for non-fixed contamination and 20 times higher for fixed contamination than the SCO-I values. On the basis of the fixed and non-fixed contamination specified above (for the SCO-I object), released fixed contamination would be $4 \times 10^9 \times 20 \times 0.2 \times 0.2 \times 10^{-2}$ Bq $= 32$ MBq, when the assumed released fraction is $10^{-2}$ and the non-fixed contamination would be $0.4 \times 10^6 \times 100$ Bq $= 40$ MBq. Thus the total released amount would be 72 MBq, which is equal to $3.6 \times 10^{-3} A_2$. This also provides type A package safety level. Such objects are transported in type IP-2 packages.

**Transport Index** (TI): The TI gives an indication of the radiation level in the vicinity of a package, over-pack, freight container, unpackaged LSA-I or SCO-I. It is the maximum radiation level at 1 m from the external surface of a package expressed in mSv.h$^{-1}$ and multiplied by 100. It is then rounded up to the first decimal place. The value of TI for a consignment, not under exclusive use, is to be limited to 10. There are additional considerations in TI calculations. For packages of large cross-sectional areas, the estimated TI value is further multiplied by a factor of 10 which depends on the largest cross sectional area of the load. This factor is in the range 1–10. It is 1 if the largest area is 1 m$^2$ or less. It is 10 if the largest area is more than 20 m$^2$.

The TI for an over-pack (either rigid or non-rigid) or for a freight container containing a number of packages can be obtained by adding the TIs of all the packages contained in the over-pack. This would be a conservative estimate as the self-shielding effect of the packages would not be taken into account.

## 19.4 Waste Packages

One of the most important pre-requisites for the safe transportation of radioactive materials is that the characteristics of the package must match the radioactive content of a package. Five types of packages have been defined by IAEA for the transport of radioactive materials and these are: **excepted package, industrial package, type A package, type B package and type C package** [4].

Small quantities of radioactive materials can be transported in simplified packages, called 'excepted packages'. Nonetheless, these packages have to meet the general safety requirements for packages such as having sufficient structural strength, attachments to facilitate lifting, quality assured structural material which is compatible with the radioactive content etc. The activity limits in 'excepted packages' are to be maintained within the values shown in Table 19.2 in order to meet the general safety standards with sufficient safety margin. The values of $A_1$ and $A_2$ for some important radionuclides are given in Table 19.1. The values of $A_2$ are generally lower than or at most equal to $A_1$. The activity concentrations of the material in these 'excepted packages' are generally limited to LLW. The radiation level at any point on the external surface of an excepted package must not exceed 5 $\mu Sv.h^{-1}$.

Industrial packages may contain LSA material or SCO. Large volumes of radioactive materials of LSA and SCO can be transported in the industrial packages. The LSA material has a limited specific activity, around low to

**Table 19.2** *Activity limits for excepted packages*

| Contents | Instrument or article* | | Material** |
|---|---|---|---|
| | Item Limit | Package limit | Package Limit |
| Solids | | | |
| special form | $1.0 \times 10^{-2} A_1$ | $A_1$ | $1.0 \times 10^{-3} A_1$ |
| other form | $1.0 \times 10^{-2} A_2$ | $A_2$ | $1.0 \times 10^{-3} A_2$ |
| Liquids | $1.0 \times 10^{-3} A_2$ | $1.0 \times 10^{-1} A_2$ | $1.0 \times 10^{-4} A_2$ |
| Gases | | | |
| tritium | $2.0 \times 10^{-2} A_2$ | $2.0 \times 10^{-1} A_2$ | $2.0 \times 10^{-2} A_2$ |
| special form | $1.0 \times 10^{-3} A_1$ | $1.0 \times 10^{-2} A_1$ | $1.0 \times 10^{-3} A_1$ |
| other forms | $1.0 \times 10^{-3} A_2$ | $1.0 \times 10^{-2} A_2$ | $1.0 \times 10^{-3} A_2$ |

\* Material either enclosed in or included as a component part of an instrument

\** Material not enclosed in nor a component part of an instrument

intermediate level. Industrial package type 1 (type IP-1) may contain LSA-I materials which are mainly natural U, natural Th, depleted U etc. Type IP-2 may contain LSA-II materials which are low activity wastes and type IP-3 may contain LSA-III materials which may include ILW immobilised in concrete, bitumen, ceramic etc. The external radiation level at 3 m from the unshielded LSA material or SCO in an industrial package or unpackaged container is limited to 10 mSv.h$^{-1}$. The activity limits for LSA material and SCO are given in Table 19.3.

Type A packages may contain radioactivity up to $A_1$, if it is in special form radioactive material, or up to $A_2$, if not in special form radioactive material. The values of $A_1$ and $A_2$ can be found in Table 1 of [4] and some of the commonly used radionuclides are quoted in Table 19.1. If a radionuclide is not specified in Table 1 of [4], then the general values of $A_1$ and $A_2$, given in Table 19.4, apply. It should be noted that a type A package may contain a single radionuclide or a mixture of radionuclides, but the limit imposed by $A_1/A_2$ system would apply as per equation (19.1).

$$\sum_{j,k}\left(\frac{A_{1j}}{A_1} + \frac{A_{2k}}{A_2}\right) \leq 1 \tag{19.1}$$

where $A_{1j}$ is the activity of a particular radionuclide, $j$ in the special form, and $A_{2k}$ is the activity of another radionuclide, $k$ in non-special form.

The type A package is intended to provide economical transport for large numbers of low activity consignments, while at the same time achieving a high level of safety. As the anticipated radiological consequences of type A packages following accidents are not considered excessive, there is no need for the regulatory approval for the design of type A packages, except when fissile material is to be transported.

Activities in excess of type A package limits are used in type B packages which are of two types: B(U) and B(M). If these packages are transported by air, there are specific restrictions on low dispersible radioactive material, special form radioactive material or for other radioactive material. As the

**Table 19.3** *Activity limits in industrial packages*

| Material | Activity limits in conveyances |
|---|---|
| LSA-1 | No limit |
| LSA-II and LSA-III non-combustible solids | No limit |
| LSA-II and LSA-III combustible solids, liquids and gases | 100 $A_2$ |
| SCO | 100 $A_2$ |

**Table 19.4** *General values for $A_1$ and $A_2$*

| Contents | $A_1$ (TBq) | $A_2$ (TBq) |
|---|---|---|
| Only β/γ-emitter | 0.2 | 0.02 |
| α-emitter along with other emitters | 0.1 | $2.0 \times 10^{-5}$ |

radiological consequences of a type B package in an accident would be more severe than a type A package, there is a need for regulatory approval of type B design.

Type C packages are primarily used for transporting fissile material and other high activity materials. Thus restrictions are imposed on these packages on radiological and criticality considerations.

## 19.4.1  General Requirements for all Packaging and Packages

There are some general requirements which apply to all types packaging and packages.

- The package must have provisions to secure it in or on the conveyance during transport.
- Any lifting attachments on the package must be fail-safe, if used properly. The design will take into account of appropriate safety factors to cover snatch lifting.
- The external surfaces of the packages should be free from protruding features and the surfaces should be easy to decontaminate.
- The outer layer of the package should be designed so as to prevent collection and retention of water.
- The package should be able to withstand the effects of any acceleration, vibration etc. during normal conditions of transport. In particular, nuts, bolts and other securing devices should not become loose or be released unintentionally.
- The packaging materials should be physically and chemically compatible with each other and with the radioactive contents.

The above requirements apply to excepted packages as well as to industrial package type 1 (type IP-1). For type IP-2, all the above requirements apply; additionally, the package should meet the free drop test and stacking test (see Section 19.4.2) without showing a loss of shielding resulting in an increase of more than 20% in the radiation level at any external surface of the package.

For industrial package type 3 (type IP-3) and type A packages, all the above requirements will apply and, in addition, the following requirements apply:

- The smallest external dimension of the package must not be less than 10 cm.
- The package must incorporate an external feature such as a seal, which will not be readily breakable and which, while intact, will be the evidence that it has not been opened.
- The package will have to withstand a temperature range from −40°C to +70°C.
- If a separate containment system is incorporated within the package, then that system must be properly secured within the package.
- If the package is designed to contain liquid radioactive material, then provisions for ullage to accommodate the temperature variations of the contents, dynamic effects etc. must be taken into account.
- If the package is designed to carry liquid radioactive materials, then there must a primary inner and a secondary outer containment to ensure retention of the liquid contents.
- The package must be able to withstand the water spray test, free drop test, stacking test and the penetration test (see Section 19.4.2) without the loss of shielding resulting in more than a 20% increase in the external radiation level.

The requirements for type B (type B(U) and type B(M)) and type C packages are progressively more stringent. All these types will have to meet all the requirements specified above and additionally meet the following conditions:

- The heat generated within the package by the radioactive contents must not adversely affect the shielding and containment provisions of the package, even if it is left unattended for one week.
- The surface temperature of a package must not exceed +50°C when the ambient temperature is assumed to be +38°C, unless the package is transported under exclusive use.
- As the package is designed to meet the water spray test, free drop test, stacking test and the penetration test, the maximum loss of radioactive content would be limited to $10^{-6} A_2$ per hour following these tests.
- Following additional tests such as the mechanical test, thermal test and water drop test simulating accident conditions, the package should retain sufficient shielding

to ensure that the radiation level at 1 m from the surface would not exceed 10 mSv.h$^{-1}$, with the maximum radioactive content.

## 19.4.2 Tests for Packages

A type A package is designed to maintain its integrity during normal transport conditions as well as in minor mishaps. In more serious transport accidents, it may release part of its contents. However, as it only contains LLWs, the consequences of any accidents are limited.

Type A packages must include a containment system which may be a special form radioactive material or a separate unit of the package. They should be able to withstand normal conditions of transport. This capability is demonstrated by the following tests: **water spray test, free drop test, stacking test** and **penetration test**. A specimen of a package is subjected to the free drop test, the stacking test and the penetration test. Each test is preceded by a water spray test. The time interval between the water spray test and the succeeding test is to be two hours if water spray is applied from four directions simultaneously. If water spray is applied from each of the four directions consecutively, then there is to be no time gap. One specimen is to be used for all the tests. For the water spray test, the package is subjected to a water spray that simulates exposure to rainfall of approximately 5 cm per hour for at least one hour. For the free drop test, the package is dropped onto a hard surface from a height of 1.2 m if the package is less than 5 te or 0.3 m if the package is heavier than 15 te. For package weights between 5 te and 15 te, the drop height would be linearly interpolated between 1.2m and 0.3m. For the penetration test, the package is placed on a rigid surface and a bar of 3.2 cm in diameter with a hemispherical end and having a mass of 6 kg is dropped from a height of 1 m. If the package is designed to carry liquid or gas, the free drop test should be carried out from a height of 9 m and the penetration test from a height of 1.7 m.

Packages of types B (B(U) and B(M)) and C are designed to conform to additional tests which simulate accident conditions. These tests are: **mechanical, thermal** and **water immersion tests**. The specimen is subjected to the cumulative effects of the mechanical and thermal tests and then the water immersion test.

The mechanical test consists of three different drop tests. The order in which the specimen is subjected to the drops is specified so that it suffers damage which will lead to the maximum damage in the thermal test which follows.

- Drop I: The specimen is dropped onto a hard surface from a height of 9 m. The height is measured from the lowest point of the specimen to the upper surface of the target.
- Drop II: The specimen is dropped from a height of 1 m to a

bar rigidly mounted perpendicularly on the target. The bar will be solid mild steel of circular cross-section, 15 cm in diameter and 20 cm long.

- Drop III: This is a crush test when the specimen is placed on a hard surface and a 500 kg steel plate 1 m × 1 m is dropped from a height of 9 m in a horizontal attitude. The height of the drop will be measured from the underside of the plate to the highest point of the specimen.

The thermal test is carried out when the specimen is in thermal equilibrium at an ambient temperature of 38°C. The specimen is then subjected to an average flame temperature of 800°C for a period of 30 min.

The water immersion test is done by subjecting the specimen to a head of water of 15 m for a period of not less than 8 h.

Type B(U) packages are qualified unilaterally whereas B(M) packages require multilateral qualification between the consignor and the consignee. Type C packages are primarily used for the transport of fissile material.

## 19.5 Transportation of Waste

### 19.5.1 Transportation of LLW

The low level wastes contain such a small amount of radioactive material that they can be packaged in normal steel drums and containers. Most operational LLW is suitable for packaging in standard 200 litre drums (0.51 m OD × 0.86 m high). The drummed waste needs to be compressed before disposal and possibly before transport. The UK nuclear waste executive body, known as Nirex (now being merged with the NDA), plans to compress each drum to about one-fifth of its size in a high force compactor. The compressed drums are then packed into one of the two standard Nirex boxes: 2 m and 4 m boxes. The details of the Nirex drum and boxes for LLW are given in Table 19.5 [10].

The LLWs arising from decommissioning operations comprise contaminated plant items and equipment, pipe-work, ducts, cables etc. which are inherently

**Table 19.5** *Nirex standard waste containers for LLW*

| Description | Application | Dimension (m) (external) | Gross wt (kg) |
|---|---|---|---|
| 2 m LLW box | For higher density operational and decom. wastes (shielded) | 2.0 × 2.4 plan × 2.2 high | 30 000 |
| 4 m LLW box | For general operational and decom. wastes (shielded) | 4.0 × 2.4 plan × 2.2 high | 30 000 |

**Fig. 19.1** *4 m LLW box with compressed 200 litre drums (Courtesy of Nirex)*

of large dimensions. These large items are placed in standard 4 m box packages which are designed to conform to the standard ISO freight container dimensions. The 4 m box, shown in Figure 19.1, although designed specifically for LLW, can also carry wastes which are at the lower end of the ILW scale. Both of these boxes (2 m and 4 m boxes) have been designed to satisfy the industrial package type 2 (type IP-2) requirements. It should be noted that the LLW specified in the UK qualifies as LSA/SCO under the IAEA transport regulations.

### 19.5.2 Transportation of ILW

For the transportation and disposal of ILW, Nirex has designed four standard waste packages (see Table 19.6). The principal waste package for ILW is a

**Table 19.6** *Nirex standard waste packages for ILW*

| Description | Application | Dimension/m (external) | Gross wt./kg |
|---|---|---|---|
| 500 l drum | For most operational and general waste (unshielded) | 0.8 OD × 1.2 high | 2 000 |
| 3 m³ ILW box | For solid wastes (unshielded) | 1.72 × 1.72 plan × 1.2 high | 12 000 |
| 3 m³ ILW drum | For solid wastes for in-drum mixing and sludge immobilisation (unshielded) | 1.72 OD × 1.2 high | 12 000 |
| 4 m ILW box | For larger items of decommissioning wastes (shielded) | 4.0 × 2.4 plan × 2.2 high | 65 000 |

**Fig. 19.2** *A 500 litre drum made of stainless steel with cement wasteform (Courtesy of Nirex)*

special 500 litre stainless steel drum. Wastes may be loaded into the drum and immobilised using cement or other suitable materials. Alternatives to the 500 litre drum are the 3 m³ drum and 3 m³ box. These drums and boxes may be used when larger containers are needed. For some wastes, even 3 m³ containers may not be sufficiently large. A standard 4 m long box may be used for such items.

All ILW packages except the 4 m box are unshielded. It may seem contradictory that, when the ILW needs to be shielded to protect the workers and the public from the effects of radiation, these packages are unshielded. The principle behind this design philosophy is that these small packages will be transported in packages known as Reusable Shielded Transport Containers

**Fig. 19.3** *A 3 m³ ILW drum with mixing mechanism (Courtesy of Nirex)*

**Fig. 19.4** *A 3 m³ ILW box (Courtesy of Nirex)*

(RSTCs) thereby saving costs and resources by not incorporating shielding material into each package. The 500 litre drum, shown in Figure 19.2, the 3 m³ box, shown in Figure 19.3 and 3 m³ drum, shown in Figure 19.4, are all disposable with the waste. However, as they are not shielded, they need to be operated and handled remotely. The RSTC, shown in Figure 19.5, can carry four 500 litre drums or one 3 m³ box or drum. Two design concepts for the transport containers are being developed by Nirex: RSTC-70 and RSTC-285, both made of steel. The RSTC-70, with a nominal shielding thickness of 70 mm, has an unladen weight of about 16 te and a maximum laden weight of 28 te. The RSTC-285, with a nominal shielding thickness of 285 mm, has an unladen weight of about 52 te and a laden weight of 64 te. These RSTCs are designed to satisfy the regulations for type B packages. It should be noted that type B packages must be designed so that they can withstand not only the

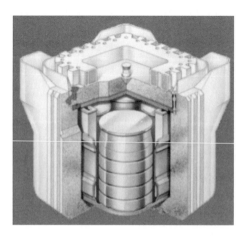

**Fig. 19.5** *An RSTC with four 500 litre drums (Courtesy of Nirex)*

**Fig. 19.6** *A 4 m ILW box on a lorry (Courtesy of Nirex)*

conditions of normal transport and minor mishaps, but also accidents including impact, fire and water immersion, while not sustaining significant loss of either shielding or containment.

The Nirex shielded 4 m ILW boxes with concrete walls have been designed to transport ILW without further requirements for shielded packages. These boxes qualify as industrial package type 2 (type IP-2) of the IAEA transport regulations.

The main use of the 4 m ILW box (shielded) is to transport large quantities of decommissioning waste which may include large items such as redundant plant items and equipment, building materials etc. The box is usually made of stainless steel and has a concrete lining that can be varied in thickness to suit the shielding requirements of the contents. Figure 19.6 shows a 4 m ILW box being transported by road.

# Revision Questions

1. What is meant by the storage of radioactive material? Explain the significance of, and differences between, interim storage and long-term interim storage.

2. What are the implications of interim storage vis-a-vis long-term storage from the safety point of view? What regulatory imposition is placed on the owner/operator of a long-term storage facility?

3. What is meant by the transportation of radioactive material? Describe the full range of activities involved in the transportation process.

4. What is the purpose of the Radiation Protection Programme (RPP) in the transportation of radioactive material? Describe the scope of the RPP in detail.

5. What is meant by Low Specific Activity (LSA) material? List the types of LSA materials and describe each type briefly.

6. Define the following terms, giving values and units, if applicable:
   - (i) $A_1$ and $A_2$ values
   - (ii) Criticality Safety Index (CSI)
   - (iii) Low dispersible radioactive material
   - (iv) Special form radioactive material
   - (v) Surface Contaminated Objects (SCO)
   - (vi) Transport Index (TI)

7. What are the various types of waste package. Describe each type, giving a broad outline of its application. Also describe the differences between B(U) and B(M) packages.

8. What are the tests for type A packages? Describe these tests fully.

9. What tests are applied to type B packages? Briefly describe these tests.

10. What are the standard waste containers for LLW in the UK? Describe these containers and give their dimensions.

11. Describe the various ILW packages that have been designed by Nirex. How are unshielded ILW packages transported in the UK?

## REFERENCES

[1] International Atomic Energy Agency, Storage of Radioactive Waste, IAEA Safety Guide No. WS-G-6.1, Vienna, Austria, 2006, http://www-pub.iaea.org/MTCD/publications/PDF/Pub1254_web.pdf.

[2] International Atomic Energy Agency, Predisposal Management of Radioactive Waste, Including Decommissioning, IAEA Safety Standards Series No. WS-R-2, IAEA, Vienna, Austria, 2000.

[3] International Atomic Energy Agency, International Basic Safety Standards for Protection against Ionising Radiation and for the Safety of Radiation Sources, Jointly sponsored by FAO, IAEA, ILO, OECD/NEA, PAHO, WHO, IAEA Safety Series No. 115, Vienna, Austria, 1996.

[4] International Atomic Energy Agency, Regulations for the Safe Transport of Radioactive Material, IAEA Safety Standards Series No. TS-R-1 (ST-1, Revised) IAEA, Vienna, Austria, 2000.

[5] EC Council Directive 96/29/Euratom of 13 May 1996, laying down basic safety standards for the protection of the health of workers and the general public against the dangers arising from ionising radiation, *Official Journal of the European Union* L159, 29 June 1996.

[6] UK Government, The Radioactive Material (Road Transport) (Great Britain) Regulations 1996, Statutory Instruments 1996 No. 1350, Reprinted in 2000.

[7] International Atomic Energy Agency, Advisory Material for the IAEA Regulations for the Safe Transport of Radioactive Material, IAEA Safety Standards Series No TS-G-1.1 (ST-2), IAEA, Vienna, Austria, 2002.

[8] UK Government, The Ionising Radiations Regulations 1999, Statutory Instruments No. 3232, HSE, 1999.

[9] UK Government, Radioactive Substances Act 1993, The Stationary Office.

[10] UK Nirex Limited, Foundations of the Phased Disposal Concept, Radioactive Waste Transport, Nirex Report No. N/022, October 2002.

# 20

# DISPOSAL OF RADIOACTIVE WASTE

## 20.1 Introduction

The back-end of the management of radioactive waste is its disposal. Disposal is defined as the emplacement of waste in an approved repository without the intention of retrieval. However, this categorical statement of 'without the intention of retrieval' has recently been slightly modified in order to accommodate a rather pragmatic approach in the light of inherent uncertainties in the likely conditions and performance of repositories far into the future. There may be situations which may necessitate wastes being retrieved, repackaged and emplaced in another facility from a potentially unsafe or failing repository. So, retrievability and reversibility are the new requirements which may need to be accommodated in the design of disposal facilities.

Thus, the disposal of waste in a repository may not be a permanent solution, as much as storage is not. The clear distinction between disposal and storage is becoming somewhat blurred. Storage may be for an extended period of time, which may be described as a long-term interim storage. Disposal at a repository may last for a short period of time requiring retrieval for safety or other reasons. The distinction lies rather in the intention and strategic view. Storage of waste for any length of time is not intended to be permanent, although it may continue for a long time. On the other hand, disposal is intended to be permanent, although retrieval may have to be carried out for safety, security and other reasons on a shorter time scale.

Before the disposal, wastes are subjected to a number of management steps such as the pre-treatment, treatment, conditioning, storage and finally transportation, which can collectively be called the 'pre-disposal management' [1]. These pre-disposal management operations are carried out in order to prepare waste for the long-term containment and isolation from human environment in a disposal facility. Treatment and conditioning of waste have been discussed in Chapter 18; whereas Chapter 19 discussed storage and transportation. This chapter deals first with the disposal concepts and systems applicable to all types of waste: VLLW, LLW, ILW and HLW of half-lives ranging from short-lived to long-lived, followed by some examples of implementation of waste disposal in some countries. Half-lives shorter than 30 years are generally considered to be short half-lives, beyond 30 years are long half-lives. This quantitative specification of 30 years is, to a large extent, arbitrary, although justification can be drawn from the fact that the half-lives of two most significant fission products, Sr-90 and Cs-137, are 29.1 y and 30.17 y respectively. Finally, the chapter considers the disposal of decommissioned nuclear submarines.

## 20.2  Pre-disposal Management

Pre-disposal management of radioactive waste comprises all management steps prior to disposal [1]. These steps include processing of operational and decommissioning waste as well as waste from cleanup activities. The safety requirements for pre-disposal management are exactly the same as those applicable to decommissioning (see Chapter 7).

In the design of facilities and the planning of activities that have the potential to generate radioactive waste, measures are taken to avoid or reduce its generation. Wastes are collected and then segregated, if necessary. They may be released from regulatory control if they require no further considerations from the radiation safety point of view. This includes the controlled discharge of effluents produced during pre-disposal operations. The reuse or recycling of materials is applied as a means of minimising waste generation. The radioactive materials that can be released from regulatory controls for reuse or recycling are based on the criteria that the radioactive concentrations are below the 'clearance levels' (see Section 17.4). The remaining waste is processed in accordance with the radioactive waste management strategy, as described in Chapter 17.

## 20.3  Safety Objectives

The two basic safety objectives for the disposal of radioactive waste are [2]

- to protect human beings and the environment from the harmful effects of wastes either radiological or non-radiological.
- to dispose of the waste in such a way that the transfer of responsibility to future generations is minimised.

A considerable amount of work has been done to translate these objectives into applicable safety standards and goals. The overall safety standards and international conventions and obligations enforcing these objectives have been described in Chapter 17.

Safety standards for the disposal of radioactive wastes can be conveniently separated into (i) pre-closure or operational phase, and (ii) post-closure phase of the disposal facility. The pre-closure phase spanning the period when wastes are received could be as long as 100 years for some disposal facilities. This phase of the disposal facility is governed by the safety standards applicable to operational nuclear facilities. The levels of safety and operational practices (as embodied in the Waste Acceptance Criteria (WAC)) in the pre-closure period will determine the safety state of the facility at the time of closure. The post-closure period extends from the time when the repository is closed until that time when it is deemed safe. Safety standards during the post-closure period are governed by risk criteria.

## 20.4 Disposal Concepts and Systems

Safe and effective disposal concepts and practices are intricately linked with the waste categories. The primary requirement is to isolate the waste from the accessible human environment over a period sufficiently long enough to allow substantial decay of radionuclides so that they pose no unacceptable risks to human beings and to the environment. To this end, there is a distinction in disposal practices applied to various categories of waste. VLLWs with short or long half-lives and LLWs with short half-lives would not require isolation for a very long period of time to reduce radiological risks to acceptable levels and hence may be disposed of at surface facilities (**surface disposal**). LLWs with long half-lives and ILWs with short half-lives, because of their higher levels of activity, would require somewhat longer period of isolation and more stringent disposal criteria which may be fulfilled by sub-surface disposal (**near-surface disposal**). These surface and near-surface disposal options are, in fact, variations of surface disposal with more stringent isolation provisions for the latter. Long-lived ILWs and HLWs with short as well as long half-lives would require long to very long periods of isolation and that can be achieved by deep underground disposal (**geological disposal**).

These are the various disposal options based on categories of waste. It is vital to have a waste categorisation scheme which facilitates proper disposal. The UK waste categorisation scheme based on activity concentrations (see Section 16.3.2) takes no account of the half-lives of radionuclides and consequently waste streams do not follow the disposal method mentioned above very well. This is due to the fact that in the UK the waste categories were developed some 60 years ago when waste disposal provision was not the prime consideration. Consequently, waste disposal provisions now being mooted in the UK, are based on somewhat conservative standards which may not be the most cost-effective. Other countries which developed nuclear technologies at a later date had the benefits of technological development and experiences of other nuclear advanced countries and addressed these problems effectively. The waste categorisation schemes presented in Section 16.3 show that the French have a more pragmatic approach and have a waste categorisation scheme which is in harmony with their disposal scheme.

## 20.5 Multi-barrier Concept

The isolation of radioactive waste from the human environment is achieved by applying the general principle of defence in depth. The higher the level of activity of waste or longer its half-life, in other words the higher the hazard level, the more rigorous is the application of defence in depth principle. The defence in depth is provided by the multi-barrier concept where a combination of different lines of defence is employed against potential challenges to the safety of the disposal system. For LLW or short-lived ILW, the hazard is low or somewhat time-limited and consequently no rigorous defence in depth is required. Rigorous defence in depth using the multi-barrier concept is applied to long-lived ILW and all types of HLW.

The multiple barriers of the geological repository consist of an Engineered Barrier System (EBS) comprising waste form, waste containers, buffer mass and backfill material; the natural barrier comprising site geology (beyond the EBS) and wider geosphere; and biosphere comprising atmosphere, soil and surface waters; all contribute to contain the hazardous radionuclides. When, some time far into the future, the EBS fails, the activity will be released into the surrounding rocks, and groundwater will carry activity, either in dissolved or in suspended form, through the geosphere into the biosphere. This transfer of activity through the natural barrier will take a very long time. The whole idea is that the engineered barrier and the natural barrier together will offer a barrier that is robust enough and will contain activity for a sufficiently long period of time during which activity will decay sufficiently to have any significant impact on either human beings or on the environment.

It should be noted here that the ICRP in ICRP-81 report [3] mentions that in a geological repository, unlike in a nuclear reactor, the multiple barriers can neither be completely independent nor would there be redundancy. However, the idea is to have a design objective which would maximise the extent to which the safety functions provided by the barriers are maintained over a long period of time (see Section 7.4.1 for a discussion of the defence in depth).

It may, additionally, be noted that the multi-barrier concept incorporates a number of unsubstantiated or untested ideas which may not be anticipated or envisioned from the present-day perspective. In addition, there may be occasions when the disposal facility may be breached either due to human intrusions, natural causes such as a seismic event, global warming, glaciation, flooding etc. All of these factors and their outcome may be put together as uncertainties in the multi-barrier concept. Thus, there cannot be any deterministic estimate that the activity will remain isolated and contained within the confines of a repository for a certain length of time so as to have an insignificant impact on human beings when released. Due to all these uncertainties, some flexibility in the management of waste is required. This flexibility may require retrieval of waste or reversal of the disposal process. These concepts will be discussed further in Section 20.8.

## 20.5.1 Engineered Barrier System

The EBS comprises conditioned waste in solid form, the waste package (consisting of a waste container and/or an over-pack), a buffer mass and backfill [4]. These barriers are made by human beings during the construction, operation and closure of a repository. The main factors involved in the design of engineered barriers are the flow characteristics and chemistry of groundwater in the vicinity of the barriers as well as mechanical, thermal and hydraulic conditions imposed on the barriers. These barriers are intended to confine the radioactive waste for as long as possible, at least for a period of a few hundred years and preferably up to a few thousand years. After that time, the barriers are most likely to be degraded sufficiently to allow penetration of groundwater to the waste material. Their role, therefore, becomes one of limiting release of dissolved radionuclides by sorption and retarding transport of radionuclides. Table 20.1 identifies the types of engineered barriers and their intended functions.

### 20.5.1.1 Waste Forms

The primary role of the waste form, which is an inert matrix in which the radionuclides are distributed, is to contain radionuclides and constrain their release. Irradiated fuel is placed in an array of metallic parts within canisters and encapsulated in cementitious material. HLW is usually vitrified in borosilicate glass or in a ceramic matrix such as Synroc. LLW and ILW are immobilised in a cement matrix in steel drums or concrete boxes. Metallic components such as pipes, tubes, metal plates etc. are size reduced and placed in metallic containers and grouted with cement. Highly activated metallic components containing long-lived radionuclides such as Ni-59, Ni-63 and Nb-94 may be

**Table 20.1** *Engineered barriers and their functions*

| Engineered barrier | Function | Waste type | |
|---|---|---|---|
| | | **HLW** | **ILW/LLW** |
| Waste form | To immobilise radionuclides | Cladding, glass, ceramic, Synroc | Cement, bitumen, resins |
| Waste package | Containment | Canister of steel, cast iron, copper, titanium, ceramic | Steel drum |
| Buffer mass or backfill | To stabilise underground openings, delay soil and ground-water ingress | Bentonite clay, crushed rock, remoulded host clay etc. | Concrete, clay, sand |

melted to ingots and then grouted in a cement mixture. Further details about the treatment and conditioning of waste can be found in Chapter 18.

The key requirements are that the stability of the radionuclides within the matrix should be high and the rate of physical and chemical degradation of the matrix should be low. The groundwater in a deep geological repository is the main mechanism which causes degradation of the matrix.

**Groundwater**: At the upper zones of the geological profile, nearer the top of the water table, fresh water flow is dynamic and significant. But at depths where the deep repositories are likely to be built, 500 m or so below the surface, groundwater flow is slow and stable. Slow groundwater movement over long periods of time results in water in chemical equilibrium with the minerals of the surrounding rocks. Consequently this groundwater is likely to become saline and reducing. The absence of oxygen causes corrosion rates of metallic components to slow down, thus slowing the degradation of the containers.

**Dissolution of Waste Forms**: Waste form whether borosilicate glass or cementitious mixture or other types dissolves in water, although very slowly, and gives rise to a mechanism of radionuclide release. The factors that are involved in the release of radionuclides from any waste form are

- surface area of waste form exposed to water
- location of radionuclides in the waste matrix
- water composition and rate of water ingress into the repository
- solubilities of radionuclides
- potential for radiolysis (particularly from $\alpha$-emitters)

All of these factors at different levels of significance determine the rate of dissolution of waste form. It should be noted that the solubility of many radionuclides in water is extremely low.

### 20.5.1.2 Waste Packages

Waste packages constitute the second level of the EBS, after the initial level of the waste form. For Spent Fuel (SF) and HLW, there may be a primary metallic container which is placed inside an over-pack; whereas LLW and ILW may be placed directly in mild steel or stainless steel or concrete containers. Waste packages, whether over-packs or canisters, are primarily designed to contribute to the containment of radioactivity in the EBS. This period of containment is determined by the resistance of the container material to corrosion and disintegration. The thickness of the metal (mild steel or cast iron) may be sufficiently large to delay container failure for some thousands of years when short-lived as well as intermediate range half-life fission products have decayed to insignificant levels. Metallic containers may also be made of corrosion resistant materials such as copper or titanium alloys that prevent water ingress for much longer (~100,000 years). For mechanical strength, the containers are usually of metallic construction; but ceramic and metal–ceramic containers are now being considered. The ceramic and metal–ceramic containers, although largely corrosion resistant, suffer from brittle fracture. The container material may additionally be chosen to provide radiation shielding during emplacement operations.

## 20.5.1.3 Backfill Materials

Backfill materials are used to fill up the void spaces in the repository. The backfill materials which are used in the immediate surrounding of the waste packages are called buffer materials; whereas those providing backfill of the void spaces in the repository such as shafts, tunnels and access ways are called the mass backfill materials. They are selected to fulfil several functions. The important ones are

- to fill any voids in the repository around the waste packages and in excavated areas, thereby enchancing structural strength and stability and reducing permeability.
- to reduce water ingress within the repository.
- to provide chemical buffering around the waste packages.
- to retard migration of the radionuclides should they be released from the waste packages.
- to provide sufficient heat conduction.

The buffer materials need to be homogeneous and provide uniform physico-chemical and hydraulic conditions to waste containers and they need to be manufactured to high QA standards. Generally, backfills are natural materials such as clay mixed with bentonite (see below) (which usually would have to be imported to the repository site) and crushed host rock (salt, granite) accumulated during excavation. Cement and concrete may also be used as mass backfill in repositories containing ILW.

The purpose of backfill materials, particularly the buffer materials, is to isolate the waste containers from the processes taking place in the geological environment. It cushions waste containers from mechanical damage due to rock movements or tectonic displacements. As bentonite clay expands when it absorbs water, the gap and voids are filled up, thereby limiting further ingress of water. Bentonite clay is chemically very stable in deep underground water, but it tends to decompose at temperatures above 100°C. So the repositories incorporating bentonites should take this temperature constraint into consideration.

**Bentonite ($(Na, Ca)_{0.33}(Al,Mg)_2 Si_4 O_{10}(OH)_2.(H_2O)_n$)**

The name bentonite was given to the absorbent clay by an American geologist after its discovery in about 1890 at Benton Formation in Eastern Wyoming, USA. There are two types of bentonites: (i) swelling bentonite, known as sodium bentonite; and (ii) non-swelling bentonite known as calcium bentonite.

Sodium bentonite expands significantly when it is wet. It can absorb water up to several times its dry weight. This ability to swell makes it useful as a sealant, especially for sealing disposal systems containing nuclear wastes and also quarantining metal pollutants of groundwater.

Calcium bentonite is a non-swelling type of bentonite. It is mainly used and sold in the alternative health market for its cleansing properties. It is usually combined with water and ingested as part of a detox diet. It is claimed that the microscopic structure of this bentonite draws impurities into it from the digestive system, which are then excreted along with bentonite. The native tribes of South America, Africa and Australia have long used bentonite clay for medicinal purposes.

The Swedish company, SKB (Swedish Radioactive Waste Company) in collaboration with the Spanish company, ENRESA (Spanish Radioactive Waste Agency) is in the process of conducting research at Äspö Hard Rock Laboratory to characterise the hydro-mechanical behaviour of the bentonite in salt-water solution. The aim was to see the variation of permeability in bentonite with salt concentration in the fluid. The bentonite material was obtained by mixing 30% of sodium bentonite with 70% of crushed granite rock by weight.

### 20.5.1.4 Transport Mechanism

Once radionuclides have been dissolved in water in and around the degraded containers, they must pass through the buffer and backfill materials before migrating to the surrounding rocks. Buffers are designed to have low permeability to water flow. There are two main mechanisms of solute flow: advective transport and diffusion. The advective transport by water of dissolved radionuclides is considerably smaller than that due to molecular diffusion. The molecular diffusion of dissolved nuclides takes place in the pore waters in the clay and shale. Clay is a good impervious material whereas shale is slightly pervious. Shale is composed of microscopic layered crystals whose main components are silica and alumina. These shale minerals may be combined with other chemical compounds such as carbonates and silicates. These solid particles leave voids between them, known as pores, which may contain interstitial fluids such as water or gases. The ratio of the voids to the total volume of the rock is called the porosity. Shale retards water ingress but cannot completely stop it.

Different radionuclides form different ionic species (cationic or anionic) or uncharged complexes and they move at different rates in the varying electro-chemical environment of the buffer and backfill material. This transport process is generally referred to as dispersion.

## 20.5.2 Natural Barrier

The natural barrier or geosphere is the undisturbed host rock which provides a stable environment for long-term isolation of activity before it eventually escapes to the biosphere. However, the intrinsic properties of the geosphere are somewhat perturbed by the excavation and presence of the repository itself. Geological sites having long-term stability (little or no seismic activity) and with desirable rock characteristics are normally chosen for the disposal of radioactive wastes as they offer good natural barriers. Geological repositories may be constructed in a variety of rock types such as crystalline rocks, clays and salt. The key characteristics of the geological environment that need to be considered in determining the efficacy of the natural barriers in isolating wastes from the biosphere are

- physical isolation and seismic stability
- hydrogeological conditions and processes
- geochemical and geomechanical conditions and processes

To ensure physical isolation for a long period of time, the disposal site needs to be in a geologically stable environment. Stability here does not mean that there will be no changes in the geological conditions. It only means that there will be no violent changes which may change the geological conditions drastically far beyond what may have been anticipated from the knowledge of geological history. Taking into account all possible Features, Events and Processes (FEPs) associated with the geological site, one can identify two broad categories of situations: (i) tectonic and magmatic activity, and (ii) weathering and natural erosion. The first category can be predicted from detailed geological studies. But the second category is very difficult to predict and estimate. The main mechanisms of natural erosion are glaciation on the one hand and global warming on the other. The effects of either phenomenon are extremely difficult to gauge.

To understand the behaviour of various FEPs in situ in geological formations, a collaborative research programme by a number of European countries, the EC and Japan was started at Grimsel, some 20 years ago [5]. Grimsel is at an altitude of 1730 m in granite rock in the Alps in central Switzerland. The Grimsel Test Site (GTS) is in a crystal rock 450 m below the surface and is used to carry out R&D on issues such as long-term diffusion, colloid formation and migration, gas migration in the shear zone, gas migration in EBS and geosphere, super-alkaline plume in fractured rock etc. It is not a waste disposal site. Figure 20.1 shows the entrance to the site.

Based on the characteristics of the host rock where waste is to be deposited, the components of the engineered barrier such as waste form, waste package and buffer mass etc. are chosen. In general, the greater the depth of the waste repository, the longer it takes for wastes to be released into the human environment. The HLW and SF are normally disposed of at sufficient depth (200–1000 m below the surface). Such repositories are normally referred to as deep geological repositories, although the word 'deep' is often omitted. The shallow repositories are those which are at some depth from

**Fig. 20.1** *Grimsel test site, Switzerland*

the surface; not as deep as deep repositories but not as shallow as surface or sub-surface repositories. The actual depth of the repository required for the projected period of isolation of activity depends on a number of factors such as the rock type giving strength and stability, physical and chemical properties of the rock, local geological and hydrogeological conditions.

### 20.5.2.1 Radionuclide Migration

Radionuclides released from the EBS as solutes in groundwater will enter the pores and fractures and be transported by advection or diffusion. In crystalline rocks such as granite, radionuclide migration is dependent on fractures in crystal boundaries. In sedimentary rocks (rock salt, soft clays etc.), radionuclide migration is through rock matrices. Rock matrices in sedimentary rock remain continuous over large distances. They are made up of small individual crystals which are often separated by microscopically small pores or cracks that form a network of pores. The porosity of the matrix can vary strongly between different rock types. In sedimentary rocks, porosity varies between a few per cent to tens of per cent of volumes. In crystalline rocks, porosity ranges from 0.1–0.5%. However, the determining factor for the transport process is the connectivity of pores rather than their porosity.

Radionuclide transport may also take place through fractures and fracture zones [4]. Consolidated hard rocks are subjected to stress fields which may induce fractures of varying magnitudes. There are enormous fractures between tectonic plates. Even in monolithic rocks, there are smaller but widespread fractures. The distances between individual fractures can range between tens of centimetres and tens of metres, depending on the rock type and location. The fractures and fracture zones, if not sealed by the deposition of secondary minerals, form conduits for water flow and solute transport. These conduits form a complex three-dimensional network of water channels through the rocks.

### 20.5.3 Biosphere

The biosphere includes the atmosphere; soils; surface waters such as ponds, lakes, rivers, seas and oceans and their sediments in which living organisms normally reside [6]. The interface between the geosphere and the biosphere is somewhat blurred, but it is normally taken to be the bottom of near surface materials such as soil, sediment and rocks that are affected by human activities such as farming.

The biosphere itself is not considered to be part of the system of natural barriers, although it is viewed as the last barrier in the multi-barrier concept. Several processes in the biosphere contribute to the retardation or dilution of radionuclides. In the safety assessment when estimating the effects of the biosphere, two types of biosphere are usually considered: (i) a temperate biosphere, which includes a flat ecosystem with lakes, streams and marshy lands where human populations live in small areas and for which there are specific exposure pathways. (ii) An arid biosphere which includes areas where the human population density is relatively high such as urban areas where the exposure pathways may differ from those in the temperate biosphere.

As transport by groundwater is the principal mechanism for the migration of radionuclides to the biosphere, the repository should be located in a geological environment where the fluxes and velocities of groundwater are sufficiently small to provide further isolation. This can be achieved by locating the repository where the rocks are of low permeability and have a very low water content.

## 20.6 Waste Emplacement Options

The options for the emplacement of encapsulated waste underground can be categorised as (i) tunnel, (ii) cavern, and (iii) deep borehole. There are considerable flexibilities in the details of these options or concepts with regard to their layout, dimensions etc. allowing them to be adapted to various categories of waste and the designated geological environment [7].

### 20.6.1 Tunnel Concept

The tunnel option is currently favoured in most countries which are planning waste disposal underground. The canisters are placed either
- horizontally, parallel to the tunnel axes, as in Belgian, Swiss and USA Yucca Mountain concepts, or
  - in boreholes drilled vertically into tunnel floors or horizontally in tunnel walls or in vaults, as in the Swedish, Finnish and French concepts.

The void spaces around the canisters are usually filled with a buffer material such as bentonite clay on its own or mixed with silica sand. The buffer material provides a low permeability hydraulic barrier and a stable geochemical environment for the waste. The choice of buffer material is dependent on the waste form and the geological environment. The Belgian disposal concept for HLW and SF in a clay formation employs cementitious backfill. Access tunnels and other openings are backfilled with a low permeability material and seals are installed. In the special USA case of disposal of relatively fresh SF in unsaturated tuff, the tunnels are left open for long periods of time in order to limit the temperatures in the repository.

## 20.6.2 Cavern Concept

Massive shielded containers containing multiple waste packages are placed in caverns. For example, in Japan it is proposed that 20 HLW packages would be placed in massive 110 te shielded transportable containers. This concept is explicitly designed for ease of retrieval and inspection during the storage period which could last for several hundred years. Ventilation, drainage and inspection facilities are needed during the storage period and after that caverns would be backfilled. An advantage of the cavern concept is that if forced ventilation is used during the storage period to limit temperatures in the repository, a high waste emplacement density is possible. Once the heat output from the waste is reduced sufficiently due to radioactive decay, the ventilation system can be turned off, and the repository can be closed off by backfilling and sealing. The cavern concept is shown in Figure 20.2.

## 20.6.3 Deep Borehole Concept

In this concept, the waste is placed in canisters surrounded by a protective clay buffer inside steel vertical shells which may be a kilometre or so deep. This concept aims to utilise the deep, stable geological environment where ground-

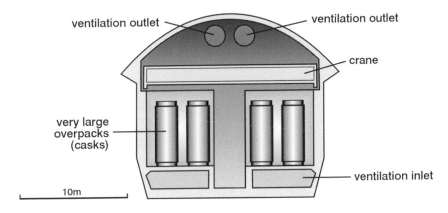

**Fig. 20.2** *A cavern concept*

water is dense and reducing. The various types of vertical boreholes are shown in Figure 20.3. Type (a) shows the vertical boreholes; type (b) shows a horizontal gallery above the disposal zones; and type (c) has two horizontal galleries, one above and one below the zones connected by a vertical shaft.

## 20.7 Disposal Practices

### 20.7.1 Disposal of Very Low Level Waste

VLLW (short or long half-lives) and LLW with short half-lives are relatively easy to dispose of. This type of waste may arise from operations and decommissioning of nuclear facilities (scrap metals, debris from demolition etc.), from the use of radionuclides in medicine, industry and research laboratories [8]. Generally, this is the type of waste whose concentrations are slightly above the free release criteria, but very low in comparison to standard packages from nuclear facilities. It may also include Naturally Occurring Radioactive Materials (NORMs). It should be noted that most NORMs are long-lived materials, for example, Ra-226 ($t_{1/2}$ =1600 y), but their activity concentrations are in the very low level range. There are, however, some restrictions on this waste type: long-lived wastes and wastes containing hazardous and toxic chemicals are generally excluded.

Very low level and low level wastes may originate from diverse sources such as nuclear power plants and various other nuclear facilities, industry, hospitals, universities etc. Irrespective of its origin, the waste must meet certain specific requirements before it is accepted by a repository for disposal. These specific requirements are called the Waste Acceptance Criteria (WAC). The criteria cover various issues such as the type of waste, physical and chemical state of the waste, the type of waste container, the total amount of radioactivity in a container, the way a container is packaged and labelled, the level of contamination at the outside surface of the container etc. It is a regulatory requirement that the management of the waste disposal facility ensures that the WAC are

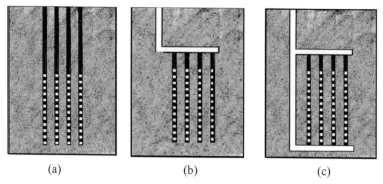

|   (a)   |   (b)   |   (c)   |

**Fig. 20.3** *Various types of boreholes*

fully met before acceptance. Also it is obligatory for the management to record and maintain an accurate inventory of all the wastes accepted for disposal at the facility. The whole purpose of the disposal facility is to provide adequate provisions for isolation for a period long enough so that the activity decays to or below the clearance level. After that the site may be cleared for unrestricted use. Some examples of VLLW disposal practices in some of the European countries are given below.

### 20.7.1.1 French Disposal Practice

The French nuclear industry, with an anticipated large decommissioning programme over the next 30 years or so, expects large volume of VLLW to arise and hence it was considered necessary to have a new disposal facility [9]. The design principle for such a facility was that it should comply not only with radiological regulations but also with non-radiological hazardous regulations. The containment of waste is ensured by a low permeability clay layer, where the repository is situated. In 2000, a site was chosen near the village of Morvilliers, about 2 km from the Centre de l'Aube facility that accommodates low and intermediate level short-lived radionuclides. After two public inquiries in 2001 and 2002, the repository started operation in October 2003. The total capacity is 650,000 m³, equivalent to about 1,000,000 te of waste. It is made up of a number of cells excavated in a layer of clay, about 15–25 m thick. The cells are 80 m long, 25 m wide and 6 m deep. The sides and the floor of the cells are covered with watertight membranes before accepting wastes, which come in large bags or in drums. Bulk items are deposited without any form of preliminary packaging. Waste treatment units comprising compactors and solidifiers are used. A mobile roof is provided to protect operations during loading and from rainwater ingress. After loading, the cells are backfilled and sealed with the same type of membrane. The repository is then covered with clay. After operations, a post-closure monitoring for a period of 30 years is scheduled. As the repository is also designed for non-radiological hazardous materials, toxic materials such as As, Zn, Pb, Cd etc. with low concentrations are also accepted. The annual inflow of waste is in the range 20,000–30,000 m³.

### 20.7.1.2 Spanish Disposal Practice

The Spanish radioactive Waste Management Agency (ENRESA), in anticipation of large volume of decommissioning waste, applied in 2003 for the construction of a VLLW disposal facility at its El Cabril site where LLW and ILW are presently disposed [10]. The advantage of the proposed site, within the existing site, is that there would be a synergy with the infrastructure and organisation of the present site. Also the monitoring and surveillance programmes may be coordinated.

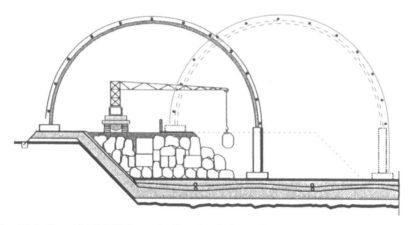

**Fig. 20.4** *Spanish VLLW disposal cell with a light roof*

The proposed disposal facility will consist of four cells, although only one will be built initially. All the basic tenets of repository design will be utilised. An artificial geological barrier made up of compacted clay with a minimum thickness of 1 m and a bentonite film providing together an equivalent thickness of 5 m thick clay with permeability, $K = 10^{-9}$ m.s$^{-1}$ will be designed. Each cell will be 4–6 m high. During operation, there would be a light cover to minimise rainwater ingress. There will be a leachate collection system. A scheme of operation of the disposal cell and the design of the light roof are shown in Figure 20.4. At the end of operation, an engineered multi-layered cap of clay and drainage layers will be constructed. A treatment building will be located nearby where waste will be received, identified, segregated and compacted.

### 20.7.2 Disposal of Low and Intermediate Level Waste

Activities for ensuring safe management of short-lived LLW and ILW have been going for many years now. These wastes are disposed of at intermediate depth in geological repositories. More than 80 near-surface disposal facilities have been built worldwide and more are under development.

### 20.7.3 Disposal of High Level Waste

Deep geological disposal of SF and HLW as well as waste with a significant amount of long-lived radionuclides is the internationally preferred option. The fundamental requirement for such a disposal facility is the long-term isolation of waste from the human environment so that the risks become insignificant when waste breaches all of the barriers.

Various countries are at different stages in the development of geological sites. They include Belgium, France, Finland, Germany, Sweden, Switzerland

and the USA. Finland has already started the construction work at Olkiluoto in Finland for SF disposal and the repository is scheduled to be operational within the next 10–15 years. Some countries such as Switzerland and Germany are considering geological disposal for long-lived LLW and ILW. The USA is operating a geological repository, called the Waste Isolation Pilot Plant (WIPP) in New Mexico for the disposal of long-lived ILW (transuranic waste) arising from military activities. It is located in a salt formation 650 m below the surface. Germany has recently licensed the Konrad mine facility for long-lived ILW. Sweden and Finland have operated shallow facilities for short-lived ILW and LLW for some time. However, these facilities are not strictly geological repositories, they are shallow vaults.

## 20.7.4  Disposal Practice in the UK

In the UK, wastes of various categories which have arisen from nuclear activities over the last 60 years or so need to be disposed of properly. The current disposal situation is shown in Table 20.2 [11]. Except for a near-surface disposal facility at Drigg for VLLW and LLW, there is no disposal facility in the UK. The present situation is to store wastes on sites. However, Nirex had proposed a phased disposal concept [7] for deep geological disposal, which is described below.

### 20.7.4.1  Phased Disposal Concept

The phased disposal concept is applicable only to a deep geological repository where discrete, and to some extent reversible, steps may be taken in the disposal process. The aim is to have a number of disposal options during the disposal process such that the rate of waste disposal can be varied, terminated or even retrieved, if necessary. The time period for this phased disposal could be several hundred years during which the waste is monitored and is readily retrievable. The main phases may be

- packaging of the waste
- surface storage period
- waste emplacement
- monitored retrievable storage
- backfilling period
- repository closure

## 20.7.5  Co-location

A concept [7] which is being actively considered and developed internationally is the so-called co-location or co-disposal where various types of HLWs are being disposed of in various compartments of the same disposal facility. The deep underground facility would have the same access but the emplacement of different wastes such as SF, HLW and long-lived ILW would be different. Thus the long-term management of the facility would be easier, would provide better

**Table 20.2** *Waste disposal practice in the UK*

| Location | Present position in the UK | Comment |
|---|---|---|
| Surface | Stored at licensed nuclear sites | Various types of wastes are stored at various licensed sites. LLWs were disposed of at Dounreay at waste pits at the surface rock |
| Near surface | Licensed repositry at Drigg, Cumbria | Solid VLLWs and LLWs are disposed of in a shallow land burial facility |
| Deep geological | No disposal facility within the UK for ILW and HLW | NDA is actively searching for an Integrated Waste Strategy (IWS) in cooperation with the CoRWM |

safety and security as well as being cost-effective. An additional advantage of this provision is that there would be a single site selection, site characterisation and development process, thereby negating the requirements for separate sites for different types of HLW/ILW.

# 20.8 Performance Assessment of Disposal System

As the behaviour and performance of geological repositories cannot be definitively assessed by deterministic methods, probabilistic methods are used to carry out 'safety assessment' and 'performance assessment' [4]. Safety assessment aims to derive the consequences of radionuclides release as a function of time and space and compares with the regulatory safety standards. Performance assessment, on the other hand, deals with the behaviour of the disposal facility in relation to stated performance criteria. These two assessment methodologies are not segregated, in fact, they are very much inter-related.

The selection of a geological repository requires a full-scale performance assessment. Performance assessment is an iterative process which is conducted to help develop a site characterisation programme to support site selection, repository design and licensing. The purpose of this performance assessment and safety assessment is to demonstrate compliance with the regulatory requirements that risks from such a facility during operations and following closure far into the future are within acceptable limits. The time span for these assessments is required to be long (~10,000 years) to very long (~1,000,000 years), as some of the radionuclides of concern such as Ni-59 ($t_{1/2}$ = 75,000 years), Nb-94 ($t_{1/2}$ = 20,000 years) and Tc-99 ($t_{1/2}$ = 213,000 years) have long half-lives.

The performance assessments of a repository must address both pre-closure and post-closure phases. The pre-closure assessment is concerned with the radiological and non-radiological safety of the repository before closure. As at this stage, the repository is deemed to be a nuclear site, all the safety require-ments associated with a nuclear site must be complied with. Following closure, post-closure assessment using predictive models that describe the behaviour of the repository over thousands or even hundreds of thousands of years must be carried out.

## 20.8.1 Elements of Performance Assessment

The general framework for performance assessment can be considered to comprise a number of inter-related and iterative steps, as shown in Figure 20.5. The main elements are given below.

**Scenario Development**: In this step, all Features, Events and Processes (FEPs) that can initiate release of radionuclides from the wastes are identi-fied. Geological stability cannot be assured over the timescale required for performance assessment. Consequently, some amount of variation from the geological stability needs to be incorporated. The probabilities of occurrence of these FEPs are then estimated.

**Model Development**: The development of databases and computer codes predicting the behaviour of the disposal facility over a long period of time is carried out here. This model may conveniently be separated into three sub-models covering three sub-systems

- the repository and its contents, i.e. waste inventory, waste form, waste container, buffer and backfill mass
- the geosphere which is composed of the host formation and surrounding rocks including fluids and gases capable of transporting radionuclides
- the biosphere which consists of soil, aquifers, surface water, atmosphere and human beings

The physical processes described in Sections 20.5.1 and 20.5.2 all come into play here. Additionally, many of these processes can influence each other and hence they need to be considered in a coupled way. All of these aspects impose enormous complications to the predictive model.

Results from the modelling of the biosphere can form the basis of the Environmental Impact Assessment (EIA) and the preparation of an Environmental Statement (ES). It should be noted that both radiological and non-radiological issues need to be addressed in the ES.

**Consequence Assessment**: Consequences arising from various FEPs are assessed using the predictive models developed above. The consequences could be assessed in terms of activity released into the biosphere as a function of time. With an assumed population distribution, the societal risks, economic consequences etc. from the disposal site can be assessed.

**Sensitivity and Uncertainty Analysis**: Using the predictive computer codes, the parameters and assumptions which contribute significantly to

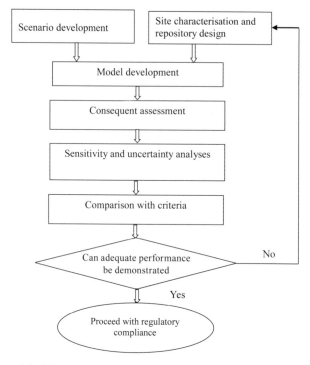

**Fig. 20.5** *Simplified flow diagram illustrating performance assessment methodology*

the variations in the consequences are estimated here. This is the sensitivity analysis. Estimations of the uncertainties in the input parameters and assumptions leading to the uncertainties in the endpoints are carried out here. This is the uncertainty analysis. Uncertainties that need to be dealt with as part of the safety case generally fall into five categories [4]

    (i)   System uncertainties: These arise from incomplete knowledge about systems, their interrelationships with each other (system coupling). They are more relevant to natural barrier processes than to the EBS.

   (ii)  Scenario uncertainties: These refer to situations when the geological environment may change either due to natural causes (tectonic behaviour, glaciation, global warming etc.) or human intervention.

 (iii)  Conceptual model uncertainties: These refer to the conceptual evolution of modelling of a system in a particular way as against other alternatives.

 (iv)  Mathematical model uncertainties: These refer to the mathematical modelling of components of systems. Various mathematical models may be available and they may be applicable in various situations.

(v)   Parametric uncertainties: These refer to values or ranges of values of parameters to be used in modelling the system.

It should be noted that these uncertainties do not contribute equally to the uncertainties of the end-point or end-points. In other words, some uncertain parameters may influence one end-point more strongly than others. Consequently uncertainty analysis should also cover the evaluation of correlations between input and output parameters. By estimating input/output correlations, one can identify those input parameters which correlate strongly with output parameters and thus identify those parameters where greater emphasis is needed and improvements need to be made to the quality of the output parameter.

## 20.9  Disposal Options for Nuclear Submarines

The disposal of a nuclear submarine is very much inter-related to the disposal options for radioactive waste. Following decommissioning of a nuclear submarine in the UK, the irradiated fuel is sent to Sellafield for long-term storage. As no submarine has gone beyond stage 3 (afloat storage) in the UK, the disposal of the Reactor Compartment (RC) has not yet been faced. The first nuclear submarine in the UK called *Dreadnought* was decommissioned in 1982 and has gone through two 10-year dry dockings. The future strategy for the decommissioning of submarines in the UK is currently under discussion.

Other countries such as USA and France have separately developed their own national decommissioning strategies. Instead of following five decommissioning stages, they adopted the strategy that after the DDLP (see Section 5.9), the RC is cut-out intact, lifted out of the submarine hull and then stored at a surface facility in preparation for disposal. The front and aft sections of the hull are then welded together and left afloat. Following the removal of the RC, there are two main options for its disposal: land burial at a shallow trench and land burial at a deep repository. The option of sea disposal is out of question following the 'Joint Convention' (see Section 17.3.1) to which both the USA and France are signatories.

Following the Radioactive Waste Management Advisory Committee (RWMAC) report in 1997 (RWMAC has now been replaced by CoRWM) which criticised the MoD for not having a long-term strategy for the disposal of decommissioned nuclear submarines and the MoD realising the likely shortage of space for 'afloat storage' of submarines beyond 2012, a study was commissioned by the Warship Support Agency (WSA). The study resulted in a report called 'The ISOLUS Investigation' and its main conclusions are as follows:

- Land storage of the separated RC of the submarine is the favoured option. There will be cost savings from its early implementation.

- Afloat storage (stage 3 and beyond) should be considered only as a 'stop-gap' measure. The shortage of storage space at the dockyard as additional submarines are decommissioned precludes this option as a viable long-term option.
- A BRDL proposal for a pilot project to store primary reactor plant components on land as unpackaged ILW is worth further investigation.

### 20.9.1 Shallow Land Burial

In this option, the intact RC is laid in a trench at a remote site. The advantage of this option is two-fold: (i) as the RC is kept intact and not cut-up, the exposure to the workforce is minimal; and (ii) the RC can be monitored and, if necessary, maintained over a long period of time.

This option has been adopted by the USA and France. A site of about 1500 km² at Hanford, Washington state is used by the USA for disposal of RCs, whereas the French submarine RCs are disposed of at Cherbourg. Figure 20.6 shows the site at Hanford where over 75 RCs are buried.

### 20.9.2 Deep Burial

This option entails burial of decommissioned nuclear submarine wastes in a deep geological repository. However, the design of a conventional waste repository for civil waste is unlikely to be large enough to allow passage of the whole intact RC. Consequently the RC would need to be cut up, packaged in approved packages and laid in the repository. The workforce exposure from this option would be higher than that of the shallow land burial. It should be noted that no country has adopted this option for the disposal of reactor compartments.

**Fig. 20.6** *The RC disposal site at Hanford, WA, USA*

# Revision Questions

1. Compare and contrast the concepts of storage and disposal of radioactive waste. Why are retrievability and reversibility important considerations in the context of waste disposal?

2. What is the pre-disposal management of waste? Describe the steps that may be taken in the pre-disposal management.

3. What are the 'pre-closure' and 'post-closure' of a disposal facility? Discuss the implications and regulatory requirements applicable to these two phases of a disposal facility.

4. Disposal of waste can be broadly separated into surface disposal and geological disposal. Identify waste types with their half-lives that are suitable for these disposal options.

5. What is the multi-barrier concept in the context of geological disposal? Briefly discuss these barriers and their intended functions.

6. What is the purpose of the Engineered Barrier System (EBS)? What are the components of this barrier system and how are they intended to provide long-term isolation of wastes?

7. What are the major transport mechanisms of radioactive material within the natural barrier? Briefly describe these mechanisms.

8. What are the possible exposure pathways to human beings once radioactivity gets into the biosphere? Briefly describe these pathways.

9. What are the various emplacement options of encapsulated waste in deep geology? Describe these options with examples of countries where they are practised.

10. What is meant by the 'phased disposal' concept? How does the phased disposal concept match up with the retrievability option?

11. Why is the performance assessment of a disposal system important? Show the methodology with a flow diagram.

12. What are the major options for the disposal of nuclear submarines in the UK? Describe these options vis-à-vis those available in other nuclear weapon states.

## REFERENCES

[1] International Atomic Energy Agency, Predisposal Management of Radioactive Waste, Including Decommissioning, Safety Requirements, No. WS-R-2, IAEA, Vienna, Austria, 2000

[2] International Atomic Energy Agency, Classification of Radioactive Waste, A Safety Guide, IAEA Safety Series No. 111-G-1.1, Vienna, Austria, 1994.

[3] International Commission on Radiological Protection, Radiation Protection Recommendations as applied to the Disposal of Long-lived Solid Radioactive Waste, ICRP 81, *Annals of the ICRP* Vol 28, No 4.

[4] International Atomic Energy Agency, Scientific and Technical Basis for the Geological Disposal of Radioactive Wastes, Technical Reports Series No. 413, Vienna, Austria, 2003.

[5] Grimsel Test Site, http://www.grimsel.com.

[6] International Atomic Energy Agency, Scientific and Technical Basis for the Near Surface Disposal of Low and Intermediate Level Waste, Technical Reports Series No. 412, Vienna, Austria, 2003.

[7] UK Nirex Limited, Repository Concept and Programme Development for UK High Level Waste and Spent Nuclear Fuel, Technical Note No. 502659, September 2005.

[8] International Atomic Energy Agency, Disposal of Low Activity Radioactive Waste, Proceedings of an International Symposium held in Cordova, Spain, 13–17 December 2004, http://www-pub.iaea.org/MTCD/publications/PDF/pub1224_Web. pdf.

[9] Dutzer M., Chastagner F., Riquart N. and Duret F., Disposal of Very Low Level Waste and Safety Assessment, Proceedings of International Symposium on Disposal of Low Activity Radioactive Waste, Cordova, Spain, 13–17 December 2004.

[10] Zuloaga P., Management of Very Low Activity Radioactive Waste in Spain, Proceedings of an International Symposium on Disposal of Low Activity Radioactive Waste, Cordova, Spain, 13–17 December 2004.

[11] UK Nirex Limited, Comparison of the Nirex Waste Package Specification with Waste Acceptance Criteria for Storage and Disposal Facilities in Other Countries, December 2002.

# Appendix 1

# Abbreviations
# and Acronyms

| | |
|---|---|
| AA | Approving Authority |
| AC | Authorisation Condition |
| ACoP | Approved Code of Practice |
| ACSNI | Advisory Committee on the Safety of Nuclear Installations |
| ADS | Accelerator Driven System |
| AGR | Advanced Gas cooled Reactor |
| AHP | Analytical Hierarchy Process |
| ALARA | As Low As Reasonably Achievable |
| ALARP | As Low As Reasonably Practicable |
| ANSTO | Australian Nuclear Science and Technology Organisation |
| AVM | Atelier de vitrification de Marcoule |
| AWE | Atomic Weapons Establishment |
| AWIJ | Abrasive Water Injection Jet |
| AWSJ | Abrasive Water Suspension Jet |
| | |
| BATNEEC | Best Available Techniques Not Entailing Excessive Cost |
| BC | Basic Clearance |
| BE | British Energy |
| BNFL | British Nuclear Fuels Limited |
| BNG | British Nuclear Group |
| BPEO | Best Practicable Environmental Option |
| BPM | Best Practicable Means |
| BPSS | Baseline Personnel Security Standard |
| Bq | Becquerel |
| BSL (LL) | Basic Safety Level (Legal Limit) |
| BSO | Basic Safety Objective |
| BSS | Basic Safety Standard |
| BWR | Boiling Water Reactor |
| | |
| C&M | Care and Maintenance |
| C&MP | Care and Maintenance Preparation |
| CAD | Computer-Aided Design |
| CAMC | Contact Arc Metal Cutting |
| CBA | Cost-Benefit Analysis |
| CBS | Cost Breakdown Structure |
| CEA | Commissariat á l'Energie Atomique (Atomic Energy Commission of France) |
| CEA | Cost-Effective Analysis |
| CERN | European Organisation for Nuclear Research (French acronym) |
| CHIP | Chemical Hazard Information and Packaging |
| Ci | Curie |
| CNNRP | Chairman, Naval Nuclear Regulatory Panel (now defunct) |
| COMAH | Control Of Major Accident Hazards |
| CoRWM | Committee on Radioactive Waste Management |
| COSHH | Control Of Substances Hazardous to Health |

| | |
|---|---|
| cpm | counts per minute |
| CPM | Critical Path Method |
| CR | Cost Reimbursable |
| CS | Carbon Steel |
| CSI | Criticality Safety Index |
| CTC | Counter Terrorist Check |
| CVF | Constant Volume Feeder |
| | |
| D&D | Decommissioning & Dismantling |
| DBA | Design Basis Accident |
| DBERR | Department for Business, Enterprise and Regulatory Reform |
| DBIF | Design Basis Initiating Fault |
| DDLP | Defuel, De-equip and Lay-up Preparation |
| DDREF | Dose and Dose Rate Effectiveness Factor |
| DEFRA | Department for Environment, Food and Rural Affairs |
| DF | Decontamination Factor |
| DFR | (Dounreay) Demonstration Fast Reactor |
| DfT | Department for Transport |
| DGD | Dangerous Goods Division |
| DNA | Deoxyribo Nucleic Acid |
| DNFR | Defence Nuclear Facilities Regulation |
| DNSR | Defence Nuclear Safety Regulator |
| DP | Decommissioning Plan |
| DQA | Data Quality Assessment |
| DQO | Data Quality Objective |
| DSC | Decommissioning Safety Case |
| DTLR | Department of Transport, Local Government and Regions |
| DTPA | Diethylene Triamine Penta-acetic Acid |
| DV | Developed Vetting |
| DVA | Defence Vetting Agency |
| DWP | Department for Work and Pensions |
| | |
| EA | Environment Agency |
| EBC | Enhanced Basic Check |
| EBS | Engineered Barrier System |
| EC | European Commission |
| EDM | Electro Discharge Machine |
| EDTA | Ethylene Diamine Tetra-acetic Acid |
| EFDA | European Fusion Development Agreement |
| EIA | Environmental Impact Assessment |
| EIAD | Environmental Impact Assessment for Decommissioning |
| EIADR | Environmental Impact Assessment Decommissioning Regulations |
| EMP | Environmental Management Plan |
| EPA | Environmental Protection Agency |
| EPS | Encapsulation Product Store |
| ES | Environmental Statement |
| EU | European Union |
| Euratom | European Atomic Energy Treaty |
| EW | Exempt Waste |
| EWN | Energie Worke Nord |
| | |
| FA | Faible Activité |
| FAO | Food and Agriculture Organisation |
| FEP | Features, Events and Processes |
| FSC | Final Site Clearance |

| | |
|---|---|
| GCR | Gas-Cooled Reactor |
| GDA | Generic Design Assessment |
| GHS | Globally Harmonised System (of Classification and Labelling of Chemicals) |
| G-M | Geiger–Müller (counter) |
| GOCO | Government Owned, Contractor Operated |
| GTS | Grimsel Test Site |
| Gy | Gray |
| | |
| H&S | Health & Safety |
| HAL | Highly Active Liquor |
| HASS | High Activity Sealed radioactive Sources |
| HAZAN | HAZard ANalysis |
| HAZOP | HAZard and OPerability |
| HEDTA | Hydroxy Ethylene Diamine Tri-acetic Acid |
| HLW | High Level Waste |
| HPA | Health Protection Agency |
| HPGe | High Purity Germanium (detector) |
| HSC | Health and Safety Commission |
| HSE | Health and Safety Executive |
| HSWA | Health and Safety at Work etc. Act |
| HVL | Half Value Layer |
| HVT | Half Value Thickness |
| | |
| IAEA | International Atomic Energy Agency |
| IATA | International Air Transport Association |
| ICAO | International Civil Aviation Organisation |
| ICDT | In-Cell Drum Transporter |
| ICPMS | Inductively Coupled Plasma Mass Spectrometry. |
| ICPP | Idaho Chemical Processing Plant |
| ICRP | International Commission on Radiological Protection |
| ICRU | International Commission on Radiation Units and Measurements |
| ILO | International Labour Organisation |
| ILW | Intermediate Level Waste |
| IMO | International Maritime Organisation |
| INEEL | Idaho National Environmental Engineering Laboratory |
| IP | Industrial Package |
| IPC | Integrated Pollution Control |
| IPRI | Industrial Pollution and Radiochemical Inspectorate |
| IRR | Ionising Radiations Regulations |
| ISN | Interim Storage North |
| ISOLUS | Interim Storage Of Laid Up Submarine |
| IWS | Integrated Waste Strategy |
| | |
| JAERI | Japan Atomic Energy Research Institute |
| JET | Joint European Torus |
| JPDR | Japan Power Demonstration Reactor |
| | |
| LANL | Los Alamos National Laboratory |
| LC | Licence Condition |
| LCBL | Life-Cycle Base Line |
| LET | Linear Energy Transfer |
| LILW | Low and Intermediate Level Waste |
| LL | Long Lived |
| LLNL | Lawrence Livermore National Laboratory |

| | |
|---|---|
| LLW | Low Level Waste |
| LLWR | Low Level Waste Repository |
| LNT | Linear No Threshold |
| LOLER | Lifting Operations and Lifting Equipment Regulations |
| LOMI | Low Oxidation state Metal Ion |
| LRGS | Low Resolution Gamma Spectrometry |
| LSA | Low Specific Activity |
| LZH | Laser Zentrum (Centre) Hannover (in Germany) |
| | |
| M&O | Management and Operation |
| MAC | Miscellaneous Activated Components |
| MCA | Multi-Criteria Analysis |
| MCDA | Multi-Criteria Decision Analysis |
| MCI | Miscellaneous Contaminated Items |
| MDA | Minimum Detectable Activity |
| MEP | Magnox Encapsulation Plant |
| MeV | Million electron Volts |
| MFP | Mean Free Path |
| MN | Mega Newton |
| MoD | Ministry of Defence |
| MODIX | Multi-stage OxiDative by Ion Exchange |
| MOX | Mixed OXide (fuel) |
| MUA | Multi-attribute Utility Analysis |
| | |
| NaK | Sodium Potassium |
| ND | Nuclear Directorate (within the HSE) |
| NDA | Nuclear Decommissioning Authority |
| NDPB | Non-Departmental Public Body |
| NEA | Nuclear Energy Agency |
| NGO | Non-Governmental Organisation |
| NHL | Nominal Hygienic air requirement Limit |
| NIA | Nuclear Installations Act |
| NII | Nuclear Installations Inspectorate (now part of the ND) |
| NIR | Nuclear Installations Regulation |
| NISR | Nuclear Industries Security Regulations |
| NNPP | Naval Nuclear Propulsion Programme |
| NORM | Naturally Occurring Radioactive Material |
| NPP | Nuclear Power Plant |
| NPT | Non-Proliferation Treaty |
| NPV | Net Present Value |
| NRC | Nuclear Regulatory Commission (of the USA) |
| NRP | Nuclear Reactor Plant |
| NRPB | National Radiological Protection Board (now reorganised as the RPD) |
| NRTE | Naval Reactor Test Establishment |
| NTWP | Near Term Work Plan |
| NuSAC | Nuclear Safety Advisory Committee |
| NWP | Nuclear Weapons Programme |
| NWR | Nuclear Weapons Regulator |
| | |
| OBS | Organisational Breakdown Structure |
| OC | Operational Circular |
| OCNS | Office for Civil Nuclear Security |
| OECD | Organisation for Economic Co-operation and Development |
| OEEC | Organisation for European Economic Cooperation |
| OK | Odourless Kerosene |

| | |
|---|---|
| OM | Organisational Manual |
| OR | Operating Rule |
| OSC | Operational Safety Case |
| OSHA | Occupational Safety and Health Administration |
| OSPAR | OSlo and PARis (Convention) |
| OWR | Outside Workers Regulations |
| | |
| PAHO | Pan-American Health Organisation |
| P&T | Partitioning and Transmutation |
| PAS | Personal Air Sampler |
| PBO | Parent Body Organisation |
| PCM | Plutonium Contaminated Material |
| PCmSR | Pre-Commissioning Safety Report |
| PCSR | Pre-Construction/Commencement Safety Report |
| PDR | Post-Decommissioning Report |
| PERT | Programme Evaluation and Review Technique |
| PFR | Prototype Fast Reactor |
| PIRER | Public Information for Radiation Emergency Regulation (replaced by REPPIR) |
| PMIS | Project Management Information Service |
| PMP | Project Management Plan |
| PMP | Plant Modification Proposal |
| PMS | Planned Maintenance Schedule |
| PMT | Photo-Multiplier Tube |
| POCO | Post-Operational Clean Out |
| POSR | Pre-Operational Safety Report |
| PPE | Personal Protective Equipment |
| PQM | Project Quality Management |
| PRM | Project Risk Management |
| PSA | Probabilistic Safety Assessment |
| PSR | Periodic Safety Review |
| PSR | Preliminary Safety Report |
| PVC | Poly Vinyl Chloride |
| PWR | Pressurised Water Reactor |
| | |
| QA | Quality Assurance |
| QC | Quality Control |
| QFD | Quartz Fibre Dosimeter |
| QMS | Quality Management System |
| | |
| R&D | Research and Development |
| R2P2 | Reducing Risks, Protecting People |
| RBE | Relative Biological Effectivness |
| RC | Reactor Compartment |
| RCEP | Royal Commission on Environmental Pollution |
| REDOX | REDucing OXidising |
| REPPIR | Radiation Emergency Preparedness and Public Information Regulations |
| RHS | Right-Hand Side |
| RMTD | Radioactive Materials Transport Division |
| RMW | Radioactive Mixed Waste |
| RPD | Radiation Protection Division |
| RPE | Respiratory Protective Equipment |
| RPP | Radiation Protection Programme |
| RPV | Reactor Pressure Vessel |

| | |
|---|---|
| RSA | Radioactive Substances Act |
| RSTC | Reusable Shielded Transport Container |
| RTD | Research and Technological Development |
| RWM | Radioactive Waste Management |
| RWMAC | Radioactive Waste Management Advisory Committee (now disbanded) |
| RWPG | Radioactive Waste Policy Group |
| | |
| SAPs | Safety Assessment Principles |
| SC | Security Check |
| SCO | Surface Contaminated Object |
| SD | Standard Deviation |
| SE | Standard Error |
| SEPA | Scottish Environment Protection Agency |
| SETP | Sellafield Effluent Treatment Plant |
| SF | Spent Fuel |
| SFAIRP | So Far As Is Reasonably Practicable |
| SGHWR | Steam Generating Heavy Water Reactor |
| SL | Sellafield Limited |
| SL | Short Lived |
| SLC | Site Licence Company |
| SMART | Simple Multi Attribute Rating Technique |
| SoLA | Substances of Low Activity |
| SPSC | Safety Principles and Safety Criteria |
| SQEP | Suitably Qualified and Experienced Person |
| SS | Stainless Steel |
| Sv | Sievert |
| | |
| T&M | Time and Material |
| TBP | Tri-Butyl Phosphate |
| TFA | Très Faible Activité |
| TI | Transport Index |
| TLD | Thermo-Luminescent Dosimeter |
| TRU | TRansUranic (waste) |
| | |
| UKAEA | United Kingdom Atomic Energy Authority |
| UKSO | UK Safeguards Office |
| UN | United Nations |
| UNCED | United Nations Conference on Environment and Development |
| UNEP | United Nations Environment Programme |
| UNSCEAR | United Nations Scientific Committee on the Effects of Atomic Radiation |
| | |
| VLLW | Very Low Level Waste |
| | |
| WAC | Waste Acceptance Criteria |
| WAGR | Windscale Advanced Gas-cooled Reactor |
| WBM | Whole Body Monitor |
| WBS | Work Breakdown Structure |
| WENRA | Western European Nuclear Regulators' Association |
| WHO | World Health Organisation |
| WIPP | Waste Isolation Pilot Plant |
| WO | Work Order |
| WP | Work Package |
| WRS | Wilcoxon Rank Sum |
| WSA | Warship Support Agency |
| WVP | Windscale Vitrification Plant |

# Appendix 2

# Physical Quantities, Symbols and Units

**Table A2.1** *Physical quantities, symbols and units (SI)*

| Physical quantity | Symbol | Unit |
|---|---|---|
| Length | l, L | m |
| Mass | m, M | kg |
| Time | t, T | s |
| Frequency | $\nu$ | $Hz = 1\ s^{-1}$ |
| Density | $\rho$ (rho) | $kg.m^{-3}$ |
| Force | f, F | $N\ (newton) = 1\ kg.m.s^{-2}$ |
| Pressure | p | $Pa\ (pascal) = 1\ N.m^{-2}$ |
| Pressure | p | $bar = 10^5\ N.m^{-2}$ |
| Energy | E | $J\ (joule) = 1\ N.m\ or\ W.s$ |
| Energy | E | $eV\ (electron\ volt) = 1.602 \times 10^{-19}\ J$ |
| Power | P | $W\ (watt) = 1\ J.s^{-1}$ |
| Activity | A | $Bq\ (becquerel)$ |
| Particle fluence rate or flux | $\phi$ | $m^{-2}.s^{-1}$ |
| Absorbed dose | D | $Gy\ (gray) = 1\ J.kg^{-1}$ |
| Equivalent dose | $H_T$ | $Sv\ (sievert) = 1\ J.kg^{-1}$ |
| Effective dose | E | $Sv\ (sievert) = 1\ J.kg^{-1}$ |
| Committed effective dose | E(t) | $Sv\ (sievert) = 1\ J.kg^{-1}$ |
| Collective effective dose | S | person.Sv or man.Sv |

**Table A2.2** *Physical constants, symbols and units (SI)*

| Physical quantity | Symbol | Unit |
|---|---|---|
| Electronic charge | e | $1.602 \times 10^{-19}\ C$ |
| Electronic rest mass | $m_e$ | $9.109 \times 10^{-31}\ kg$ |
| Proton rest mass | $m_p$ | $1.673 \times 10^{-27}\ kg$ |
| Neutron rest mass | $m_n$ | $1.675 \times 10^{-27}\ kg$ |
| Atomic mass unit | AMU | $1.67 \times 10^{-27}\ kg$ |
| Planck constant | h | $6.626 \times 10^{-34}\ J.s$ |
| Avogadro's constant | $L, N_A$ | $6.023 \times 10^{23}\ mol^{-1}$ |
| Speed of light | c | $3.0 \times 10^8\ m.s^{-1}$ |
| Acceleration due to gravity | g | $9.8\ m.s^{-2}$ |

**Table A2.3** *SI prefixes*

| Prefix | Symbol | Quantity |
|--------|--------|----------|
| peta | P | 1.0E+15 |
| tera | T | 1.0E+12 |
| giga | G | 1.0E+09 |
| mega | M | 1.0E+06 |
| kilo | k | 1.0E+03 |
| hecto | h | 1.0E+02 |
| deca | da | 1.0E+01 |
| unit | unit | 1.0E00 |
| deci | d | 1.0E−01 |
| centi | c | 1.0E−02 |
| milli | m | 1.0E−03 |
| micro | μ | 1.0E−06 |
| nano | n | 1.0E−09 |
| pico | p | 1.0E−12 |
| femto | f | 1.0E−15 |

# Appendix 3

# Properties of Elements and Radionuclides

**Table A3.1** *Properties of elements [A3.1, A3.2]*

| Element | Symbol | Atomic number, $Z$ | Nominal density kg.m$^{-3}$($\times 10^3$) | Atomic mass* | No. of atoms per m$^3$($\times 10^{28}$) |
|---------|--------|--------|--------|--------|--------|
| Hydrogen | H (8) | 1 | Gas | 1.00795 | – |
| Helium | He (10) | 2 | Gas | 4.0026 | – |
| Lithium | Li (1) | 3 | 0.534 | 6.941 | 4.630E+00 |
| Beryllium | Be (2) | 4 | 1.85 | 9.012 | 1.236E+01 |
| Boron | B (7) | 5 | 2.34 | 10.811 | 1.304E+01 |
| Carbon | C (8) | 6 | 2.267 | 12.0108 | 1.137E+01 |
| Nitrogen | N (8) | 7 | Gas | 14.0067 | – |
| Oxygen | O (8) | 8 | Gas | 15.9994 | – |
| Fluorine | F (9) | 9 | Gas | 18.9984 | – |
| Neon | Ne (10) | 10 | Gas | 20.1797 | – |
| Sodium | Na (1) | 11 | 0.968 | 22.9898 | 2.536E+00 |
| Magnesium | Mg (2) | 12 | 1.738 | 24.305 | 4.307E+00 |
| Aluminium | Al (6) | 13 | 2.70 | 26.9815 | 6.027E+00 |
| Silicon | Si (7) | 14 | 2.33 | 28.0855 | 5.000E+00 |
| Phosphorus | P (8) | 15 | 1.823 | 30.9738 | 3.545E+00 |
| Sulphur | S (8) | 16 | 2.07 | 32.065 | 3.888E+00 |
| Chlorine | Cl (9) | 17 | Gas | 35.453 | – |
| Argon | A (10) | 18 | Gas | 39.948 | – |
| Potassium | K (1) | 19 | 0.89 | 39.098 | 1.371E+00 |
| Calcium | Ca (2) | 20 | 1.55 | 40.078 | 2.329E+00 |
| Scandium | Sc (5) | 21 | 2.985 | 44.956 | 3.999E+00 |
| Titanium | Ti (5) | 22 | 4.506 | 47.867 | 5.670E+00 |
| Vanadium | V (5) | 23 | 6.0 | 50.942 | 7.094E+00 |
| Chromium | Cr (5) | 24 | 7.15 | 51.996 | 8.282E+00 |
| Manganese | Mn (5) | 25 | 7.21 | 54.938 | 7.905E+00 |
| Iron | Fe (5) | 26 | 7.86 | 55.845 | 8.477E+00 |
| Cobalt | Co (5) | 27 | 8.90 | 58.933 | 9.096E+00 |
| Nickel | Ni (5) | 28 | 8.908 | 58.693 | 9.141E+00 |
| Copper | Cu (5) | 29 | 8.96 | 63.546 | 8.492E+00 |
| Zinc | Zn (5) | 30 | 7.14 | 65.409 | 6.575E+00 |
| Gallium | Ga (6) | 31 | 5.91 | 69.723 | 5.105E+00 |
| Germanium | Ge (7) | 32 | 5.323 | 72.64 | 4.414E+00 |
| Arsenic | As (7) | 33 | 5.727 | 74.922 | 4.604E+00 |

**Table A3.1** *Properties of elements [A3.1, A3.2] (continued)*

| Element | Symbol | Atomic number, $Z$ | Nominal density kg.m$^{-3}$($\times 10^3$) | Atomic mass* | No. of atoms per m$^3$($\times 10^{28}$) |
|---|---|---|---|---|---|
| Selenium | Se (8) | 34 | 4.81 | 78.96 | 3.669E+00 |
| Bromine | Br (9) | 35 | 3.103 | 79.904 | 2.339E+00 |
| Krypton | Kr (10) | 36 | Gas | 83.798 | – |
| Rubidium | Rb (1) | 37 | 1.532 | 85.468 | 1.080E+00 |
| Strontium | Sr (2) | 38 | 2.64 | 87.62 | 1.815E+00 |
| Yttrium | Y (5) | 39 | 4.472 | 88.906 | 3.030E+00 |
| Zirconium | Zr (5) | 40 | 6.52 | 91.224 | 4.305E+00 |
| Niobium | Nb (5) | 41 | 8.57 | 92.906 | 5.556E+00 |
| Molybdenum | Mo (5) | 42 | 10.28 | 95.942 | 6.454E+00 |
| Technetium | Tc (5) | 43 | 11.0 | [98.906] | 6.699E+00 |
| Ruthenium | Ru (5) | 44 | 12.45 | 101.07 | 7.419E+00 |
| Rhodium | Rh (5) | 45 | 12.41 | 102.906 | 7.264E+00 |
| Palladium | Pd (5) | 46 | 12.023 | 106.42 | 6.805E+00 |
| Silver | Ag (5) | 47 | 10.49 | 107.868 | 5.857E+00 |
| Cadmium | Cd (5) | 48 | 8.65 | 112.41 | 4.635E+00 |
| Indium | In (6) | 49 | 7.31 | 114.82 | 3,835E+00 |
| Tin | Sn (6) | 50 | 7.265 | 118.71 | 3.686E+00 |
| Antimony | Sb (7) | 51 | 6.697 | 121.76 | 3.313E+00 |
| Tellurium | Te (7) | 52 | 6.24 | 127.60 | 2.945E+00 |
| Iodine | I (9) | 53 | 4.933 | 126.904 | 2.341E+00 |
| Xenon | Xe (10) | 54 | Gas | 131.293 | – |
| Caesium | Cs (1) | 55 | 1.93 | 132.905 | 8.746E-01 |
| Barium | Ba (2) | 56 | 3.51 | 137.33 | 1.539E+00 |
| Lanthanum | La (3) | 57 | 6.162 | 138.905 | 2.672E+00 |
| Cerium | Ce (3) | 58 | 6.77 | 140.12 | 2.910E+00 |
| Praseodymium | Pr (3) | 59 | 6.77 | 140.908 | 2.894E+00 |
| Neodymium | Nd (3) | 60 | 7.01 | 144.24 | 2.927E+00 |
| Promethium | Pm (3) | 61 | 7.26 | [146.92] | 2.976E+00 |
| Samarium | Sm (3) | 62 | 7.52 | 150.36 | 3.012E+00 |
| Europium | Eu (3) | 63 | 5.264 | 151.964 | 2.086E+00 |
| Gadolinium | Gd (3) | 64 | 7.90 | 157.25 | 3.026E+00 |
| Terbium | Tb (3) | 65 | 8.23 | 158.925 | 3.119E+00 |
| Dysprosium | Dy (3) | 66 | 8.54 | 162.50 | 3.165E+00 |
| Holmium | Ho (3) | 67 | 8.79 | 164.93 | 3.210E+00 |
| Erbium | Er (3) | 68 | 9.066 | 167.26 | 3.265E+00 |
| Thulium | Tm (3) | 69 | 9.32 | 168.934 | 3.323E+00 |
| Ytterbium | Yb (3) | 70 | 6.90 | 173.04 | 2.402E+00 |
| Lutetium | Lu (3) | 71 | 9.84 | 174.97 | 3.387E+00 |
| Hafnium | Hf (5) | 72 | 13.31 | 178.49 | 4.491E+00 |
| Tantalum | Ta (5) | 73 | 16.69 | 180.95 | 5.555E+00 |
| Tungsten | W (5) | 74 | 19.25 | 183.84 | 6.307E+00 |

**Table A3.1** *Properties of elements [A3.1, A3.2] (continued)*

| Element | Symbol | Atomic number, $Z$ | Nominal density kg.m$^{-3}$($\times 10^3$) | Atomic mass* | No. of atoms per m$^3$($\times 10^{28}$) |
|---|---|---|---|---|---|
| Rhenium | Re (5) | 75 | 21.02 | 186.21 | 6.799E+00 |
| Osmium | Os (5) | 76 | 22.61 | 190.23 | 7.159E+00 |
| Iridium | Ir (5) | 77 | 22.65 | 192.22 | 7.097E+00 |
| Platinum | Pt (5) | 78 | 21.45 | 195.08 | 6.623E+00 |
| Gold | Au (5) | 79 | 19.3 | 196.97 | 5.902E+00 |
| Mercury | Hg (5) | 80 | 13.53 | 200.59 | 4.063E+00 |
| Thallium | Tl (6) | 81 | 11.85 | 204.38 | 3.492E+00 |
| Lead | Pb (6) | 82 | 11.34 | 207.2 | 3.296E+00 |
| Bismuth | Bi (6) | 83 | 9.78 | 208.98 | 2.819E+00 |
| Polonium | Po (7) | 84 | 9.196 | [208.98] | 2.650E+00 |
| Astatine | At (9) | 85 | – | [209.99] | – |
| Radon | Rn (10) | 86 | Gas | [222.02] | – |
| Francium | Fr (1) | 87 | – | [223.02] | – |
| Radium | Ra (2) | 88 | 5.5 | [226.03] | 1.466E+00 |
| Actinium | Ac (4) | 89 | 10.0 | [227.03] | 2.653E+00 |
| Thorium | Th (4) | 90 | 11.7 | 232.04 | 3.037E+00 |
| Protactinium | Pa (4) | 91 | 15.37 | 231.04 | 4.007E+00 |
| Uranium | U (4) | 92 | 19.1 | 238.03 | 4.833E+00 |
| Neptunium | Np (4) | 93 | 20.2 | [237.05] | 5.132E+00 |
| Plutonium | Pu (4) | 94 | 19.816 | 244.06 | 4.890E+00 |
| Americium | Am (4) | 95 | 12.0 | [243.06] | 2.974E+00 |
| Curium | Cu (4) | 96 | 13.51 | [247.07] | 3.293E+00 |
| Berkelium | Bk (4) | 97 | 14.78 | [247.07] | 3.603E+00 |
| Californium | Cf (4) | 98 | 15.1 | [251.08] | 3.622E+00 |
| Einsteinium | Es (4) | 99 | 8.84 | [252.08] | 2.112E+00 |
| Fermium | Fm (4) | 100 | – | [257.09] | – |
| Mendelevium | Md (4) | 101 | – | [258.09] | |
| Nobelium | No (4) | 102 | – | [259.10] | |
| Lawrencium | Lr (4) | 103 | – | [260.11] | – |
| Rutherfordium | Rf (5) | 104 | – | [261.11] | – |
| Dubnium | Db (5) | 105 | – | [262.11] | – |
| Seaborgium | Sg (5) | 106 | – | [263.12] | – |
| Bohrium | Bh (5) | 107 | – | [262.12] | – |
| Hassium | Hs (5) | 108 | – | [265] | – |
| Meitnerium | Mt (5) | 109 | – | [266] | – |
| Darmstadtium | Ds (5) | 110 | – | [269] | – |
| Roentgenium | Rg (5) | 111 | – | [272] | – |
| Unstable nuclides | Uub, Uut, Uuq,Uup, Uuh, Uuo | 112,113, 114,115, 116,118 | – | – | – |

\*    Relative atomic mass based on the mass of C-12 as 12.0000

(1)   Alkali metal
(2)   Alkaline earth metal
(3)   Lanthanide
(4)   Actinide
(5)   Transition metal
(6)   Post-transition metal
(7)   Metalloid
(8)   Non-metal
(9)   Halogen
(10)  Noble gas

An atomic mass within a parenthesis in column 5 signifies that either the element does not occur in nature or its half-life is so short that it cannot be found in nature

**Notes**

- A transition metal, according to IUPAC, is one whose atom has an incomplete d sub-shell, or which can give rise to cations with an incomplete d sub-shell. There are about 40 transition metals.
- A post-transition metal is a metallic element, similar to a transition metal, but it is more electro-negative.
- The term metalloid is used in chemistry to classify chemical elements. Although elements can generally be classified as either a metal or a non-metal, there are a few elements which exhibit intermediate properties and are referred to as metalloids. These behave as semiconductors (B, Si, Ge) to semi-metals (Sb).
- Lanthanides are also known as 'rare earth elements'.
- Alkali metals, alkaline earth metals, transition metals, post-transition metals, lanthanides and actinides are collectively known as metals.
- Atomic mass calculation is based on the aggregation of the natural abundance of the isotopes of an element. For example, chlorine occurs in nature as Cl-35 (75.78%) and Cl-37 (24.22%). The atomic mass of chlorine is calculated as 34.9688 (atomic mass of Cl-35) × 0.7578 + 36.9659 (atomic mass of Cl-37) × 0.2422 = 35.453.
- Number of atoms per unit volume, $D_v$ is calculated as

$$D_v = \frac{\rho N_A}{M}$$

where $\rho$ is the mass density, $N_A$ is Avogadro's number and M is the atomic or molecular weight.

---

Example: The number of atoms per m³ of Li is calculated as:

$$D_v \text{ for Li} = \frac{0.534 \times 10^3 \left(\frac{kg}{m^3}\right) \times 6.023 \times 10^{23}}{6.941 \times 10^{-3} (kg)} = 4.63 \times 10^{28} \, m^{-3}$$

---

**Table A3.2** *Half-lives and specific activities of some significant radionuclides [A3.3, A3.4]*

| Radionuclide | Element(atomic number) | Decay mode | Half-life | Specific activity (Bq.g$^{-1}$) |
|---|---|---|---|---|
| Be-7 | Beryllium (4) | ε | 53.3 d | 1.297E+16 |
| C-14 | Carbon (6) | β$^-$ | 5730 y | 1.652E+11 |
| F-18 | Fluorine (9) | β$^+$, ε | 1.83 h | 3.526E+18 |
| Na-22 | Sodium (11) | β$^+$ | 2.602 y | 2.315E+14 |
| Na-24 | Sodium (11) | β$^-$; γ | 15 h | 3.225E+17 |
| P-32 | Phosphorus (15) | β$^-$ | 14.29 d | 1.058E+16 |
| Cl-36 | Chlorine (17) | β$^-$, ε | 3.01E+05 y | 1.223E+9 |
| Ar-41 | Argon (18) | β$^-$; γ | 1.827 h | 1.550E+18 |
| K-40 | Potassium (19) | β$^-$; γ | 1.28E+09 y | 2.589E+05 |
| K-43 | Potassium (19) | β$^-$; γ | 22.6 h | 1.195E+17 |
| Sc-44 | Scandium (21) | IT, ε; γ | 3.927 d | 6.720E+17 |
| Sc-46 | Scandium (21) | β$^-$; γ | 83.83 d | 1.255E+15 |
| V-48 | Vanadium (23) | β$^+$; γ | 16.238 d | 6.207E+15 |
| Mn-52 | Manganese (25) | β$^+$, ε; γ | 5.591 d | 1.664E+16 |
| Mn-56 | Manganese (25) | β$^-$; γ | 2.578 h | 8.041E+17 |
| Fe-55 | Iron (26) | ε | 2.7 y | 8.926E+13 |
| Fe-59 | Iron (26) | β$^-$; γ | 44.529 d | 1.841E+15 |
| Co-58 | Cobalt (27) | β$^+$, ε; γ | 70.8 d | 1.178E+15 |
| Co-58m | Cobalt (27) | IT | 9.1 h | 2.188E+17 |
| Co-60 | Cobalt (27) | β$^-$; γ | 5.271 y | 4.191E+13 |
| Ni-59 | Nickel (28) | ε | 7.5E+04 y | 2.995E+09 |
| Ni-63 | Nickel (28) | β$^-$ | 96 y | 2.192E+12 |
| Cu-64 | Copper (29) | β$^-$, ε; γ | 12.701 h | 1.428E+17 |
| Zn-65 | Zinc (30) | β$^+$, ε; γ | 243.9 d | 3.052E+14 |
| Ga-68 | Gallium (31) | β$^+$, ε; γ | 1.13 h | 1.507E+18 |
| Ga-72 | Gallium (31) | β$^-$; γ | 14.1 h | 1.144E+17 |
| As-72 | Arsenic (33) | β$^+$, ε; γ | 26.0 h | 6.203E+16 |
| As-74 | Arsenic (33) | β$^+$, ε; γ β$^-$, ε; γ | 17.76 d | 3.681E+15 |
| As-76 | Arsenic (33) | β$^-$; γ | 26.32 h | 5.805E+16 |
| Br-76 | Bromine (35) | β$^+$; γ | 16.2 h | 9.431E+16 |
| Br-82 | Bromine (35) | β$^-$; γ | 35.3 h | 4.011E+16 |
| Kr-85 | Krypton (36) | β$^-$; γ | 10.72 y | 1.455E+13 |
| Rb-81 | Rubidium (37) | β$^+$, ε; γ | 4.58 h | 3.130E+17 |
| Rb-83 | Rubidium (37) | ε; γ | 86.2 d | 6.762E+14 |
| Rb-84 | Rubidium (37) | β$^+$, ε; γ β$^-$, ε; γ | 32.77 d | 1.758E+15 |
| Rb-86 | Rubidium (37) | β$^-$; γ | 18.66 d | 3.015E+15 |
| Sr-89 | Strontium (38) | β$^-$; γ | 50.5 d | 1.076E+15 |
| Sr-90 | Strontium (38) | β$^-$; γ | 29.12 y | 5.057E+12 |
| Sr-91 | Strontium (38) | β$^-$; γ | 9.5 h | 1.343E+17 |

**Table A3.2** *Half-lives and specific activities of some significant radionuclides [A3.3, A3.4] (continued)*

| Radionuclide | Element (atomic number) | Decay mode | Half-life | Specific activity (Bq.g$^{-1}$) |
|---|---|---|---|---|
| Sr-92 | Strontium (38) | $\beta^-; \gamma$ | 2.71 h | 4.657E+17 |
| Y-88 | Yttrium (39) | $\beta^+, \varepsilon; \gamma$ | 106.64 d | 5.155E+14 |
| Y-90 | Yttrium (39) | $\beta^-; \gamma$ | 64.0 h | 2.016E+16 |
| Y-91 | Yttrium (39) | $\beta^-; \gamma$ | 58.51 d | 9.086E+14 |
| Y-91m | Yttrium (39) | IT | 49.71 min | 1.540E+18 |
| Y-92 | Yttrium (39) | $\beta^-; \gamma$ | 3.54 h | 3.565E17 |
| Y-93 | Yttrium (39) | $\beta^-; \gamma$ | 10.1 h | 1.236E+17 |
| Zr-93 | Zirconium (40) | $\beta^-; \gamma$ | 1.53E+6 y | 9.315E+07 |
| Zr-95 | Zirconium (40) | $\beta^-; \gamma$ | 63.98 d | 7.960E+14 |
| Zr-97 | Zirconium (40) | $\beta^-; \gamma$ | 16.9 h | 7.083E+16 |
| Nb-94 | Niobium (41) | $\beta^-; \gamma$ | 2.03E+04 y | 6.946E+09 |
| Nb-95 | Niobium (41) | $\beta^-; \gamma$ | 35.15 d | 1.449E+15 |
| Nb-97 | Niobium (41) | $\beta^-; \gamma$ | 1.23 h | 9.961E+17 |
| Mo-93 | Molybdenum(42) | $\varepsilon; \gamma$ | 3500 y | 4.072E+10 |
| Mo-99 | Molybdenum(42) | $\beta^-; \gamma$ | 66 h | 1.777E+16 |
| Tc-95m | Technetium(43) | $\varepsilon; \gamma$ | 61 d | 8.349E+14 |
| Tc-96 | Technetium(43) | $\varepsilon; \gamma$ | 4.28 d | 1.177E+16 |
| Tc-96m | Technetium(43) | IT, $\beta^+, \varepsilon; \gamma$ | 51.5 min | 1.409E+18 |
| Tc-99 | Technetium(43) | $\beta^-; \gamma$ | 2.13E+05 | 6.286E+08 |
| Tc-99m | Technetium(43) | IT, $\beta^-; \gamma$ | 6.02 h | 1.948E+17 |
| Ru-103 | Ruthenium (44) | $\beta^-; \gamma$ | 39.28 d | 1.196E+15 |
| Ru-105 | Ruthenium (44) | $\beta^-; \gamma$ | 4.44 h | 2.491E+17 |
| Ru-106 | Ruthenium (44) | $\beta^-$ | 368.2 d | 1.240E+14 |
| Rh-99 | Rhodium (45) | $\beta^+, \varepsilon; \gamma$ | 16 d | 3.054E+15 |
| Rh-102 | Rhodium (45) | $\varepsilon; \gamma$ | 2.9 y | 4.481E+13 |
| Rh-102m | Rhodium (45) | $\beta^-, \beta^+, \varepsilon; \gamma$ | 207 d | 2.291E+14 |
| Rh-105 | Rhodium (45) | $\beta^-; \gamma$ | 35.36 h | 3.127E+16 |
| Ag-108m | Silver (47) | IT, $\varepsilon; \gamma$ | 127 y | 9.664E+11 |
| Ag-110m | Silver (47) | IT, $\beta^-, \varepsilon; \gamma$ | 249.9 d | 1.760E+14 |
| Cd-115 | Cadmium (48) | $\beta^-; \gamma$ | 53.46 h | 1.889E+16 |
| Cd-115m | Cadmium (48) | $\beta^-; \gamma$ | 44.6 d | 9.433E+14 |
| In-114m | Indium (49) | IT, $\varepsilon; \gamma$ | 49.51 d | 8.572E+14 |
| In-115m | Indium (49) | IT, $\beta^-; \gamma$ | 4.486 h | 2.251E+17 |
| Sn-123 | Tin (50) | $\beta^-; \gamma$ | 129.2 d | 3.044E+14 |
| Sn-125 | Tin (50) | $\beta^-; \gamma$ | 9.64 d | 4.015E+15 |
| Sb-122 | Antimony (51) | $\beta^-; \gamma$ | 2.7 d | 1.469E+16 |
| Sb-124 | Antimony (51) | $\beta^-; \gamma$ | 60.2 d | 6.481E+14 |
| Sb-125 | Antimony (51) | $\beta^-; \gamma$ | 2.77 y | 3.828E+13 |
| Sb-126 | Antimony (51) | $\beta^-; \gamma$ | 12.4 d | 3.096E+15 |
| Te-121 | Tellurium (52) | $\varepsilon; \gamma$ | 17 d | 2.352E+15 |
| Te-121m | Tellurium (52) | IT, $\varepsilon; \gamma$ | 154 d | 2.596E+14 |

**Table A3.2** *Half-lives and specific activities of some significant radionuclides [A3.3, A3.4]* *(continued)*

| Radionuclide | Element (atomic umber) | Decay mode | Half-life | Specific activity (Bq.g$^{-1}$) |
|---|---|---|---|---|
| Te-123m | Tellurium (52) | IT, ε; γ | 119.7 d | 3.286E+14 |
| Te-127 | Tellurium (52) | β$^-$; γ | 9.35 h | 9.778E+16 |
| Te-129 | Tellurium (52) | β$^-$; γ | 69.6 min | 7.759E+17 |
| Te-129m | Tellurium (52) | IT, β$^-$; γ | 33.6 d | 1.116E+15 |
| Te-131m | Tellurium (52) | IT, β$^-$; γ | 30 h | 2.954E+16 |
| Te-132 | Tellurium (52) | β$^-$; γ | 78.2 h | 1.125E+16 |
| I-124 | Iodine (53) | ε; γ | 4.18 d | 9.334E+15 |
| I-126 | Iodine (53) | β$^-$, β$^+$, ε; γ | 13.02 d | 2.949E+15 |
| I-131 | Iodine (53) | β$^-$; γ | 8.04 d | 4.593E+15 |
| I-132 | Iodine (53) | β$^-$; γ | 2.3 h | 3.824E+17 |
| I-133 | Iodine (53) | β$^-$; γ | 20.8 h | 4.197E+16 |
| I-134 | Iodine (53) | β$^-$; γ | 52.6 min | 9.884E+17 |
| I-135 | Iodine (53) | β$^-$; γ | 6.61 h | 1.301E+17 |
| Xe-122 | Xenon (54) | ε; γ | 20.1 h | 4.735E+16 |
| Xe-123 | Xenon (54) | ε; γ | 2.08 h | 4.538E+17 |
| Xe-127 | Xenon (54) | ε; γ | 36.41 d | 1.046E+15 |
| Xe-131m | Xenon (54) | IT | 11.9 d | 3.103E+15 |
| Xe-133 | Xenon (54) | β$^-$; γ | 5.245 d | 6.935E+15 |
| Xe-135 | Xenon (54) | β$^-$; γ | 9.09 h | 9.462E+16 |
| Cs-129 | Caesium (55) | β$^+$, ε; γ | 32.06 h | 2.808E+16 |
| Cs-134 | Caesium (55) | β$^-$; γ | 2.062 y | 4.797E+13 |
| Cs-134m | Caesium (55) | IT; γ | 2.9 h | 2.988E+17 |
| Cs-135 | Caesium (55) | β$^-$; γ | 2.3E+06 y | 4.269E+07 |
| Cs-136 | Caesium (55) | β$^-$; γ | 13.1 d | 2.716E+15 |
| Cs-137 | Caesium (55) | β$^-$; γ | 30 y | 3.225E+12 |
| Ba-131 | Barium (56) | β$^+$, ε; γ | 11.8 d | 3.130E+15 |
| Ba-133 | Barium (56) | ε; γ | 10.74 y | 9.279E+12 |
| Ba-140 | Barium (56) | β$^-$; γ | 12.74 d | 2.712E+15 |
| La-140 | Lanthanum (57) | β$^-$; γ | 40.27 h | 2.059E+16 |
| Ce-141 | Cerium (58) | β$^-$; γ | 32.50 d | 1.056E+15 |
| Ce-143 | Cerium (58) | β$^-$; γ | 33 h | 2.461E+16 |
| Ce-144 | Cerium (58) | β$^-$; γ | 284.3 d | 1.182E+14 |
| Pr-142 | Praseodymium (59) | β$^-$; γ | 19.13 h | 4.274E+16 |
| Pr-143 | Praseodymium (59) | β$^-$; γ | 13.56 d | 2.495E+15 |
| Nd-147 | Neodymium(60) | β$^-$; γ | 10.98 d | 2.997E+15 |
| Nd-149 | Neodymium(60) | β$^-$; γ | 1.73 h | 4.504E+17 |
| Pm-143 | Promethium(61) | ε; γ | 265 d | 1.277E+14 |

**Table A3.2**  *Half-lives and specific activities of some significant radionuclides [A3.3, A3.4]  (continued)*

| Radionuclide | Element (atomic number) | Decay mode | Half-life | Specific activity (Bq.g$^{-1}$) |
|---|---|---|---|---|
| Pm-144 | Promethium(61) | ε; γ | 363 d | 9.255E+13 |
| Pm-146 | Promethium(61) | β⁻, ε; γ | 5.53 y | 1.639E+13 |
| Pm-148m | Promethium(61) | IT, β⁻; γ | 41.3 d | 7.915E+14 |
| Pm-149 | Promethium(61) | β⁻; γ | 53.08 h | 1.468E+16 |
| Pm-151 | Promethium(61) | β⁻; γ | 28.4 h | 2.708E+16 |
| Sm-153 | Samarium (62) | β⁻; γ | 46.7 h | 1.625E+16 |
| Eu-147 | Europium (63) | β⁺, ε; γ | 24 d | 1.371E+15 |
| Eu-148 | Europium (63) | β⁺, ε; γ | 54.5 d | 5.998E+14 |
| Eu-149 | Europium (63) | ε; γ | 93.1 d | 3.488E+14 |
| Eu-150 (SL) | Europium (63) | β⁻, ε; γ | 12.62 h | 6.134E+16 |
| Eu-150 (LL) | Europium (63) | ε; γ | 34.2 y | 2.584E+12 |
| Eu-152 | Europium (63) | β⁻, ε; γ | 13.33 y | 6.542E+12 |
| Eu-152m | Europium (63) | IT,β⁻; γ | 9.32 h | 8.196E+16 |
| Eu-154 | Europium (63) | β⁻; γ | 8.8 y | 9.781E+12 |
| Eu-156 | Europium (63) | β⁻; γ | 15.19 d | 2.042E+15 |
| Gd-146 | Gadolinium(64) | ε; γ | 48.3 d | 6.861E+14 |
| Gd-149 | Gadolinium(64) | ε; γ | 9.3 d | 3.486E+15 |
| Gd-159 | Gadolinium(64) | β⁻; γ | 18.56 h | 3.935E+16 |
| Tb-158 | Terbium (65) | β⁻; γ | 150 y | 5.593E+11 |
| Tb-160 | Terbium (65) | β⁻; γ | 72.3 d | 4.182E+14 |
| Dy-165 | Dysprosium(66) | β⁻; γ | 2.334 h | 3.015E+17 |
| Dy-166 | Dysprosium(66) | β⁻; γ | 81.6 h | 8.572E+15 |
| Ho-166 | Holmium (67) | β⁻; γ | 26.8 h | 2.610E+16 |
| Ho-166m | Holmium (67) | β⁻; γ | 1200 y | 6.654E+10 |
| Er-171 | Erbium (68) | β⁻; γ | 7.52 h | 9.029E+16 |
| Tm-170 | Thulium (69) | β⁻; γ | 128.6 d | 2.213E+14 |
| Tm-171 | Thulium (69) | β⁻; γ | 1.92 y | 4.037E+13 |
| Yb-175 | Ytterbium(70) | β⁻; γ | 4.19 d | 6.598E+15 |
| Lu-172 | Lutetium (71) | ε; γ | 6.7 d | 4.198E+15 |
| Lu-173 | Lutetium (71) | ε; γ | 1.37 y | 5.592E+13 |
| Lu-174 | Lutetium (71) | ε; γ | 3.31 y | 2.301E+13 |
| Lu-174m | Lutetium (71) | IT,ε; γ | 142 d | 1.958E+14 |
| Lu-177 | Lutetium (71) | β⁻; γ | 6.71 d | 4.073E+15 |
| Hf-175 | Hafnium (72) | ε; γ | 70 d | 3.949E+14 |
| Hf-181 | Hafnium (72) | β⁻; γ | 42.4 d | 6.304E+14 |
| Hf-182 | Hafnium (72) | β⁻; γ | 9.0E+06 | 8.092E+06 |
| Ta-178 | Tantalum (73) | ε; γ | 2.2 h | 2.965E+17 |
| Ta-182 | Tantalum (73) | β⁻; γ | 115 d | 2.311E+14 |
| W-185 | Tungsten (74) | β⁻; γ | 75.1 d | 3.482E+14 |
| W-187 | Tungsten (74) | β⁻; γ | 23.9 h | 2.598E+16 |
| W-188 | Tungsten (74) | β⁻; γ | 69.4 d | 3.708E+14 |

**Table A3.2** *Half-lives and specific activities of some significant radionuclides [A3.3, A3.4] (continued)*

| Radionuclide | Element (atomic number) | Decay mode | Half-life | Specific activity (Bq.g⁻¹) |
|---|---|---|---|---|
| Re-184 | Rhenium (75) | $\varepsilon; \gamma$ | 38 d | 6.919E+14 |
| Re-184m | Rhenium (75) | IT, $\varepsilon; \gamma$ | 165 d | 1.594E+14 |
| Re-186 | Rhenium (75) | $\beta^-; \gamma$ | 90.64 h | 6.887E+15 |
| Re-188 | Rhenium (75) | $\beta^-; \gamma$ | 16.98 h | 3.637E+16 |
| Re-189 | Rhenium (75) | $\beta^-; \gamma$ | 24.3 h | 2.528E+16 |
| Os-185 | Osmium (76) | $\varepsilon; \gamma$ | 94 d | 2.782E+14 |
| Os-193 | Osmium (76) | $\beta^-; \gamma$ | 30 h | 2.005E+16 |
| Ir-190 | Iridium (77) | $\varepsilon; \gamma$ | 12.1 d | 2.104E+15 |
| Ir-192 | Iridium (77) | $\beta^-, \varepsilon; \gamma$ | 74.02 d | 3.404E+14 |
| Ir-194 | Iridium (77) | $\beta^-; \gamma$ | 19.15 h | 3.125E+16 |
| Pt-191 | Platinum (78) | $\varepsilon; \gamma$ | 2.8 d | 9.046E+15 |
| Pt-193m | Platinum (78) | IT; $\gamma$ | 4.33 d | 5.789EE+15 |
| Pt-195m | Platinum (78) | IT; $\gamma$ | 4.02 d | 6.172E+15 |
| Pt-197 | Platinum (78) | $\beta^-; \gamma$ | 18.3 h | 3.221E+16 |
| Pt-197m | Platinum (78) | IT,$\beta^-; \gamma$ | 94.4 min | 3.746E+17 |
| Au-193 | Gold (79) | $\varepsilon; \gamma$ | 17.65 h | 3.409E+16 |
| Au-194 | Gold (79) | $\beta^+; \gamma$ | 39.5 h | 1.515E+16 |
| Au-198 | Gold (79) | $\beta^-; \gamma$ | 2.696 d | 9.063E+15 |
| Au-199 | Gold (79) | $\beta^-; \gamma$ | 3.139 d | 7.745E+15 |
| Hg-195m | Mercury (80) | IT,$\varepsilon; \gamma$ | 41.6 h | 1.431E+16 |
| Hg-197m | Mercury (80) | IT,$\varepsilon; \gamma$ | 23.8 h | 2.476E+16 |
| Hg-203 | Mercury (80) | $\beta^-; \gamma$ | 46.6 d | 5.114E+14 |
| Tl-200 | Thallium (81) | $\beta^+; \gamma$ | 26.1 h | 2.224E+16 |
| Tl-202 | Thallium (81) | $\beta^+; \gamma$ | 12.23 d | 1.958E+15 |
| Tl-204 | Thallium (81) | $\beta^-$ | 3.779 y | 1.719E+14 |
| Pb-201 | Lead (82) | $\beta^+; \gamma$ | 9.4 h | 6.145E+16 |
| Pb-203 | Lead (82) | $\varepsilon; \gamma$ | 52.05 h | 1.099E+16 |
| Bi-205 | Bismuth (83) | $\beta^+; \gamma$ | 15.31 d | 1.541E+15 |
| Bi-206 | Bismuth (83) | $\beta^+; \gamma$ | 6.243 d | 3.762E+15 |
| Bi-207 | Bismuth (83) | $\beta^+; \gamma$ | 38 y | 1.685E+12 |
| Bi-210 | Bismuth (83) | $\alpha,\beta^-; \gamma$ | 5.012 d | 4.597E+15 |
| Bi-210m | Bismuth (83) | $\alpha; \gamma$ | 3.0E+6 y | 2.104E+7 |
| Bi-212 | Bismuth (83) | $\alpha,\beta^-; \gamma$ | 60.55 min | 5.427E+17 |
| Po-210 | Polonium (84) | $\alpha; \gamma$ | 138.38 d | 1.665E+14 |
| At-211 | Astatine (85) | $\alpha; \gamma$ | 7.214 h | 7.628E+16 |
| Rn-222 | Radon (86) | $\alpha; \gamma$ | 3.8235 d | 5.70E+15 |
| Ra-223 | Radium (88) | $\alpha; \gamma$ | 11.434 d | 1.897E+15 |
| Ra-224 | Radium (88) | $\alpha; \gamma$ | 3.66 d | 5.901E+15 |
| Ra-225 | Radium (88) | $\beta^-; \gamma$ | 14.8 d | 1.453E+15 |
| Ra-226 | Radium (88) | $\alpha; \gamma$ | 1600 y | 3.666E+10 |
| Ra-228 | Radium (88) | $\beta^-; \gamma$ | 5.75 y | 1.011E+13 |

**Table A3.2** *Half-lives and specific activities of some significant radionuclides [A3.3, A3.4] (continued)*

| Radionuclide | Element (atomic number) | Decay mode | Half-life | Specific activity (Bq.g$^{-1}$) |
|---|---|---|---|---|
| Ac-225 | Actinium (89) | $\alpha$; $\gamma$ | 10 d | 2.150E+15 |
| Ac-227 | Actinium (89) | $\alpha$,$\beta^-$; $\gamma$ | 21.773 y | 2.682E+12 |
| Ac-228 | Actinium (89) | $\alpha$,$\beta^-$; $\gamma$ | 6.13 h | 8.308E+16 |
| Th-227 | Thorium (90) | $\alpha$; $\gamma$ | 18.72 d | 1.139E+15 |
| Th-228 | Thorium (90) | $\alpha$; $\gamma$ | 1.913 y | 3.039E+13 |
| Th-229 | Thorium (90) | $\alpha$; $\gamma$ | 7340 y | 7.886E+09 |
| Th-230 | Thorium (90) | $\alpha$; $\gamma$ | 7.7E+4 | 7.484E+08 |
| Th-231 | Thorium (90) | $\beta^-$; $\gamma$ | 25.52 h | 1.970E+16 |
| Th-232 | Thorium (90) | $\alpha$; $\gamma$ | 1.405E+10 y | 4.066E+03 |
| U-230 | Uranium (92) | $\alpha$; $\gamma$ | 20.8 d | 1.011E+15 |
| U-232 | Uranium (92) | $\alpha$; $\gamma$; SF | 72 y | 7.935E+11 |
| U-233 | Uranium (92) | $\alpha$; $\gamma$; SF | 1.585E+5 y | 3.589E+08 |
| U-234 | Uranium (92) | $\alpha$; $\gamma$; SF | 2.445E+5 y | 2.317E+08 |
| U-235 | Uranium (92) | $\alpha$; $\gamma$; SF | 7.038E+8 y | 8.014E+04 |
| U-236 | Uranium (92) | $\alpha$; $\gamma$; SF | 2.342E+7 y | 2.399E+06 |
| U-238 | Uranium (92) | $\alpha$; $\gamma$; SF | 4.468E+9 y | 1.246E+04 |
| Np-235 | Neptunium (93) | $\alpha$; $\gamma$ | 396.1 d | 5.197E+13 |
| Np-236(LL) | Neptunium (93) | $\beta^-$, $\varepsilon$; $\gamma$ | 1.15E+5 y | 4.884E+08 |
| Np-236(SL) | Neptunium (93) | $\beta^-$, $\varepsilon$; $\gamma$ | 22.5 h | 2.187E+16 |
| Np-237 | Neptunium (93) | $\alpha$; $\gamma$ | 2.14E+6 y | 2.613E+07 |
| Np-239 | Neptunium (93) | $\beta^-$; $\gamma$ | 2.355 d | 8.596E+15 |
| Pu-236 | Plutonium (94) | $\alpha$; $\gamma$; SF | 2.851 y | 1.970E+13 |
| Pu-237 | Plutonium (94) | $\alpha$; $\gamma$ | 45.3 d | 4.506E+14 |
| Pu-238 | Plutonium (94) | $\alpha$; $\gamma$; SF | 87.74 y | 6.347E+11 |
| Pu-239 | Plutonium (94) | $\alpha$; $\gamma$; SF | 24065 y | 2.305E+09 |
| Pu-240 | Plutonium (94) | $\alpha$; $\gamma$; SF | 6537 y | 8.449E+09 |
| Pu-241 | Plutonium (94) | $\beta^-$; $\gamma$ | 14.4 y | 3.819E+12 |
| Pu-242 | Plutonium (94) | $\alpha$; $\gamma$; SF | 8.26E+7 y | 6.577E+05 |

## Notes

$\alpha$     alpha particle
$\beta^-$     electron
$\beta^+$     positron
$\gamma$     gamma ray
$\varepsilon$     electron capture
IT     isomeric transition
LL     long-lived
m     metastable state
SF     spontaneous fission
SL     short-lived

## REFERENCES

[A3.1] Table of Chemical Elements, Wikipedia the free encyclopedia. http://en.wikipedia.org/wiki/List_of_elements_by_atomic_number.

[A3.2] Lamarsh J.R., *Introduction to Nuclear Engineering*, 3rd edn., Addison-Wesley, Reading, MA, USA, 1975.

[A3.3] International Atomic Energy Agency, Advisory Material for the IAEA Regulations for the Safe Transport of Radioactive Material, Safety Guide, No. TS-G-1.1 (ST-2), IAEA, Vienna, Austria, 2002.

[A3.4] General Electric, Chart of the Nuclides, 14th edn.

# Appendix 4

# NDA Management of Decommissioning Activities in the UK

## A4.1 NDA Operating Regime

Under the terms of the Energy Act 2004 [A4.1], the NDA became responsible for managing the programme of decommissioning and restoration of all legacy nuclear sites in the UK. Prior to the formation of the NDA, BNFL, Magnox Electric and UKAEA were the main site licence holders who owned and operated all the major sites directly, except for Springfields which was operated by Westinghouse. Altogether there were 20 sites which were: Berkeley, Bradwell, Calder Hall, Capenhurst, Chapelcross, Culham JET, Dounreay, Dungeness A, Harwell, Hinkley Point A, Hunterston A, Low Level Waste Repository at Drigg, Oldbury, Sellafield, Sizewell A, Springfields, Transfynydd, Windscale, Winfrith and Wylfa. Figure A4.1 shows the location of these nuclear sites. Table 5.1 of Chapter 5 also shows the present site licence holders of these sites and gives a short description of the sites. On the vesting day for the NDA on 1 April 2005, the ownership of the sites was transferred to the NDA, but the site licences (with all their conditions and restrictions) remained with the Site Licence Companies (SLCs). The range of activities the present site licensees undertake include plant operations, reprocessing, radioactive waste retrieval, radioactive waste storage and conditioning, spent fuel management, waste transportation etc. The remit of the NDA is to decommission and clean up the sites safely, securely and cost-effectively, with a strategic focus on the competitiveness of the incumbents.

In preparation for the new operating regime, the BNFL had carved out of the parent company a group, called the British Nuclear Group (BNG), to function as a Management and Operations (M&O) organisation which would run nuclear sites. However, in 2007, due to the changed operating model of the NDA and due to the privatisation pressure from the stakeholder, the BNG was broken into three separate companies: The Project Services, Sellafield Ltd and the Magnox Electric Ltd. The BNFL announced in August 2007 that the Project Services will be privatised as a separate entity. Sellafield Ltd having responsibilities at the moment for Sellafield site operations, Capenhurst site operations and the international nuclear services operations may operate as a Site Licence Company. Magnox Electric Ltd will consist of Magnox North (Chapelcross, Hunterston A, Oldbury, Trawsfynydd and Wylfa magnox

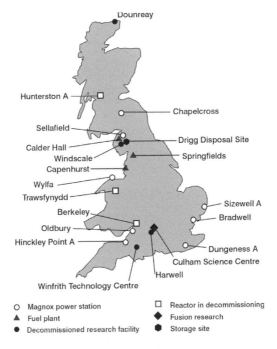

Fig. A4.1 *Location of nuclear legacy sites in the UK*

reactors) and Magnox South (Berkeley, Bradwell, Dungeness A, Hinkeley Point A and Sizewll A magnox reactors). At that point BNFL will effectively cease to exist.

## A4.2 NDA Operating Model

As stated above, the focus of the NDA operating model is competitiveness. That does not preclude, in any way, the basic requirements of safety, security and environmental acceptability. The key organisations in this competitive strategy are the Parent Body Organisations (PBOs) and the SLC [A4.2]. Figure A4.2 shows the key relationships between the NDA and these organisations.

The SLC is the bedrock of this process to deliver the NDA strategy. It will be the legal entity with responsibility for operating one or more nuclear licensed sites. As the holder of the site licence, the SLC must have all the attributes of a licensee and it will be the prime contractor to the NDA via an M&O contract. The PBO will provide management, leadership and strategic direction to the SLC to ensure effective delivery of the NDA contract in return for the ownership of the shares of the SLC. The PBO will also provide the financial guarantee required by the NDA. The PBO will have no operational responsibilities, but will have a limited number of key personnel who will

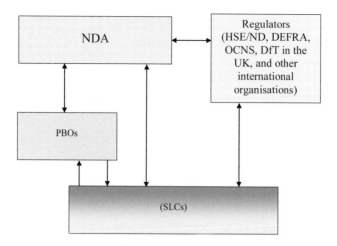

**Fig. A4.2** *NDA operating model*

work alongside the SLC personnel. A number of organisations may form a consortium to be the PBO of an SLC.

The competition will be directed towards the selection of the PBO which will hold the shares of the SLC for the duration of the NDA contract. The NDA will award the contract to the SLC and sign a parent body agreement with the PBO. The PBO receives income via fees from the SLC on the successful delivery of the NDA contract. At the appropriate time, another competition will be initiated and if a different PBO wins the competition and after an appropriate transition period, the shares of the SLC will be transferred to the new PBO [A4.2].

It should be noted that the SLC itself will not be subjected to the competitive process, as the primary requirements of the SLC are technical competence, knowledge and skill which must be retained to carry out the technical work. Moreover, an organisation which holds the site licence, radioactive waste disposal authorisation, a security plan and possible transport approvals cannot be thrown out at regular intervals and replaced by another organisation with similar attributes. The prerequisites of the SLC and its relationship with the PBO and sub-contractors are shown in Figure A4.3.

In order to implement the operating model and facilitate competition, the NDA is proposing to restructure the present 20 site licensees into a more manageable seven site licensees. The proposed SLCs are as follows:

- Existing Sellafield site, Windscale site and Capenhurst site will be put together into a single nuclear SLC. Sellafield Ltd. will be the SLC.
- LLW disposal site near Drigg will have the LLW Repository Company as the SLC.

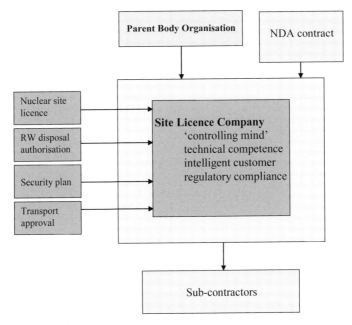

**Fig. A4.3** *SLC model*

- Dounreay site will have the Dounreay SLC.
- A new Magnox South SLC will have five nuclear licensed sites at Berkeley, Bradwell, Dungeness A, Hinkley Point A and Sizewell A.
- A new Harwell and Winfrith SLC will have two licensed sites at Harwell and Winfrith.
- A new Magnox North SLC will have six sites at Chapelcross, Hunterston A, Oldbury, Trawsfynydd and Wylfa.
- Springfields Fuels Limited will be the SLC with no changes.

It should be noted that the last remaining site under NDA ownership, Culham JET Plant, will remain with the UKAEA as a Non-Departmental Public Body (NDPB) and will not be subject to the SLC provisions.

## A4.3 Supply Chain

The operating practice under the NDA regime is that the NDA awards an M&O contract to a SLC who would effectively be a tier 1 company. The tier 1 company would be the 'intelligent customer' who would operate the site. In order to do so, it may award contracts, when necessary, to tier 2 companies who may, in turn, award contracts to tier 3 companies. This hierarchy of the supply chain is shown in Figure A4.4 [A4.3]. The NDA will not normally award contracts directly to the lower tier subcontractors, but will maintain an oversight through the review of annual procurement plans of tier 1 companies.

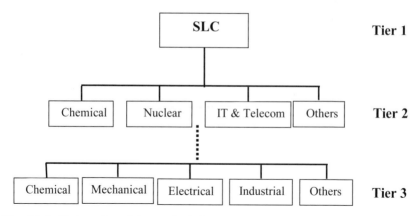

**Fig. A4.4** *Various tiers of the supply chain.*

The existing SLCs will draw up Life-Cycle Base Lines (LCBLs) and Near Term Work Plans (NTWPs) for every site for scrutiny and approval by the NDA. The NDA will then consolidate the individual LCBLs into a national LCBL for the UK. The NDA would also produce annual plans for each year, initially a draft plan for public consultation followed by a final plan. The annual plan would be the expenditure plan on a site-by-site basis and on work categories. This plan along with the NTWP would give a fair view of decommissioning activities and expenditure for every site for the year in question.

### A4.3.1 Tier 1 Organisations

Tier 1 organisations will be responsible for managing operations at a site or sites contracted to it under the M&O contract from the NDA. Tier 1 companies will earn fees from delivering agreed deliverables to NDA under the NTWP. To do so, they will have to liaise effectively with tier 2 companies and manage staff on a day-to-day basis.

### A4.3.2 Tier 2 Organisations

Tier 2 organisations may be either junior partners of a tier 1 company or a main supplier to the tier 1 company. They will supply solutions, rather than services and goods, to their customers. They will also manage the tier 3 companies who will be supplying goods and services to their customers. Tier 2 companies will offer more than excellence in skills and capabilities. They will need to be able to 'buy intelligently' and integrate the services bought, in order to satisfy their customers. By doing so, they will be assisting tier 1 companies in completing the contract to time and cost. Thus, the tier 2 companies will share some of the financial risk with tier 1 companies.

### A4.3.3 Tier 3 Organisations

Tier 3 organisations will supply goods and services to tier 2 organisations. These organisations will be specialist suppliers of goods and services with little or no management responsibilities for the contract. However, tier 2 organisation may delegate some responsibilities and commensurate rewards to these companies for satisfactory performance and delivery of services and products to time and cost. These companies will probably be small, lean and fit and responsive to the demands from higher tier companies.

### REFERENCES

[A4.1] UK Government, Energy Act 2004, The Stationary Office.

[A4.2] Nuclear Decommissioning Authority, NDA Operating Model: The Roles of the Site Licence Company and the Parent Body Organisations, Doc No. NSG 31, Revision 2, 15 January 2007.

[A4.3] Nuclear Industry Association, The impact of the Nuclear Decommissioning Authority on the Supply Chain, February 2005.

# Appendix 5

# Institutional Framework and Regulatory Standards in Decommissioning in Selected EU Member States

Article 4 of the Convention on Nuclear Safety (Convention on Nuclear Safety, INFCIRC/449, IAEA, Vienna, 1994) states, 'Each Contracting Party shall take, within the framework of its national law, the legislative, regulatory and administrative measures and other steps necessary to implement its obligations under the Convention'.

**Table A5.1** *Overview of decommissioning regulatory issues in selected EU Member States*

| Question | Belgium | Bulgaria | France |
|---|---|---|---|
| What is the name of the national regulatory body empowered to issue site licence? | Federal Agency for Nuclear Control (FANC/AFCN) | Nuclear Regulatory Agency (NRA) | French Nuclear Safety Authority, ASN (Autorité de Sureté Nucléaire) |
| What is the primary legislation under which the nuclear site licence is issued? | Royal Decree of 28 February 1963 gives general regulations. Under the Royal Decree of 20 July 2001, the king gives authorization for Class I facilities | 'Act on the Safe Use of Nuclear Energy', promulgated in the State Gazette No. 63 of 28 June 2002 | Law 61-842 of 2 August 1961 on Atmospheric Pollution and its Application Decree 63-1228 of 11 December 1963 are the primary legislation |
| Is EC Directive 96/29/Euratom of 13 May 1996 on BSS followed? | Yes. Royal Decree of 20 July 2001 | Yes. 'Regulation for the Basic Norms for Radiation Protection', promulgated in State Gazette No. 73, 2004 | Yes |
| Is EC Directive 97/11/EC of 3 March 1997 on the EIA on decommissioning implemented? | Yes. Royal Decree of 20 July 2001, C-2001/00726 enforces EC Directive | Yes. Act on Environmental Protection' came into force in September 2002. It was last amended in January 2005 | Yes |
| Is there a requirement for a separate licence for decommissioning work? | Yes, Article 17 of the Royal Decree of 20 July 2001. Licence period is specified in the licence | Yes. Decommissioning permit from the NRA chairman is required. Permit for each stage has terms and conditions attached to it | Yes. A new administrative note dated 17 February 2003 requires a separate licence |
| Are there defined stages of decommissioning? | No. Approval is given on a case-by-case basis | No, there are no nationally defined stages of decommissioning. Decommissioning plan specifies the stages | Three stages of decommissioning are followed without any delay time between stages |
| What is the name of the national regulatory body overseeing radiological protection? | FANC/AFCN | NRA | ASN draws on the expertise of DGSNR |
| Is IAEA waste categorisation scheme followed? | For long-term management (>30 y), IAEA scheme is followed | Yes | Primarily yes. But there is no exempt waste category in France |
| What are the criteria for delicensing a site? | Free release certificate based on clearance levels is required | No delicensing provision exists. However, radiological criteria of $10\ \mu Sv.y^{-1}$ public dose applies | Delicensing is done on a case by case basis, taking into account that there are no exemption and clearance criteria in France |
| Is there a regulatory body overseeing industrial safety? | Federal Ministry of Employment and Labour | General Labour Inspectorate of the Executive Agency | |

**Table A5.1** *Overview of decommissioning regulatory issues in selected EU member states (continued)*

| Question | Germany | Italy | Lithuania |
|---|---|---|---|
| What is the name of the national regulatory body empowered to issue site licence? | BMU (Federal Agency) is the regulatory body. The länder (federal states) are empowered to execute administrative duties (licensing and supervision) | Ministry of Productive Activities (MAP) in consultation with APAT issues operational and decommissioning licence | State Nuclear Power Industry Safety Inspectorate (VATESI) is empowered to issue site licence |
| What is the primary legislation under which the nuclear site licence is issued? | Atomic Energy Act of 23 December 1959, amended on 27 April 2002, prohibits any further issue of licences for nuclear power generation | Law No. 1860 of 1962, amended by the Presidential Decree No. 519 of 1975. Nuclear power plants are governed by the Legislative Decree 230/95 | The Law on Nuclear Energy (1996, No.1-1613, last amended 2004) and the Government Resolution No. 103 |
| Is EC Directive 96/29/Euratom of 13 May 1996 on BSS followed? | Radiation Protection Ordinance (RPO) of July 2001 implements BSS | Yes. Legislative Decree No. 241 of 2000 implements the BSS | Yes. Lithuanian hygiene standard is HN 73:2001 'Basic Standards of Radiation Protection' |
| Is EC Directive 97/11/EC of 3 March 1997 on the EIA on decommissioning implemented? | Yes. The EIA Act of September 2001 requires implementation of this EC Directive | No. The Ministry of Environment uses this EC Directive directly with the advice of the EIA Commission | Yes. The Law on Environmental Impact Assessment of the Proposed Economic Activity (2000, last amended in 2005) implements EC EIA |
| Is there a requirement for a separate licence for decommissioning work? | Yes. The länder regulatory body issues the decommissioning licence | Yes. The licences are granted for specific phases of work by MAP and regulated by APAT | Yes. VATESI in consultation with Ministry of Environment (RSC) issues licence |
| Are there defined stages of decommissioning? | No. Operator decides on either immediate or deferred dismantling on a case-by-case basis | No. Waste management problem dictates this issue | Operator chooses on a case by case basis, which may not agree with the IAEA stages |
| What is the name of the national regulatory body overseeing radiological protection? | Federal Office for Radiation Protection (BfS) implements on behalf of the BMU | APAT is responsible for the regulation and supervision of nuclear safety and radiation protection | Radiation Protection Centre (RSC) of the Ministry of Health is responsible for radiation protection of workers and the public |
| Is IAEA waste categorisation scheme followed? | Yes, but with minor variations | No. Technical Guide No. 26 is followed | Yes, with some variation |
| What are the criteria for delicensing a site? | Radiation Protection Ordinance (RPO) of July 2001 defines clearance levels for delicensing | Delicensing requires compliance with clearance criteria | Primarily radiological protection (<10 µSv individual and <1 manSv collective dose for the critical group) |
| Is there a regulatory body overseeing industrial safety? | Federal Ministry for Labour and Social Affairs (BMAS) | | Ministry of Social Security and Labour |

**Table A5.1** *Overview of decommissioning regulatory issues in selected EU member states (continued)*

| Question | The Netherlands | Spain | UK |
|---|---|---|---|
| What is the name of the national regulatory body empowered to issue site licence? | Directorate for Chemicals, Waste, Radiation Protection of the Ministry of Housing, Spatial Planning and the Environment is the licensing authority | Ministry of Industry, Tourism and Trade on the basis of assessment by the Nuclear Safety Council (CSN) | Health and Safety Executive (HSE)/ Nuclear Directorate (ND) |
| What is the primary legislation under which the nuclear site licence is issued? | Nuclear Energy Act: Bulletin of Acts, Orders and Decrees, 82, 1963, as amended 2004 | Nuclear Energy Law L 25/1964 and then the Royal Decree as 'Nuclear Installations Regulations' issued in December 1999 | Nuclear Installations Act 1965 (as amended) (NIA65) |
| Is EC Directive 96/29/Euratom of 13 May 1996 on BSS followed? | Nuclear Safety Service (KFD) of the Ministry of Housing, Spatial Planning and the Environment (VI: VROM Inspectorate) implements it | Yes, by the the Royal Decree 1836/1999 of 3 December defined as "Nuclear Installations Regulations" | Yes. IRR99 implements this directive |
| Is EC Directive 97/11/EC of 3 March 1997 on the EIA on decommissioning implemented? | Yes. The Environmental Protection Act (1979, as amended 2002) | Yes, implemented by the Royal Legislative Decrees RLD 1306/1986 of 26 June and RLD 9/2000 of 7 October 2000 | Yes. Nuclear Reactors (Environmental Impact Assessment for Decommissioning) Regulations 1999 |
| Is there a requirement for a separate licence for decommissioning work? | Yes. The Nuclear Energy Act (Section 15b) defines licence requirements | No | No, operational licence is carried through to decommissioning operation |
| Are there defined stages of decommissioning? | No. Operator identifies the stages on a case by case basis | No, but decommissioning and dismantling is to be initiated three years after the shutdown following the removal of spent fuels | There are no nationally defined stages of decommissioning |
| What is the name of the national regulatory body overseeing radiological protection? | Nuclear Safety Service (KFD) of the Ministry of Housing, Spatial Planning and the Environment | CSN is the sole competent organisation | HSE/ND is the national regulatory body. EA and SEPA are the environmental agencies |
| Is IAEA waste categorisation scheme followed? | No. It has its own categorisation scheme | No. It has its own categorisation scheme | No. UK has its own waste categorisation scheme |
| What are the criteria for delicensing a site? | Criteria are not explicit in the Dutch National Report | CSN defines criteria on a case by case basis and the Ministry of Industry, Tourism and Trade delicenses the site | There is no longer any danger from ionising radiation, generally dose $< 10\ \mu Sv.y^{-1}$ |
| Is there a regulatory body overseeing industrial safety? | Ministry of Housing, Spatial Planning and the Environment | Ministry of Labour and Social Affairs | The Health and Safety Executive (HSE) under the HSWA74 |

# APPENDIX 6

# Multi-Attribute Utility Analysis: A Major Decision-aid Technique

## A6.1  Introduction

The Multi-attribute Utility Analysis (MUA) is an advanced decision aid technique within the broad category of Multi-Criteria Analysis (MCA). The MCA technique is used for the appraisal of decision alternatives or options for programmes, projects or policies with reference to multiple criteria or objectives. In projects or policies where multiple criteria, often competing or conflicting, are to be accommodated and assessed, informal or intuitive decision making processes had been found to be unsatisfactory or lacking in transparency. To overcome such shortcomings, a rational, logical, systematic and coherent method of appraisal of options has been devised under the framework of the MCA.

The MCA covers a wide variety of methods of differing levels of sophistication: from a simple generation of a performance matrix (as used in the consumer magazine 'Which?') to the full-blown multi-criteria scoring and weighting methods as practised in the Multi-Criteria Decision Analysis (MCDA), Analytical Hierarchy Process (AHP) and MUA. The MUA technique is the most sophisticated and mathematically most demanding of the MCA techniques. An essential feature of MCA is the generation of a performance matrix. The performance matrix is essentially a two-dimensional table where decision alternatives are placed in rows and criteria in columns. Each of the decision alternatives in the performance matrix is assessed against each of the stated criteria and some sort of objective judgement is made showing the extent to which the criteria have been met. In simple performance matrices, stars (*) or ticks (√) may be sufficient. However, in MCDA, MUA etc., numerical values (scores) are assigned on the strength of performance scales. Numerical weights are then assigned to each criterion to define relative valuations or preferences based on the decision makers' attitudes to risk. These scores and weights are then multiplied and aggregated to estimate the final indices which are used to rank the decision alternatives from the most preferred to the least preferred for easier assimilation by the decision makers.

The MUA technique derives its theoretical basis from the work of Von Neumann and Morgenstern [A6.1] and of Savage [A6.2] in the late 1940s and early 1950s, respectively. However, the practical application of this theoretical formalism had to wait until the work by Keeney and Raiffa [A6.3] in the late 1970s. That work demonstrated the numerical technique for using the methodology and, at the same time, retained the essential features of the original theoretical formalism which drew extensively from disciplines such as statistical decision theory, system engineering, management science, economics and psychology. Because of extensive interplay of various disciplines requiring a high level of technical expertise on the part of the decision analysts, the MUA technique has, so far, been limited to large projects. But the technique is amenable to simplification and is thus useful in smaller and simpler projects. In the USA, the technique is used widely at local, state and federal levels. The US Nuclear Regulatory Commission (NRC) has used the technique on various occasions. The most widely publicised and widely known use of this technique was for selecting potential nuclear waste repository site(s) in the USA in the mid-1980s [A6.4]. In the UK, the technique is also gaining acceptance in both private and public sectors. Realising the importance of this technique in the decision making process, the Department for Transport, Local Government and the Regions (DTLR) (which has now been reorganized as the Department for Communities and Local Geovernment) has produced a manual on MCA [A6.5] which is available on the website: www.communities.gov.uk.

There are other decision-aid tools besides the MCA, such as the Cost Effective Analysis (CEA) or the Cost Benefit Analysis (CBA). Whereas the CEA or CBA relies exclusively on economic considerations, MUA tackles economic as well as non-economic aspects such as health consequences to the workers and to the public, environmental impacts, aesthetic impacts, public perception, technical feasibility etc. The description and methodology given in this appendix closely follows that of Covello [A6.6].

## A6.2 MUA Methodology

The MUA technique is carried out in a rational, logical and systematic way to analyse all the identified decision alternatives or options. The objectives are called here attributes or criteria and hence the name multi-attribute or multi-criteria. However, there is a subtle difference between an attribute and a criterion: an attribute is defined as a measurable criterion. Consequently if a criterion cannot be measured by standard units or scales, a constructed scale needs to be devised for the MUA. The MUA involves assessing the performance of each decision alternative with reference to the stated attributes. The attributes are quantified by performance measures. The final outcome of the analysis is the performance of each decision alternative quantified by a 'utility index' which reflects the degree of achievement against stated attributes under the defined preferences or trade-offs. The decision makers may then look at the basis and outcome of the analysis before finally making their choice.

# A6.3  Steps in the MUA

The decision analysis technique consists basically of eight steps

(1)  Identify decision alternatives and specifying decision parameters.
(2)  Define attributes or criteria.
(3)  Define performance measures or variables.
(4)  Assess probability distributions for the performance measures.
(5)  Specify preferences and value trade-offs.
(6)  Evaluate decision alternatives.
(7)  Conduct sensitivity analysis.
(8)  Identify critical uncertain variables.

Each of these steps is described below.

## A6.3.1 Identifying Decision Alternatives and Specifying Decision Parameters

The first task in the decision analysis is to identify broadly the various decision alternatives. Depending on the project or programme, the decision alternatives may be straightforward and well defined; or they may be interlinked, layered and multi-dimensional.

A small project such as the decontamination of a radioactive cell may involve identifying some discrete operational steps, some of which may run sequentially while others may run in parallel. Once the various steps have been identified, a number of distinct work programmes, each achieving the final end-point of decontamination of the cell, may need to be specified. These distinct work programmes are the decision alternatives. An example of a large complex project requiring identification of decision alternatives is the decommissioning of a nuclear power plant, where a decision alternative may include completion of phase 1 of the decommissioning operation, e.g. defuelling the reactor but deferring phases 2 and 3. Another decision alternative may include completing phases 1 and 2 in sequence within a specified timescale and deferring phase 3. A decision alternative may also include not taking any immediate action, but keeping the plant under maintenance and surveillance for a specified period of time, and then implementing a work programme which completes the project. There may, in fact, be a multitude of decision alternatives. All stakeholders such as plant owners/operators, plant managers, decision makers/analysts, experts and even regulatory bodies may become involved in this exercise. An initial sifting of a multitude of alternatives may be carried out in-house to shortlist the alternatives which deserve further consideration. In some cases, however, the alternatives may be quite straightforward. For example, when selecting a deep underground repository for the disposal of radioactive waste in the USA, decision alternatives were five potential sites as identified by the US Department of Energy.

Once all the decision alternatives have been identified, the individual alternatives are then carefully decomposed, probed and structured. All the parameters which are considered to be important for the performance of the project are clearly and explicitly defined here and these are the decision parameters. In the case of the design of a repository, health and safety of workers, health and safety of the public, environmental impacts, aesthetic effects, economic implications etc. may be all specified. These are the decision parameters.

## A6.3.2 Defining Attributes or Objectives

The next step is the clear specification of the attributes (criteria). These attributes are used as objectives on which the decision will be made. They may include, in the case of the design of a waste repository: minimising adverse health and safety impacts to workers, minimising adverse environmental impacts, minimising adverse socio-economic impacts, minimising construction costs, maximising economic benefits etc. In the case of motorway construction, the attributes may include: maximising diversion of traffic from built-up areas, minimising traffic accidents on the motorway, minimising injury or fatality to workers, and maximising economic benefits to the local community. All the attributes on which the decision will be based are clearly identified and defined here. The purpose of this multi-attribute analysis is to include all of these attributes in the final solution. Prioritisation or preference is not specified at this stage.

It should, however, be noted that some of the attributes may compete or conflict with each other. For example, minimising adverse health and safety impacts would require additional expenditure, which is counter to minimising the economic costs. The aim, therefore, should be to identify, as far as possible, a minimal set of independent attributes. However, even if the attributes are perceived not to be absolutely independent, they still merit being specified here. If an attribute is omitted at this stage, it will not be reflected in the final outcome. However, one should be careful not to incorporate attributes which are directly related or feed into each other, so as to avoid any double counting of attributes.

## A6.3.3 Defining Performance Measures

The next step is to find a way of measuring the defined attributes. A defined attribute may be minimising adverse health impacts on workers. However, this adverse health impact needs to be quantified. One way of quantifying this impact may be the expected number of fatalities or morbidities or the loss of life expectancy of workers. Any one of these may be considered as the performance measure. Adverse industrial impacts may be measured by the number of work-days lost or the number of expected mortalities or morbidities. Economic costs may be taken to be the actual expenditure. These are the performance measures or variables and they are directly quantifiable in standard units.

In cases of decision parameters such as environmental impacts, socio-economic impacts, aesthetic impacts etc., natural scales for performance measures do not exist. To overcome this lack of scale, constructed scales may be devised by specifying a scale in terms of specified levels of impact. Devising such a scale would require careful consideration and appreciation of the full extent of the impacts. An expert elicitation of the levels of impact and the relationships between various levels could be sought. An example of a performance measure for adverse impacts on a site of outstanding natural beauty is shown in Table A6.1.

The impact levels 0–6 may cover the whole range of impacts of the proposed project. The relationship between the minor and major effects would be subject to discussion and agreement among the decision makers and/or decision analysts. Expert advice and public views may be considered when resolving such contentious issues. Once agreement has been reached, it has to be adhered to throughout the course of the analysis.

## A6.3.4 Assessing Probability Distributions

Many performance measures or variables, however, cannot be quantified definitively. For example, expected radiological fatality or morbidity of workers arising from adverse health impacts of radiation cannot be specified without an element of uncertainty. These uncertainties are specified in terms of probability distributions. However, these probability distributions are not easy to specify. It requires a very detailed understanding of the way doses are likely to be accrued by workers, the variations in work patterns and practices,

**Table A6.1**  *Simplified description of performance measure for adverse impacts on a site of outstanding natural beauty*

| Impact Level | Impact on the enviroment* |
|---|---|
| 0 | None |
| 1 | One minor effect |
| 2 | Two minor effects |
| 3 | Three minor effects |
| 4 | One major effect |
| 5 | Two major effects |
| 6 | Three major effects |

\* Impacts may be considered to be adverse if there are visual impacts, damage to flora and fauna, noise degradation, asethetic disturbance etc.

levels of hazard present etc. Once all these parameters have been considered, the probability distribution for radiological consequences may be specified. Similarly other performance measures may also be specified in terms of probability distributions. Again expert advice or elicitation may be sought to define these distributions. These distributions may be of various types: they may be uniform over the whole of the range or they may follow some standard statistical distributions such as normal, log-uniform, log-normal etc.

## A6.3.5 Specifying Preferences and Value Trade-offs

At this stage, value judgements, trade-offs and preferences for each of the performance measures are specified quantitatively in equivalent monetary terms. This information is normally provided by the decision makers, although the decision analysts may help in formulating such views. This step is important as the values assigned here are crucial to the whole of this analysis process. It should be noted that, although specific values are assigned to performance measures, these values should be regarded as notional or as trade-offs. The relative weightings of the value trade-offs confer the preferences.

## A6.3.6 Evaluating decision alternatives

Here the decision analyst integrates the various components of the analysis to evaluate each decision alternative in terms of an utility index. To do so, each performance measure needs to be estimated. Initially, best estimate values may be evaluated and these values are used to estimate the utility index. A detailed evaluation of the utility indices, taking account of the probability distributions, is then undertaken.

The mathematical formalism of defining the overall utility function, $U$ of a programme or project may be delineated into a number of sub-utility functions, $U_i$. Let us consider that $U$ is composed only of $U_1$ and $U_2$ where

$$U(x_1, x_2,...,x_n ; y_1, y_2...y_n) = p_1 U_1(x_1, x_2, ..., x_n) + p_2 U_2(y_1, y_2, ..., y_n) \qquad \text{(A6.1)}$$

where $x_1, x_2, ..., x_n$ are the performance measures indicating the health and safety of workers, health and safety of the public, environmental impact, economic costs etc. during the phase of the sub-utility function $U_1$; $y_1, y_2, ..., y_n$ are the same performance measures during the phase of the sub-utility function $U_2$; $p_1$ and $p_2$ are the positive scaling factors or probabilities that sum to 1.

The analysis may then be conducted for both of these sub-utility functions. However, let us now concentrate on one of the sub-utility functions, $U_1$ which may be written as

$$U_1(x_1, x_2,...x_n) = a - b \sum_i^n k_i C_i(x_i) \qquad \text{(A6.2)}$$

where $k_i$ is the value trade-off for a performance measure; $C_i$ is the component disutility function for performance measure $x_i$; (The parameter $C_i$ is called the component disutility function as it reduces the value of the utility index) $a$ and $b$ are factors necessary to scale the sub-utility function $U_1$ in the range 0 to100. The values of $a$ and $b$ need to be estimated initially taking into account the lowest and highest levels of impact in all decision alternatives.

The weighted component disutility function, $k_i C_i$ in equation (A6.2) is called the monetary equivalent consequence function. It should be noted from equation (A6.2) that the equivalent consequence functions are aggregated by the linear additive method. This may not always be the case. In complex situations, non-linear addition of the consequence functions may be carried out. However, for the sake of simplicity, we will restrict ourselves to the simple linear additive model.

To carry out further analysis, one must have $k_i$ and $C_i$ values for each of the performance measures. The value trade-off, $k_i$ for each performance measure is assigned by decision makers with the help of experts or simply by 'expert elicitation' so that the various trade-offs for various performance measures bear some justifiable relationship to each other as well as to a fixed performance measure such as the construction cost. For example, if the value trade-off for the fatality of a worker is assigned £1,000,000, then the corresponding value for the fatality of a member of the public may be assigned a higher value, such as £3,000,000 or £4,000,000. Such assignment of values is subjective, but it does reflect the relative weighting of one performance measure against another. This relative weighting may be termed a preference. The justification for a higher weighting for a member of the public is that whereas the worker benefits from the employment, and any fatality may be considered as occupational hazard, members of the public receive no such benefits and so any fatality must be compensated at a higher rate. It should be noted that these assigned values are based on notional statistical individuals. The parameter $C_i$ is the component disutility function which is a function of the performance measure, $x_i$. For example, if $x_1$ is the number of worker fatalities, then the disutility function may be considered to be proportional to the number of worker fatalities. In other words, $C_i (x_i) = x_i$. So, if there are more worker fatalities, i.e. $x_1$ is higher, then the function $C_1$ would be higher and consequently the utility index would be lower. However, on constructed scales, the function $C_1$ may not be equal to $x_1$ and there may be some form of functional relationship between $C_1$ and $x_1$.

Taking the lowest and highest values of the component disutility functions and multiplying by the corresponding $k_i$, one can evaluate the values of $a$ and $b$ such that sub-utility function, $U_1$ is in the range 0–100. This is achieved by equating the RHS of equation (A6.2) to zero when maximum component disutility functions are considered and to 100 when minimum component disutility functions are considered. The outcome of each decision alternative is then estimated using these numerical values of *a* and *b*. This value is then the 'utility index' for that decision alternative.

### A6.3.7 Conducting Sensitivity Analysis

Sensitivity analysis is performed in order to evaluate the variations of the utility indices by systematically changing the values of the performance measures (variables) over the range of their variations. A small change in the utility index signifies a low sensitivity whereas a substantial change indicates a high sensitivity. Those variables with high sensitivity are then noted.

The analysis then re-examines these sensitive variables. For example, if public fatality is found to be a sensitive variable, then the analyst must re-visit the initial evaluation of public fatality, its assigned value trade-off and preference. The attempt here is not to alter the sensitivity of the variable but to find a better, more realistic, value of the variable so that the final outcome reflects its significance more accurately.

### A6.3.8 Identifying Critical Uncertain Variables

Critical uncertain variables are those variables whose variations across the range of uncertainty may result in a significant change in the outcome of the analysis, so much so that the decision indices may change places. Whereas sensitivity analysis brings out those variables which are sensitive to the utility index, critical uncertain variables are those variables which are critical to the decision process. The critical uncertain variables are more significant for the decision outcome than the sensitive variables.

A full range of uncertainties as well as the modes of variation within the limits for these critical uncertain variables is re-examined. This attempt to better define the probability distributions of these variables may require further consultations with experts or, in some cases, it may be achieved by experimental evaluation of the values. The MUA analyses may then be repeated with selected sample values from the distributions. A sampling procedure needs to be specified at this point. Stratified random sampling is a good method for selecting a random sample from each of the stratified areas of equal probabilities. Having done that, some correlation between the variables and the derived utility indices may be carried out. The mathematical formalism for carrying out this task is outside the scope of this appendix.

## A6.4 Outcome of the Analysis

The results of analyses ranking the decision alternatives in descending orders of utility indices are then presented to the decision makers to assimilate and finalise the decision. The higher the utility index, the better or more desirable is that decision alternative. Along with this final ranking of decision alternatives, additional information may also be presented showing monetary equivalent costs, $k_i C_i$, for various aggregates of the performance measures for each decision alternative. For example, the equivalent costs of radiological consequences to both workers and public may be grouped together or equivalent costs of radiological consequences to the workers may be grouped separately from those to the public. Other aggregations of equivalent costs may include all performance measures contributing to the environmental impacts or socio-economic impacts etc. Finally, correlations between the critical uncertain parameters and utility indices are also presented to give indications of the levels of the significance of these parameters to the estimated values of utility indices. The MUA process is shown diagrammatically in Figure A6.1.

It should be noted that some simpler versions of the MUA technique are also available. Edwards produced a variation of this technique called the Simple Multi-Attribute Rating Technique (SMART) [A6.7] which is geared mainly to economic considerations. However, it lacks the mathematical rigour of the full-blown technique described here.

## A6.5. Discussion and Conclusions

The MUA is a very powerful decision-aid technique where complex programmes, projects or policies are initially decomposed and segregated into various decision alternatives and then the various components of the analysis are processed to produce a single quantity called the 'utility index' for each decision alternative. Finally, these decision alternatives are ranked in descending orders of utility indices for easier assimilation by decision makers.

The analysis may also provide the degree of association or correlation between the derived utility indices and the assessed uncertainties of decision parameters. The application of the critical uncertainty analysis enables the robustness of the overall process to be established. For some of the very complicated and controversial decisions that have to be made in the nuclear industry such an assurance that the overall decision-making process is robust would seem to be of great value.

Although the mathematical formalism for estimating the utility index is robust and technically sound, the method of setting up value trade-offs, preferences etc.,which is crucial to the analysis, is judgemental and subjective. This subjectivity at the heart of the analysis is of concern as it may be questioned and contested. However, it should be emphasised that the numerical values of the trade-offs are not assigned by decision analysts. At the start of the analysis process they are asked to assign values to these parameters based on their preferences, value judgements and other considerations. They can do so individually, in groups, or even collectively. They may also call on expert advice or seek assistance from decision analysts. When these values are assigned, they remain fixed for the rest of the analysis, unless prompted by

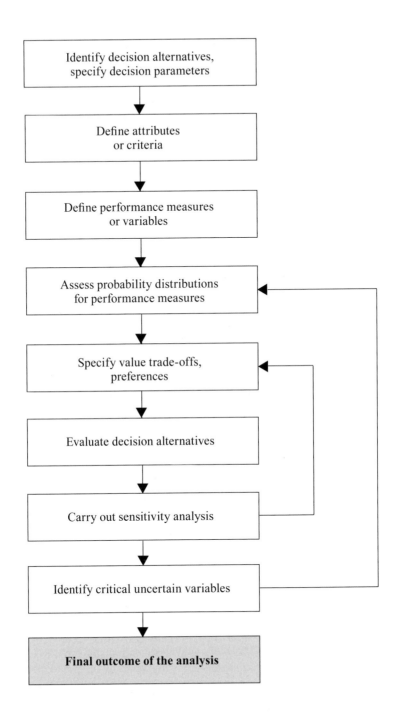

**Figure A6.1**. *Schematic representation of MUA methodology*

the sensitivity or uncertainty analysis. In other words, value trade-offs are not 'rigged' during the analysis to alter the outcome. Thus the analysis is done on a 'level playing field' where the rules of the game have been fixed at the outset without knowing the outcome.

The technique is transparent in the sense that all the decision alternatives, attributes, performance measures, preferences, value trade-offs and all other internal workings which go into the decision-making process are documented and explicit. This is the strength of the technique. However, it can also be viewed as a drawback of the system. Once the decision alternatives have been ranked, it may be difficult for the decision makers to choose anything other than the top ranking one, unless very good justification can be made to do otherwise. The technique may, thus, be viewed as imposing somewhat on the decision makers.

## A.6.6 Glossary of Terms

**Attributes**  Attributes are measurable criteria or objectives. In the case of a nuclear waste repository, attributes may include minimising adverse health effects to workers, minimising environmental impacts, minimising economic costs etc. As attributes are, by definition, measurable, they must be quantified using natural or constructed scales.

**Decision alternatives**  Options to achieve specified end-points.

**Decision parameters**  The parameters that are considered important for project performance. In the case of a nuclear waste disposal facility, health and safety of workers, health and safety of the public, environmental impacts, aesthetic impacts, economic costs etc. may be considered to be decision parameters.

**Performance measures**  The parameters which specify numerical values to attributes. For example, adverse health effects may be measured by the number of mortalities or morbidities, economic costs may be measured in pounds sterling or in other currencies. Performance measures here serve as variables and are shown as $x_i$ in equation (A6.2).

**Preferences**  These are the relative weightings of value trade-offs of various performance measures. If the value trade-off for the fatality of a member of the public is assigned £1,000,000 as against £500,000 for a worker, then the ratio indicates the level of preference for the public.

**Utility index**
The end product of the analysis, which indicates the worth or desirability of a decision alternative. This value is quoted as a percentage.

**Value trade-off**
The monetary equivalent cost for a specified performance measure and has been shown as $k_i$ in equation (A6.2). For example, the fatality of a notional worker may be assigned a value trade-off of £500,000 or £1,000,000 or some other figure.

## REFERENCES

[A6.1] Von Neumann J. and Morgenstern O., *The Theory of Games and Economic Behaviour*, 2nd Edition, Princeton University Press, 1947

[A6.2] Savage L J, *The Foundation of Statistics*, Wiley, New York, 1954

[A6.3] Keeney R L and Raiffa H, *Decisions with Multiples Objectives: Performance and Value Trade-Offs*, Wiley, New York, 1976.

[A6.4] Merkhofer M W and Keeney R L, 'A Multi-attribute Utility Analysis of Alternative Sites for the disposal of Nuclear Waste', *Risk Analysis*, Vol. 7, No. 2, pp. 173–194, 1987.

[A6.5] Department for Transport, Local Government and the Regions, 'Multi Criteria Analysis: A Manual', February 2001.

[A6.6] Covello V T, 'Decision Analysis and Risk Management Decision Making: Issues and Methods, *Risk Analysis*, Vol. 7, No. 2, pp. 131–139, 1987.

[A6.7] Edwards W, 'Social Utilities', *Engineering Economist*, Summer Symposium Series 6, pp. 119-129, 1971.

# ΛTKINS

**Atkins** is pleased to sponsor this book as a demonstration of its commitment to improving the knowledge and skill base of our people in the fields of decommissioning and radioactive waste management. We consider this book to be both timely and relevant.

Nuclear energy is a very topical issue. It is considered to be the most viable source of energy, not only to limit further global warming caused by the emission of carbon dioxide from the progressively increasing demand on fossil fuels, but also by ensuring uninterrupted supplies of energy from a large reserve of raw materials. The availability of large reserves of nuclear fuel would cater for any elasticity in demand and thereby remove market volatility – contrary to the current trends with oil and gas energies.

However, the nuclear industry needs to overcome the public fear and opposition due to the unresolved problem of disposal of radioactive waste in a safe, secure and environmentally acceptable way. Alongside this issue comes the question of capability and technical expertise of the nuclear community to be able to carry out decommissioning operations safely, to time and cost. Resolving these two major back-end fuel cycle problems satisfactorily will not only help the nuclear industry address the overall challenges, but also create enough public confidence to revitalise the nuclear industry at the front end.

Another major stumbling block to the revival of the nuclear industry, both at home and abroad is the chronic skills shortage. During the past few decades when the nuclear industry was in the doldrums, no new blood was coming into the industry. Consequently the age profile of the nuclear industry's technical workforce is now very much skewed towards towards professionals in their 50s and 60s. The skills and experience of this group are soon to disappear unless provision is made to capture and retain the expertise. We see this book as a worthwhile cause as it consolidates academic knowledge and industrial skills in a way that is transferable to future generations.

## About Atkins

Atkins is the largest engineering consultancy in the UK and the largest multidisciplinary consultancy in Europe. We are one of the world's leading providers of professional, technology-based consultancy and support services and deliver total solutions for public and private sector clients in a number of key markets, including:

| | | |
|---|---|---|
| Aerospace | Health | Buildings |
| Industry | Defence | Oil & Gas |
| Education | Telecoms | Energy |
| Transport | Environment | Water |
| Urban Development | | |

We bring significant value to our clients, harnessing an unrivalled pool of creative, professional people to produce outstanding solutions to challenging problems.

For further information, please visit www.atkinsglobal.com

# Meeting the nuclear challenge

To a certain extent the public has considered nuclear power generation to be a dangerous technology. But is it necessarily true? Incidents such as the partial core meltdown at Three Mile Island in 1979 and the explosion of the Chernobyl Nuclear Power Plant in 1986 did not help the nuclear cause and after these events the industry fell into decline and European production slowed to a crawl.

It should not be forgotten that nuclear power is now a mature technology that draws on 50 years – and over 12,000 reactor-years – of civil operational experience. There has never been major public harm from any Western-type power reactor.

There are now encouraging signs that the nuclear industry's place in the panoply of energy options is growing. Currently, some 430 nuclear power plants in 30 countries produce 16 per cent of the world's electricity; in OECD countries alone, they produce a quarter of the power. New plants are planned or in construction in 15 countries, representing well over half the world's population.

More pointedly, the latest report by the United Nations' Inter-governmental Panel on Climate Change (IPCC) warns that a huge increase in nuclear power will be required worldwide by 2030 to meet the rapidly-expanding electricity demand without exacerbating global warming.

As an industry it is incumbent upon us to demonstrate that nuclear power is safe and that we adopt the highest levels of responsibility in managing the industry. In doing this, we must show that we not only have the skills to operate nuclear sites safely, but importantly the capability to clean-up the many legacy plants around the world, restoring the local environment to a standard expected of a responsible industry.

However, this is not a challenge to be taken on lightly. Unlike the latest design of nuclear plants, the first generation of nuclear installations were neither designed nor constructed with decommissioning in mind. The resultant decommissioning projects are not only complex but often unique in their nature, requiring solutions that are innovative, efficient and cost effective.

Decommissioning of the legacy plants demands that the resources and capabilities of those who sustain the business be pushed further than ever before. The prize for successful delivery is not only a restored local environment for future generations to enjoy, but also a re-invigorated nuclear industry that, with public support, forms a major piece of the jigsaw in solving the world's energy challenge for generations to come.

**Chris Ball**
Chris Ball is director of nuclear at Atkins. Formed in 1938, Atkins has provided innovative engineering services to the world's nuclear industry for over 40 years. Today, Atkins has over 500 staff working throughout the UK nuclear industry and provides decommissioning services to the very sites the company helped construct during the 1960s.

# Index

A$_1$ and A$_2$ values 361–2
Abrasive water injection jet 258
Abrasive water suspension jet 258
Absorbed dose 44
Activity 5–6
Acute exposure 29
ALARA/ALARP 123, 125–6, 130–9, 401
Alkali metals 144
Arc saw cutting 262
Article 37 of the EURATOM Treaty 315–17
Asbestos 144
Asbestos Regulations 111
Asphyxiant 143
AT1 reprocessing plant 269–72

Backfill materials 384–5
Basel Convention 313
Basic Safety Limit (BSL) 131–4
Basic Safety Objective (BSO) 131–4
Bentonite 384–5
Best Practicable Environmental Option
    (BPEO) 113–14, 218, 235–7, 401
Binomial distribution 62–3
Biological damage 36–40
Biological effects of radiation 28–42
Biosphere 388
Bituminisation 345–6
Bottom-up technique of cost estimation 154
Brown field 113, 158, 198
Build-up factor 15–20

Calder Hall in the UK 282–7
Category A worker 129–30
Category B worker 129
Cavern concept 389
Cell membrane 36–7
Cementation 346
Central Limit Theorem 70
Characterisation 200–4
Chemical decontamination 240–4
Chemical hazards 140–5
Chromosomes 36–8
Chronic exposure 29
Clearance levels 321–8
Coefficient of variation 59–60
Collective effective dose 49

Collective equivalent dose 49
Co-location 393–4
Committed effective dose 48–9
Committed equivalent dose 48
Committee of Regulation of Waste
    Management (CoRWM) 86, 94, 401
Compaction 334–8
Comparison technique of cost estimation 154
Compton effect 9–10
Conditioning of waste 345–56
Contract 184–9
Control of Substances Hazardous to Health
    (COSHH) Regulations 110–11, 401
Controlled area 128–9
Cost estimates 158
Cost management 178–80
Cost of decommissioning 152
Cost Reimbursable (CR) contract 188–9
Criticality Safety Index (CSI) 362–3, 402
Crown censure 118
Cutting tools 252–6
Cytoplasm 36–9

Decay constant 5
Decommissioning and Dismantling (D&D)
    83–4
Decommissioning of nuclear facilities 79–95
Decommissioning Plan (DP) 191–3, 196–7
Decommissioning, regulatory aspects 96–121
Decontamination 240–51
Deep borehole concept 389–90
Defence in depth 124
Defence Nuclear Safety Regulator (DNSR)
    117, 402
Deferred dismantling 83
Defuel, De-equip and Lay-up Preparation
    (DDLP) 92
Delayed neutron 23
Delicensing 198–9
Deoxyribo Nucleic Acid (DNA) 37–9
Design Basis Accident (DBA) analysis 402,
    125–6
Deterministic effects 31–2
Direct measurement 204, 208–9
Directly ionising radiation 7–8
Discounting technique 154, 163–7
Dismantling 252–67

Disposal 378–98
Diversity 125
Dose and Dose Rate Effectiveness Factor (DDREF) 31
Dose calculations 12–15
Dosimetric quantities 44

Effective dose 48
Effective half life 33–4
Elastic scattering 9, 21, 23
Electro Discharge Machining (EDM) 262
Electromagnetic radiation 1–2
Electro-polishing 249
Energy fluence rate 6–7
Engineered Barrier System (EBS) 381–3, 402
Entomb 83
Environment Act 1995 107
Environmental baseline 227–9
Environmental Impact Assessment Decommissioning Regulations (EIADR) 108–9, 219 402
Environmental Statement (ES) 233–4
Equivalent dose 46
European Atomic Energy Community (EURATOM) 100, 402
European Commission (EC) 100
Evaporation 342–5
Excepted package 361, 365–6, 368
Excitation 4
Exclusion of sources 317
Exempt waste (EW) 290, 402
Exemption levels 318–20
Expert elicitation of cost estimation 154, 237, 434, 436
External exposure 28

Fast neutron dose calculation 23
Final radiation survey 193, 196, 198–9
Fixed price contract 189
Flame cutting 256
Flux 6
Frequency distribution 57–8
Fuel cycle 80-1
Funding mechanism 167

Gas-filled detectors 204–5
Geological disposal 380, 392–3
Germ cell 36
Globally Harmonised System of Classification and Labelling of Chemicals (GHS) 100, 141–2

Government Owned, Contractor Operated (GOCO) 117
Green field 158, 196, 198

Half value thickness or layer 14
Haphazard sampling 70–1
Hazard function 32
Health and Safety at work etc Act 102–3, 403
Health and Safety Commission (HSC)/ Health and Safety Executive (HSE) 97
High Activity Sealed radioactive Sources (HASS) 110, 403
High Level Waste (HLW) European Union (EU) 294–5
High Level Waste (HLW) International Atomic Energy Agency (IAEA) 290
High Level Waste (HLW) or Heat Generating Waste (HGW) 290–1
High pressure water jetting 245, 247
Human resource management 181–2

Impact assessment 229–33
Incineration 338–41
Independence 125
Indirectly ionising radiation 8
Industrial hazards 145–6
Industrial package 363–72, 376, 403
Inelastic scattering 21
Interim Storage of Laid Up Submarine (ISOLUS) 94, 403
Intermediate Level Waste (ILW) 289–91
Internal exposure 28
International Atomic Energy Agency (IAEA) 98–9
International Commission on Radiological Protection (ICRP) 99
Intervention 49
Ionisation 1–4, 7–8
Ionising radiation 3, 8
Ionising Radiations Regulations 1999 52, 105, 107, 403
Isotope 4–5
Joint Convention 311–14
Justification of a practice 50

KGR Greifswald, Germany 277–82
KRB-A plant in Germany 272–7

Laser cutting 263–5
Latency period 29–30

Lifting Operations and Lifting Equipment Regulations 1998 (LOLER) 112
Light ablation 249
Linear Energy Transfer (LET) 8
Linear No-Threshold (LNT) 43–4
Log-normal distribution 68–9
Low and Intermediate Level Waste (LILW) International Atomic Energy Agency (IAEA) 290
Low and Intermediate Level Water (LILW) European Union (EU) 295
Low dispersible radioactive material 363
Low Level Waste (LLW) 291–2
Low Specific Activity (LSA) material 363

Macroscopic cross section$\Sigma$ 22
Magnox Encapsulation Plant (MEP) 346–7, 404
Mass absorption coefficient, $\mu_a/\rho$ 10–12, 19
Mass attenuation coefficient, $\mu/\rho$ 10–11, 17
Mean 57–9
Mechanical decontamination 244–9
Mechanical saws 254
Median 57–9
Meiosis 38
Microscopic cross section, $\sigma$ 21
Mitochondria 37
Mitosis 38
Mixed nuclides 327–8
Mode 57–9
Multi-attribute Utility Analysis (MUA) 404, 430–40
Multi-barrier concept 381–2
Multi-barrier protection 124–5
Multi-Criteria Analysis (MCA) 136, 404, 430
Multi-phase treatment process 243–4

Natural barrier 385–7
Neutron dose calculation 23
Nuclear Energy Agency (NEA) of the Organisation for Economic Co-operation and Development (OECD) 97, 101, 405
Near surface disposal 380, 392
Net present value of money 163–5
Normal distribution 65–6
Nuclear Decommissioning Authority (NDA) 85, 404, 420–5
Nuclear Installations Act (NIA) 103–7, 404
Nuclear Installations Inspectorate (NII) 103, 404
Nuclear Directorate (ND) 102–3

Nuclear submarine decommissioning 92–4
Nuclear submarine disposal 397–8
Nuclide 4–5
Null hypothesis 73–5

Orbital cutters 254–5
Organelles 37
Office for Civil Nuclear Security (OCNS) 102–3, 404
OSPAR Convention 314–15
Overpack 364

'Polluter Pays' principle 310
Pair production 9–10, 15
Parametric technique of cost estimation 154
Particle fluence rate 6
Particulate radiation 2–3
Partitioning and Transmutation (P&T) 304–5
Percentile 60–1
Performance assessment 394–5
Personal Protective Equipment (PPE) 130, 146, 406
PERT chart 176–7
Phased disposal 393
Photoelectric effect 8–10, 15
Photon 1–4, 8–10, 12–13, 15–19, 25
Plasma arc cutting 256
Poisson's distribution 63–4
Polymerisation 355–6
Post-Operational Clean Out (POCO) 84, 86, 91, 122, 160, 195, 405
Practice (definition) 49
Pre-disposal 379
Primary waste 331
Probabilistic Safety Assessment (PSA) 126–7
Probability distributions 62–70
Probability sampling 70–1
Project management 171–3
Project Management Plan (PMP) 173–4
Prompt neutron 23
Pucks 334
Pyrolysis 341

Quality assurance 148
Quality management 180–1
Quantile 60
Quartile 60–1

Radiation Protection Programme 360
Radiation weighting factor 45–6
Radiation, effects of 28

Radiation, types 1–4
Radiative capture 21
Radioactive material 288
Radioactive mixed waste 289–90
Radioactive Substances Act 1993 107–8
Radioactive waste 288–9
Radioactivity 5–6
Radiological protection 43–56
Radionuclide 4–6
Radiotoxicity 35
Reducing Risks, Protecting People (R2P2)
    132, 406
Redundancy 125
Respiratory protective equipment (RPE) 130
Reusable Shielded Transport Containers
    (RSTC) 373–4, 406
Risk management 184

Safe-store 83
Safety 122–50
Safety case 146–7
Safety documentation 146–8
Safety Principles and Safety Criteria (SPSC)
    53
Sampling 70–6, 203–4, 211–13
Scanning 204, 209–11
Schedule management 175–8
Scintillation detectors 206
Scoping (environmental) 224–7
Screening (environmental) 220–4
Search sampling 70–2
Secondary waste 331–2
Sensitivity analysis 438, 139–40
Shaped explosive 255–6
Shears 253–4
Solid state detectors 206–8
Somatic cell 36, 39–40
Special form radioactive material 364
Specific ionisation 7
Standard deviation 59–60
Standardised list of decommissioning costs
    158–63
Standardised normal variable 67
Statistical methods 57–78
Statutory dose limit 52
Steam cleaning 246
Stochastic effects 30–1
Stopping power 7
Storage of waste 358–9
Strippable coating 246–7
Student's t-test 74–5

Subjective sampling 70–1
Super-compaction 334–8
Supervised area 128–9
Surface Contaminated Objects (SCO) 333
Surface disposal 380
Sustainable development 309–10
Synroc process 354–5

Taylor form of build-up factor 20–1
Test for packages 370–1
Test of hypothesis 74–5
Thermal neutron dose calculation 25
Thermalisation 3
Threshold level 31
Time and Material (T&M) contract 189
Tissue weighting factor 46–8
Tolerability of risk 131–4
Total absorption coefficient, $\mu_a$ 10
Total attenuation coefficient, $\mu$ 10
Transition radioactive waste 295
Transport Index (TI) 365–6, 406
Transportation of radioactive materials 361–70
Transportation of waste 371–5
Treatment of waste 333–45
Tunnel concepts 388–9
Type A, B, C packages 366–8

Ultrasonic cleaning 249
Uncertainty analysis 139–40
United Nations Conference on Environment
    and Development (UNCED) 97, 100, 406
United Nations Scientific Committee on the
    Effects of Atomic Radiation (UNSCEAR)
    97, 100, 406

Variance 59–60
Vitrification 334, 351–3
Very Low Level Waste (VLLW) 155, 289–92,
    406

Wait and see option 83
Waste classification 289–95
Waste forms 382–3
Waste management 308–30
Waste packages 383–4
Waste substitution 304
Windscale Vitrification Plant (WVP) 352–3,
    406
Work Breakdown Structure 174–5